AF588331

Neuropsychiatric Symptoms of Neurological Disease

Series editor

José M. Ferro, Serviço de Neurologia Stroke Unit, Hospital de Santa Maria, Lisboa, Portugal

More information about this series at http://www.springer.com/series/10501

Bruno Brochet
Editor

Neuropsychiatric Symptoms of Inflammatory Demyelinating Diseases

Springer

Editor
Bruno Brochet
Department of Neurology
Centre Hospitalier Universitaire Pellegrin,
INSERM U 862, Université de Bordeaux
Bordeaux, Gironde, France

ISSN 2196-2898 ISSN 2196-2901 (electronic)
Neuropsychiatric Symptoms of Neurological Disease
ISBN 978-3-319-18463-0 ISBN 978-3-319-18464-7 (eBook)
DOI 10.1007/978-3-319-18464-7

Library of Congress Control Number: 2015944986

Springer Cham Heidelberg New York Dordrecht London

Printed on acid-free paper

Springer International Publishing AG Switzerland is part of Springer Science+Business Media (www.springer.com)

Preface

Neuropsychiatric Symptoms of Inflammatory Demyelinating Diseases is the third volume of a series of volumes on the psychiatric aspects of common neurological diseases, to be published by Springer, with José M. Ferro as Series Editor.

The first volume was dedicated to the neuropsychiatric symptoms of stroke and cerebrovascular diseases, and the second to the neuropsychiatric symptoms of movement disorders.

Multiple sclerosis (MS) and other inflammatory demyelinating diseases are the commonest causes of long-standing disability in young adults due to diseases of the central nervous system. In MS, psychiatric symptoms and comorbidities could occur at onset or during the disease course. Diagnosis could be difficult and management is not easy. Some psychiatric comorbidities like depression are very frequent. In other demyelinating diseases like acute disseminated encephalomyelitis and neuromyelitis optica, this topic has been rarely addressed.

This book is an up-to-date, comprehensive review of the neuropsychiatry of MS and related diseases by active authorities in the field, with an emphasis on diagnostic and management issues.

The 19 chapters of this book cover not only the main psychiatric disorders associated with MS, their mechanisms and their management but also related and important psychological aspects including coping strategies, fatigue and behavioural therapies. There are many interactions between psychiatric comorbidities occurring in MS patients and their neurocognitive status. The last part of this book covers these aspects and also a new era of research about social cognition in MS.

This book includes critical appraisal of the methodological aspects and limitations of the current research on the neuropsychiatry of demyelinating diseases and on unanswered questions/controversies. Pharmacological and behavioural aspects of management are discussed, to provide robust information in order to enable the reader to better manage these patients.

Neuropsychiatric Symptoms of Inflammatory Demyelinating Diseases is aimed at neurologists, other multiple sclerosis specialists and psychiatrists, but will also be of interest to intensive care doctors, psychologists and neuropsychologists, research and specialist nurses and clinical researchers.

We hope this book will become a standard reference for clinicians of several specialities.

Bordeaux, France Bruno Brochet

Contents

Contributors

Peter A. Arnett, Ph.D. Department of Psychology, College of the Liberal Arts, Penn State University, University Park, PA, USA

Bertrand Audoin, MD, Ph.D. Pôle de Neurosciences Cliniques, Service de Neurologie, APHM, Hôpital de la Timone, Marseille, France

CNRS, CRMBM CEMEREM UMR 7739, Aix-Marseille Université, Marseille, France

Vikram Bhise, MD Department of Pediatrics, Neurology, Rutgers – Robert Wood Johnson Medical School, Child Health Institute of New Jersey, New Brunswick, NJ, USA

Valentina Bianchi, PsyD MS Centre, S. Andrea Hospital, Neurological Sciences, Sapienza University, Rome, Italy

Centro Interdipartimentale Medicina Sociale (CIMS), Sapienza University, Rome, Italy

Frédéric Blanc, MD, Ph.D., HdR CMRR (Memory Resources and Research Centre), University Hospital of Strasbourg, Strasbourg, France

University of Strasbourg and CNRS, ICube Laboratory UMR 7357 and FMTS (Fédération de Médecine Translationnelle de Strasbourg), Team IMIS/Neurocrypto, Strasbourg, France

Pierre Branger, MD Department of Neurology, University Hospital Center of Caen, Caen Cedex 9, France

Bruno Brochet, MD Department of Neurology, Centre Hospitalier Universitaire Pellegrin, Bordeaux, France

Margaret Cadden, B.S. Department of Psychology, Penn State University, University Park, PA, USA

Mauro Giovanni Carta, MD Department of Public Health and Clinical and Molecular Medicine, University of Cagliari, Cagliari, Italy

Consultation Liaison Psychiatric Unit at the University Hospital of Cagliari, University of Cagliari and AOU Cagliari, Cagliari, Italy

Khadija Chahraoui, Ph.D. Laboratoire de Psychopathologie et de Psychologie Médicale, Université de Bourgogne, Dijon, France

Eleonora Cocco, MD Department of Public Health and Clinical and Molecular Medicine, University of Cagliari, Cagliari, Italy

Centro Sclerosi Multipla, Cagliari, Italy

Gilles Defer, MD Department of Neurology, University Hospital Center of Caen, Caen Cedex 9, France

Emmanuelle Dieu, Ph.D. Laboratoire De Psychopathologie Et Psychologie Médicale, Université De Bourgogne, Dijon, France

Centre Hospitalier Universitaire de Dijon, Service De Psychiatrie Adulte Et Addictologie – Service De Neurologie, Dijon, France

Cécile Dulau, MD Centre Hospitalier Universitaire Pellegrin, Bordeaux, France

Department of Neurology, Université de Bordeaux, Bordeaux, France

Anthony Feinstein, MD, M.Phil., Ph.D., FRCP Department of Psychiatry, University of Toronto and Sunnybrook Health Sciences Centre, Toronto, ON, Canada

Frederick W. Foley, Ph.D. Ferkauf Graduate School of Psychology, Yeshiva University, Bronx, NY, USA

The Multiple Sclerosis Center, Holy Name Medical Center, Teaneck, NJ, USA

Olivier Heinzlef, MD Department of Neurology, Poissy-Saint-Germain Hospital, Poissy, France

Lauren B. Krupp, MD Department of Neurology, Stony Brook Medicine, Stony Brook, NY, USA

Pierre-Michel Llorca, MD, Ph.D. Department of Psychiatry B, CHU Clermont-Ferrand, University of Auvergne, EA 7280, Clermont-Ferrand, France

Lorena Lorefice, MD Department of Public Health and Clinical and Molecular Medicine, University of Cagliari, Cagliari, Italy

Centro Sclerosi Multipla, Cagliari, Italy

Giuseppe Magistrale, MSc Neurology and NeuroRehabilitation Unit, I.R.C.C.S. "Santa Lucia" Foundation, Rome, Italy

Maria Giovanna Marrosu, MD Department of Public Health and Clinical and Molecular Medicine, University of Cagliari, Cagliari, Italy

Centro Sclerosi Multipla, Cagliari, Italy

Thibault Moreau, MD, Ph.D. Department of Neurology, Centre of Epidemiology of the Populations EA 4184, University Hospital of Dijon, Dijon, France

Maria Francesca Moro, MD Department of Public Health and Clinical and Molecular Medicine, University of Cagliari, Cagliari, Italy

Consultation Liaison Psychiatric Unit at the University Hospital of Cagliari, University of Cagliari and AOU Cagliari, Cagliari, Italy

Ugo Nocentini, MD Neurology and NeuroRehabilitation Unit, I.R.C.C.S. "Santa Lucia" Foundation, Rome, Italy

Department of Systems Medicine, University of Rome "Tor Vergata", Rome, Italy

Jean Pelletier, MD, Ph.D. Pôle de Neurosciences Cliniques, Service de Neurologie, APHM, Hôpital de la Timone, Marseille, France

CNRS, CRMBM CEMEREM UMR 7739, Aix-Marseille Université, Marseille, France

Carlo Pozzilli, MD, Ph.D. MS Centre, S. Andrea Hospital, Neurological Sciences, Sapienza University, Rome, Italy

Department of Neurology and Psychiatry, Sapienza University, Rome, Italy

Audrey Rico, MD, Ph.D. Pôle de Neurosciences Cliniques, Service de Neurologie, APHM, Hôpital de la Timone, Marseille, France

CNRS, CRMBM CEMEREM UMR 7739, Aix-Marseille Université, Marseille, France

Kristoffer Romero, Ph.D. Department of Psychiatry, University of Toronto and Sunnybrook Health Sciences Centre, Toronto, ON, Canada

Aurélie Ruet, MD, Ph.D. Centre Hospitalier Universitaire Pellegrin, Bordeaux, France

Department of Neurology, Université de Bordeaux, Bordeaux, France

Ludovic Samalin, MD Department of Psychiatry B, CHU Clermont-Ferrand, University of Auvergne, EA 7280, Clermont-Ferrand, France

Marie Théaudin, MD, Ph.D. Neurology Department, CHU Bicêtre, Le Kremlin Bicêtre, France

Giuseppina Trincas, MD Department of Public Health and Clinical and Molecular Medicine, University of Cagliari, Cagliari, Italy

Consultation Liaison Psychiatric Unit at the University Hospital of Cagliari, University of Cagliari and AOU Cagliari, Cagliari, Italy

Dede Ukueberuwa, M.S. Department of Psychology, Penn State University, University Park, PA, USA

Part I
Psychiatry of Demyelinating Diseases

Chapter 1
The Spectrum of Demyelinating Inflammatory Diseases of the Central Nervous System

Bruno Brochet

Abstract Multiple sclerosis (MS) is, by far, the most common inflammatory demyelinating disease of the central nervous system (CNS), but it is not the only CNS inflammatory demyelinating disease and there is a broad spectrum of disorders with varied clinical course, imaging features, epidemiological characteristics, regional distribution of pathological abnormalities, and pathology. Other syndromes associated to MS are clinically isolated syndromes and radiologically isolated syndromes. MS variants included active MS (Marburg type), Schilder's disease, and Baló lesions. Neuromyelitis optica spectrum, idiopathic acute transverse myelitis, acute disseminated encephalomyelitis, chronic inflammatory myelopathy, and chronic relapsing inflammatory optic neuritis must be distinguished from MS.

Keywords Multiple sclerosis • Clinically isolated syndromes • Radiologically isolated syndromes • Neuromyelitis optica • Baló concentric sclerosis • Schilder's disease • Transverse myelitis • Chronic relapsing inflammatory optic neuritis • Marburg disease

Introduction

Multiple sclerosis (MS) is, by far, the most common inflammatory demyelinating disease of the central nervous system (CNS) affecting probably more than two million people worldwide [1]. However, MS is not the only CNS inflammatory demyelinating disease, and there is a broad spectrum of disorders with varied clinical course, imaging features, epidemiological characteristics, regional distribution of pathological abnormalities, and pathology.

Table 1.1 presents a classification of these disorders based on lesion location and clinical course.

B. Brochet, MD (✉)
Department of Neurology, Centre Hospitalier Universitaire Pellegrin,
INSERM U 862, Université de Bordeaux, Place Amélie Raba Léon, Bordeaux 33076, France
e-mail: bruno.brochet@chu-bordeaux.fr

B. Brochet (ed.), *Neuropsychiatric Symptoms of Inflammatory Demyelinating Diseases*, Neuropsychiatric Symptoms of Neurological Disease,
DOI 10.1007/978-3-319-18464-7_1

Table 1.1 Classification of inflammatory (not infectious) demyelinating diseases

Name	Lesion distribution	Typical course	Comment
Multiple sclerosis (MS) spectrum			
Clinically isolated syndromes (CIS) of the type seen in MS without dissemination in time			
Optic neuritis	Optic nerve	Monophasic, recovery is usual, with or without sequelae	Idiopathic, presumably autoimmune
			Absence of dissemination in time. Limited dissemination in space possible
Acute partial myelitis	Spinal cord	Monophasic, recovery is usual, with or without sequelae	Idiopathic, presumably autoimmune
			Absence of dissemination in time. Limited dissemination in space possible
Other CIS	Brainstem, cerebellum, brain hemispheres, uni- or multifocal	Monophasic, recovery is usual, with or without sequelae	Idiopathic, presumably autoimmune
			Absence of dissemination in time. Limited dissemination in space possible
Progressive syndrome mimicking primary progressive MS			
Chronic progressive inflammatory myelopathy	Spinal cord	Chronic, progressive	Inflammatory CSF usual and lesions of the inflammatory type on cord MRI. No dissemination in space outside the cord
Multiple sclerosis			
McDonald MS (or "active CIS")	Clinically CIS (optic neuritis, partial acute myelitis, brainstem, or others) but multifocal lesions (dissemination in space)	One episode and imaging evidence of dissemination in time	Clinical presentation indistinguishable from idiopathic CIS
			Usually evolves toward RRMS
Relapsing-remitting MS	Multifocal, disseminated	Relapsing-remitting	According to clinical (relapses) and imaging activity, it could be classified as active RRMS or inactive RRMS
			Frequently evolves toward SPMS

(continued)

Table 1.1 (continued)

Name	Lesion distribution	Typical course	Comment
Secondary progressive MS	Multifocal, disseminated	Progressive accumulation of disability after initial relapsing course	According to clinical (relapses) and imaging activity, it could be classified as active SPMS or inactive SPMS and with or without progression
Primary progressive MS	Multifocal, disseminated	Progressive accumulation of disability from onset	According to clinical (relapses) and imaging activity, it could be classified as active PPMS or inactive PPMS and with or without progression
Other syndromes associated with MS			
Radiologically isolated syndromes	Multifocal, disseminated	Clinically silent	Imaging abnormalities typical of MS
Marburg variant	Multifocal, disseminated, extensive lesions	Rapidly evolving course. Fatal.	Very rare
Schilder's disease			
Neuromyelitis optica (NMO) spectrum			
NMO (Devic's disease)	Mainly optic nerve and spinal cord (LETM), possible brainstem and diencephalon involvement	Relapsing (very rare secondary progression has been described)	Positive anti-aquaporin-4 antibodies in the majority of cases
Seropositive isolated LETM or optic neuritis	Cord (LETM) or optic nerve	One episode. High risk of evolution toward relapsing NMO	Positive anti-aquaporin-4 antibodies required
Unusual high-risk syndromes	Brain, infra- or supratentorial	One episode. High risk of evolution toward relapsing NMO	Positive anti-aquaporin-4 antibodies required
Recurrent syndromes	Recurrent myelitis (usually LETM) or optic neuritis	Several episodes in the same location. High risk of evolution toward relapsing NMO	Positive anti-aquaporin-4 antibodies required
Other inflammatory demyelinating diseases			
Isolated monophasic syndromes with clinical presentation usually not mimicking MS			
Acute disseminated encephalomyelitis	Brain, cord, multifocal, disseminated	Monophasic (rarely multiphasic and rare relapsing forms reported), recovery is usual, with or without sequelae	Frequently postinfectious; frequent in children; frequent gray matter involvement

(continued)

Table 1.1 (continued)

Name	Lesion distribution	Typical course	Comment
Acute hemorrhagic leukoencephalomyelitis (Hurst type)	Brain, cord, multifocal, disseminated, occasionally focal	Monophasic, hyperacute, frequently fatal	Extremely rare
Idiopathic transverse myelitis	Spinal cord (usually longitudinally extensive transverse myelitis (LETM))	Monophasic	Severe course. Negative anti-aquoporin-4 serology required
Other relapsing inflammatory diseases not mimicking MS			
Chronic relapsing inflammatory optic neuritis (CRION)	Optic nerve	Recurrent	Negative anti-aquoporin-4 serology required. Response to steroids and relapse on withdrawal or dose reduction

Multiple Sclerosis Spectrum

In many countries, MS is the most frequent disabling neurological condition that affects young adults [2]. About 5 % of MS patients are diagnosed before the age of 18 [3]. Its prevalence varies according to geographical location, with a relative north-south gradient in the north hemisphere, and ranges from 40 to 300 cases per 100,000 inhabitants [2, 4]. In nearly 85 % of patients, the disease initially follows a relapsing-remitting (RR) course and is called RRMS [5]. The remainder 15 % of cases has primary progressive MS (PPMS), which is characterized by a gradually progressive clinical course from the onset [5, 6]. In many cases, patients with RRMS experience after several years a gradual worsening after an initial relapsing disease course, with or without acute exacerbations, and are diagnosed as having secondary progressive MS (SPMS) [5, 7, 8]. A consensual classification of these clinical phenotypes has been established [9] and has been recently updated [10] (Table 1.2).

Relapsing-Onset MS

RRMS typically begins in the second or third decade of life and has a female predominance [2, 4, 5]. Some evidence has shown that the proportion of women in newly diagnosed RRMS patients increased in the last 20 years reaching more than 70 % in recent studies [11, 12]. RRMS is characterized by acute neurological episodes typically evolving over a period of several days, stabilizing, and then often improving, spontaneously or in response to corticosteroids, within weeks [13, 14].

Table 1.2 Clinical phenotypes of MS [9, 10]

1996 MS clinical subtypes [9]		2014 MS phenotypes [10]	
Relapsing disease			
RRMS	With full recovery from relapses	CIS	Not active[a]
			Active[a,b]
	With sequelae/residual deficits after incomplete recovery	RRMS	Not active[a]
			Active[a]
Progressive disease			
Progressive disease	Progressive accumulation of disability from onset with or without temporary plateaus, minor remissions and improvements (PPMS)	Progressive accumulation of disability from onset (PP)	Active[a] and with progression[c]
			Active[a] and without progression[c]
			Not active and with progression[c]
			Not active[a] and without progression[c] (stable disease)
	Progressive accumulation of disability after initial relapsing course with or without occasional relapses and minor remissions (SPMS)	Progressive accumulation of disability after initial relapsing course (SP)	Active[a] and with progression[c]
			Active[a] and without progression[c]
	Progressive accumulation of disability from onset but clear acute clinical attacks with or without full recovery		Not active[a] and with progression[c]
			Not active[a] and without progression[c] (stable disease)

[a]Activity determined by clinical relapses and/or MRI activity (contrast-enhancing lesions; new or unequivocally enlarging T2 lesions assessed at least annually)
[b]CIS active: CIS, if subsequently clinically active and fulfilling current multiple sclerosis (MS) diagnostic criteria, becomes relapsing-remitting MS (RRMS)
[c]Progression measured by clinical evaluation, assessed at least annually. If assessments are not available, activity and progression are "indeterminate"

The usual mode of presentation of RRMS is with a first episode of focal neurologic symptoms which is called a clinically isolated syndrome (CIS) in the absence of previous documented symptoms [13, 14]. A CIS is therefore defined as an acute or subacute episode of neurological dysfunction due to inflammatory demyelination that lasts more than 24 h and occurs in the absence of fever, infection, or encephalopathy [14]. The later criterion is useful to distinguish CIS from acute disseminated encephalomyelitis (ADEM) [15], but encephalopathic signs rarely occur in CIS [14]. A CIS is clinically indistinguishable from relapses that are typically associated with RRMS, except that they are isolated in time clinically (first known episode) [13]. While CIS is closely related to MS, a significant proportion of patients presenting with a CIS typical for MS have a monophasic illness without clinical and imaging dissemination in time even with long-term follow-up, and the diagnosis remains CIS or "isolated idiopathic demyelinating event" [13]. These syndromes account for approximately one third of patients with CIS although the number varies depending on the phenotype, the highest proportion having been observed in patients with optic neuritis (ON) [13]. ON, partial myelitis, and brainstem syndromes are the most common type of CIS and also of MS relapses [13, 14]. However, cerebellar, hemispheric, and multifocal episodes are possible. In large cohort studies, the proportion of patients with a progressive course among relapsing-onset MS patients is variable from 30 to 60 % [5, 7, 8, 16, 17], the proportion being higher in studies with a longer follow-up. In the British Columbia cohort, about 75 % of patients with more than 30 years of disease duration were classified as having SPMS [7]. Relapses could occur during the secondary progressive phase but their frequency is usually low. SPMS is usually characterized by progressive motor pyramidal or cerebellar impairment leading to a severe disability [5, 7, 8].

Progressive-Onset MS

Contrary to RRMS, there is no female predominance in PPMS (ratio about 1/1) [6, 18]. Usually PPMS begins in the fourth decade and is characterized in more than 80 % of cases by a progressive paraparesis due to spinal pyramidal involvement [8]. Bladder dysfunction is usual. In other cases, progressive hemiparesis or cerebellar dysfunction could occur. Very rare pure optic or cognitive forms of PPMS have been described [8].

Other Signs and Symptoms

Patients with MS frequently experience signs and symptoms that are not usually due to a relapse or the progressive stage, like fatigue, chronic pain, and urogenital dysfunction [19–21]. Psychiatric and cognitive impairment could also frequently occur and will be addressed in other chapters of this book.

Diagnostic Criteria

The diagnostic criteria of MS have been a matter of debates since many years. Various sets of criteria have been proposed on the basis of the principles of dissemination in space and time established by Schumacher [22]. Three versions of the more recent criteria, the so-called McDonald criteria, named after Pr William Ian McDonald, have been published [23–25]. Table 1.3 summarizes the current criteria published in 2011 [25].

These criteria are based on clinical and imaging evidence, the latter being obtained by magnetic resonance imaging (MRI) of the brain and the spinal cord. In relapsing-onset MS, the main challenge is to differentiate RRMS from CIS without dissemination in time as discussed above (Table 1.1). In PPMS, the main differential diagnoses are other causes of progressive myelopathy, in particular hereditary

Table 1.3 Multiple sclerosis diagnostic criteria [25]

Clinical presentation	Clinical evidence	Information needed for MS diagnosis
At least two attacks[a]	Objective clinical evidence of at least 2 lesions	None
		In these cases, MRI is not required for time dissemination but is necessary for differential diagnosis
At least two attacks[a]	Objective clinical evidence of 1 lesion and reasonable[b] historical evidence of a prior attack without other explanation	None
		In these cases, MRI is not required for time dissemination but is necessary for differential diagnosis
At least two attacks[a]	Objective clinical evidence of 1 lesion	Dissemination in space demonstrated by MRI[c]
One attack[a]	Objective clinical evidence of at least 2 lesions	Dissemination in time demonstrated by MRI[c]
One attack[a]	Objective clinical evidence of 1 lesion	Dissemination in space and time demonstrated by MRI[c]
Insidious neurological progression suggestive of MS (PPMS)	One year of disease progression (retrospectively or prospectively determined)	2 of 3 of the following criteria:
		1. Dissemination in space demonstrated by brain MRI[d]
		2. Dissemination in space demonstrated by spinal cord MRI (at least two lesions)
		3. Positive CSF[e]

MS multiple sclerosis, *MRI* magnetic resonance imaging, *PPMS* primary progressive multiple sclerosis, *CSF* cerebrospinal fluid

Notes

[a]Criteria have been proposed for patients presenting with a typical clinically isolated syndrome, including optic neuritis, myelitis, and brainstem syndromes with duration of at least 24 h, in the absence of fever or infection

[b]Some historical events with symptoms and evolution characteristic for MS, but for which no objective neurological findings are documented, can provide reasonable evidence of a prior demyelinating event, but clinical diagnosis based on objective clinical findings for 2 attacks is most secure

[c]Dissemination in space on MRI requires at least one T2 lesion in at least 2 of 4 MS-typical regions of the central nervous system (periventricular, juxtacortical, infratentorial, or spinal cord), but symptomatic lesions are excluded from consideration in subjects with brainstem or spinal cord syndromes; dissemination in time demonstrated by MRI requires simultaneous presence of asymptomatic gadolinium-enhancing and non-enhancing lesions at any time or a new T2 and/or gadolinium-enhancing lesion(s) on follow-up MRI, irrespective of its timing with reference to a baseline scan

[d]Dissemination in space demonstrated by brain MRI for PPMS requires at least one T2 lesion in the MS-characteristic (periventricular, juxtacortical, or infratentorial) regions, but symptomatic lesions are excluded from consideration in subjects with brainstem syndromes

[e]Positive CSF means isoelectric focusing evidence of oligoclonal bands and/or elevated immunoglobulin G index

spastic paraplegia and, in rare instances, chronic progressive inflammatory myelopathies which are clinically similar to PPMS and associated with evidence of inflammation on cerebrospinal fluid examination or MRI but without clinical or imaging evidence of dissemination in space [6].

Other Syndromes Associated with MS

The concept of radiological isolated syndromes (RIS) has emerged recently [26, 27]. Patients with subclinical lesions discovered on brain or cord MRI that fulfill the criteria for inflammatory demyelinating lesions suggestive of MS but with the primary reason for the acquired MRI resulting from an evaluation of a condition other than MS could be diagnosed as RIS. It is important to differentiate the lesions from nonspecific white matter hyperintense abnormalities associated with small-vessel pathology (microangiopathy), which are very frequent in patients with high blood pressure and migraine for instance. About 34 % of patients diagnosed with RIS developed neurological clinical signs suggestive of a CIS or, rarely, of PPMS, within a 5-year period from the first brain MRI study [27].

Several variants of MS have been described, including acute MS (Marburg type), Baló concentric sclerosis, and Schilder's disease (diffuse sclerosis). These conditions are very rare, and there are still nosological debates about them. Marburg MS is characterized by rapid progression of a severe demyelinating disease leading usually to death within a short period of time (a few months or a few years) and associated with destructive lesions, sometimes very similar to those found in other MS cases, but sometimes with very extensive widespread areas of demyelination [28]. These latter cases are similar to what have been described as Schilder's disease [29], a rare variant of MS predominating in children and characterized by focal neurological abnormalities, which are atypical for MS, in conjunction with tumor-like white matter lesions on MRI [29]. Peripheral demyelination could be present in acute MS. Baló concentric sclerosis lesions are characterized by alternated rims of demyelination and myelin preservation [30]. The first described cases were observed at autopsy on brain of patients who died from acute or subacute diseases, but concentric lesions have been observed on MRI of patients with typical MS. It is considered nowadays that about 50 % of Baló cases seen on MRI have typical MS lesions, and typical relapses may occur [30]. The typical syndrome is characteristic of intracerebral mass lesions including headache, cognitive impairment, seizures, aphasia, and hemiparesis [30].

Neuromyelitis Optica (NMO) Spectrum

The term neuromyelitis optica was coined by Devic and Gault in 1894 and refers to the co-occurrence of ON and myelitis [31]. NMO was regarded for many years as a clinical variant of MS and has only recently been individualized, when a highly specific antibody has been discovered [32]. NMO is a rare disease. Its prevalence is estimated to range from less than 1 to 4.4/100.000 [33]. The age at onset is usually during the fourth decade, but onset during childhood or after 60 is possible [34, 35]. Typically, NMO is characterized by episodes of severe ON, frequently bilaterally at the same time or during subsequent relapses, and severe longitudinal extensive transverse myelitis (LETM), leading to severe disability in a few years (blindness and tetraplegia) in the absence of treatment [31, 33–35]. Since the demonstration of the presence of anti-aquaporin-4 antibodies (AQP4 IgG) in the majority of cases of NMO and the absence of this antibody in MS, the diagnostic criteria of NMO has evolved. According to the 2006 criteria, a definite diagnosis can be made when ON, myelitis, and at least two of three supportive criteria (MRI evidence of a contiguous spinal cord lesion in ≥3 segments, brain MRI not diagnostic of multiple sclerosis, and AQP4-IgG seropositivity) are present [36]. The description of new syndromes associated with AQP4 IgG positivity has led to the concept of NMO spectrum disorder (NMOSD) [37]. New diagnostic criteria for NMOSD have been proposed and presented in 2014 at the American Academy of Neurology [38] which include six different core characteristics: ON, acute myelitis, area postrema syndrome (nausea, vomiting, and hiccups), other brainstem syndromes, symptomatic narcolepsy or acute diencephalic syndrome with MRI findings, and symptomatic cerebral syndrome with MRI findings. AQP4 IgG-positive patients need to show at least one of these core characteristics, with no other better explanation for their symptoms. AQP4 IgG-negative patients need to show at least two of the core characteristics, meeting the following requirements: at least one of the core symptoms must be ON, myelitis, or area postrema syndrome; the core characteristics must be disseminated in space and the MRI findings must distinguish NMOSD from MS or other demyelinating disorders.

Patients presenting with a syndrome typical of NMO, such as LETM or severe bilateral ON, and with positive AQP4 IgG but without dissemination in space and time are considered as having a high-risk syndrome [39]. Relapses could occur in patients with high-risk syndromes (recurrent severe ON, recurrent LETM). An immune-mediated optic neuropathy considered distinct from NMO and MS characterized by a recurrent or chronic unilateral or bilateral vision loss has been described under the acronym CRION (chronic relapsing inflammatory optic neuritis) [40]. Steroid responsiveness with a risk of a relapse on withdrawal of steroids is considered as a key diagnostic criterion [40].

Clinical and spinal cord imaging characteristics of patients presenting with LETM associated with NMO and patients presenting an isolated episode of LETM are similar. They are usually characterized by a severe clinical dysfunction attributable to the spinal cord with usually bilateral sensory, motor, and bladder dysfunction, reaching maximal deficit between 4 h and 21 days, with evidence of inflammation on CSF or MRI [39]. Other causes of LETM are idiopathic transverse myelitis and secondary transverse myelitis (caused by infections or associated with a connective tissue disease like systemic lupus erythematosus or Sjögren's syndrome). Acute idiopathic transverse myelitis is characterized by a typical transverse myelitis (LETM) seronegative for AQP4 IgG and without evidence of other cause of myelopathy (vascular, postradiation, MS, connective tissue disease, and NMO) [41].

Acute Disseminated Encephalomyelitis (ADEM)

ADEM is defined as a monophasic ("acute"), multifocal (dissemination of space of the lesions), inflammatory demyelinating disease [15]. ADEM is frequently secondary to infectious events or vaccinations. The absence of dissemination in time is one the main differences with MS, but the clinical characteristics (encephalitic presentation) and the MRI features could help to distinguish the two entities [15, 42, 43]. Indeed, in most cases, ADEM is monophasic, but in some patients, relapses may occur immediately after the onset of the disease. If these relapses are considered to represent part of the same acute immune process, with similar symptoms to those at onset (encephalopathic episodes), the term multiphasic disseminated encephalomyelitis (MDEM) can be used [42, 43]. However, in about 30 % of cases, a typical first episode of ADEM could subsequently evolve to typical MS, with a clear demonstration of dissemination in space and time [43]. In these cases, the relapses are clinically similar to those seen in typical MS and different from the inaugural episode. ADEM is more frequent in children but could be a mode of onset of pediatric MS [42, 44]. On a pathological point of view, ADEM can be distinguished from MS by the presence of sparse demyelination restricted to narrow perivenous sleeves widespread throughout the nervous system, giving rise to large and diffuse or multifocal lesions in MRI [45]. These features differentiate ADEM from MS where inflammation is associated with focal confluent plaques of primary demyelination showing variable degrees of axonal injury and loss. Diagnostic criteria have been proposed to distinguish ADEM from MS in adults and children [42, 43]. In adults, "ADEM corresponds to patients with at least 2 of the following 3 criteria: (1) Clinical atypical symptoms of MS. One or more of the following: consciousness alteration, hypersomnia, seizures, cognitive impairment, hemiplegia, tetraplegia, aphasia, or bilateral optic neuritis; (2) absence of oligoclonal bands in the cerebrospinal fluid; (3) MRI: Grey matter involvement (basal ganglia or cortical lesions)" [43]. Diagnostic criteria for pediatric ADEM have been recently updated [42] in which all the following items are required: "(1) a first polyfocal, clinical CNS event with presumed inflammatory demyelinating cause; (2) encephalopathy that cannot be explained by fever; (3) no new clinical and MRI findings emerge 3 months or more after the onset; (4) brain MRI is abnormal during the acute (3-month) phase; (5) typically on brain MRI: diffuse, poorly demarcated, large (>1–2 cm) lesions involving predominantly the cerebral white matter; T1 hypointense lesions in the white matter are rare; deep grey matter lesions (e.g. thalamus or basal ganglia) can be present."

A rare severe variant has been described by Hurst with perivascular hemorrhages and severe brain edema (acute hemorrhagic leukoencephalitis) [46].

Other Inflammatory Disorders of the CNS

Some other inflammatory diseases of the CNS exist and have to be taken into account for the differential diagnosis of MS including infectious diseases, paraneoplastic disorders, and other autoimmune encephalitis. The latter will be detailed in

Chap. 6 of this book. A new syndrome has been recently described with chronic lymphocytic inflammation with pontine perivascular enhancement responsive to steroids (CLIPPERS) [47]. It is characterized by episodic brainstem symptoms, characteristic punctuate and curvilinear gadolinium-enhancing lesions peppering the brainstem (mainly in the pons) on MRI, responsiveness to steroids, and T-lymphocytic infiltrate with perivascular predominance in brain biopsies.

References

1. World Health Organization. Atlas multiple sclerosis resources in the world 2008. Geneva: WHO Press; 2008.
2. Kingwell E, Marriott JJ, Jetté N, Pringsheim T, Makhani N, Morrow SA, et al. Incidence and prevalence of multiple sclerosis in Europe: a systematic review. BMC Neurol. 2013;13:128.
3. Yeh EA, Chitnis T, Krupp L, Ness J, Chabas D, Kuntz N, et al. Pediatric multiple sclerosis. Nat Rev Neurol. 2009;5:621–31.
4. Simpson Jr S, Blizzard L, Otahal P, Van der Mei I, Taylor B. Latitude is significantly associated with the prevalence of multiple sclerosis: a meta-analysis. J Neurol Neurosurg Psychiatry. 2011;82:1132–41.
5. Confavreux C, Vukusic S, Moreau T, Adeleine P. Relapses and progression of disability in multiple sclerosis. N Engl J Med. 2000;343:1430–8.
6. Thompson AJ, Polman CH, Miller DH, McDonald WI, Brochet B, Filippi M, et al. Primary progressive multiple sclerosis. Brain. 1997;120:1085–96.
7. Tremlett H, Yinshan Zhao, Devonshire V. Natural history of secondary-progressive multiple sclerosis. Mult Scler. 2008;14:314–24.
8. Koch M, Kingwell E, Rieckmann P, Tremlett H, UBC MS Clinic Neurologists. The natural history of secondary progressive multiple sclerosis. J Neurol Neurosurg Psychiatry. 2010;81:1039–43.
9. Lublin FD, Reingold SC. Defining the clinical course of multiple sclerosis: results of an international survey. National Multiple Sclerosis Society (USA) Advisory Committee on Clinical Trials of New Agents in Multiple Sclerosis. Neurology. 1996;46:907–11.
10. Lublin FD, Reingold SC, Cohen JA, Cutter GR, Sørensen PS, Thompson AJ, et al. Defining the clinical course of multiple sclerosis: the 2013 revisions. Neurology. 2014;83:278–86.
11. Trojano M, Lucchese G, Graziano G, Taylor BV, Simpson Jr S, Lepore V, et al. Geographical variations in sex ratio trends over time in multiple sclerosis. PLoS One. 2012;7(10), e48078.
12. Debouverie M. Gender as a prognostic factor and its impact on the incidence of multiple sclerosis in Lorraine, France. J Neurol Sci. 2009;286:14–7.
13. Brownlee WJ, Miller DH. Clinically isolated syndromes and the relationship to multiple sclerosis. J Clin Neurosci. 2014. doi:10.1016/j.jocn.2014.02.026. [Epub ahead of print].
14. Miller DH, Chard DT, Ciccarelli O. Clinically isolated syndromes. Lancet Neurol. 2012;11:157–69.
15. Palace J. Acute disseminated encephalomyelitis and its place amongst other acute inflammatory demyelinating CNS disorders. J Neurol Sci. 2011;306:188–91.
16. Leray E, Yaouanq J, Le Page E, Coustans M, Laplaud D, Oger J, et al. Evidence for a two-stage disability progression in multiple sclerosis. Brain. 2010;133:1900–13.
17. Scalfari A, Neuhaus A, Daumer M, Muraro PA, Ebers GC. Onset of secondary progressive phase and long-term evolution of multiple sclerosis. J Neurol Neurosurg Psychiatry. 2014;85:67–75.
18. Harding KE, Wardle M, Moore P, Tomassini V, Pickersgill T, Ben-Shlomo Y, et al. Modelling the natural history of primary progressive multiple sclerosis. J Neurol Neurosurg Psychiatry. 2014. doi:10.1136/jnnp-2014-307791. [Epub ahead of print].

19. Brochet B, Deloire MS, Ouallet JC, Salort E, Bonnet M, Jové J, et al. Pain and quality of life in the early stages after multiple sclerosis diagnosis: a 2-year longitudinal study. Clin J Pain. 2009;25:211–7.
20. Krupp LB, Serafin DJ, Christodoulou C. Multiple sclerosis-associated fatigue. Expert Rev Neurother. 2010;10:1437–47.
21. McCombe PA, Gordon TP, Jackson MW. Bladder dysfunction in multiple sclerosis. Expert Rev Neurother. 2009;9:331–40.
22. Schumacher G, Beebe G, Kibler R, Kurland L, Kurtzke J, McDowell F. Problems of experimental trials of therapy in multiple sclerosis. Ann N Y Acad Sci. 1965;122:552–68.
23. McDonald WI, Compston A, Edan G, Goodkin D, Hartung HP, Lublin FD, et al. Recommended diagnostic criteria for multiple sclerosis: guidelines from the International Panel on the diagnosis of multiple sclerosis. Ann Neurol. 2001;50:121–7.
24. Polman CH, Reingold SC, Edan G, Filippi M, Hartung HP, Kappos L, et al. Diagnostic criteria for multiple sclerosis: 2005 revisions to the "McDonald Criteria". Ann Neurol. 2005;58:840–6.
25. Polman CH, Reingold SC, Banwell B, Clanet M, Cohen JA, Filippi M, et al. Diagnostic criteria for multiple sclerosis: 2010 revisions to the McDonald criteria. Ann Neurol. 2011;69:292–302.
26. Lebrun C, Bensa C, Debouverie M, De Seze J, Wiertlievski S, Brochet B, et al. Unexpected multiple sclerosis: follow-up of 30 patients with magnetic resonance imaging and clinical conversion profile. J Neurol Neurosurg Psychiatry. 2008;79:195–8.
27. Okuda DT, Siva A, Kantarci O, Inglese M, Katz I. Tutuncu M, et al; Radiologically isolated syndrome: 5-year risk for an initial clinical event. PLoS One. 2014;9, e90509.
28. Suzuki M, Kawasaki H, Masaki K, Suzuki SO, Terada T, Tsuchida T, et al. An autopsy case of the Marburg variant of multiple sclerosis (acute multiple sclerosis). Intern Med. 2013;52:1825–32.
29. Kraus D, Konen O, Straussberg R. Schilder's disease: non-invasive diagnosis and successful treatment with human immunoglobulins. Eur J Paediatr Neurol. 2012;16:206–8.
30. Hardy TA, Miller DH. Baló's concentric sclerosis. Lancet Neurol. 2014;13:740–6.
31. Cabre P, Bonnan M, Olindo S, Brochet B, Samdja D. Neuromyélites optiques. Encclopédie Médico-Chirurgicale. 2007:17-066-A-57:1-13.
32. Lennon VA, Wingerchuk DM, Kryzer TJ, Pittock SJ, Lucchinetti CF, Fujihara K, et al. A serum autoantibody marker of neuromyelitis optica: distinction from multiple sclerosis. Lancet. 2004;364:2106–12.
33. Jacob A, McKeon A, Nakashima I, Sato DK, Elsone L, Fujihara K, et al. Current concept of neuromyelitis optica (NMO) and NMO spectrum disorders. J Neurol Neurosurg Psychiatry. 2013;84:922–30.
34. Collongues N, Marignier R, Zéphir H, Papeix C, Fontaine B, Blanc F, et al. Long-term follow-up of neuromyelitis optica with a pediatric onset. Neurology. 2010;75:1084–8.
35. Collongues N, Marignier R, Zéphir H, Papeix C, Blanc F, Ritleng C, et al. Neuromyelitis optica in France: a multicenter study of 125 patients. Neurology. 2010;74:736–42.
36. Wingerchuk DM, Lennon VA, Pittock SJ, Lucchinetti CF, Weinshenker BG. Revised diagnostic criteria for neuromyelitis optica. Neurology. 2006;66:1485–9.
37. Wingerchuk DM, Lennon VA, Lucchinetti CF, Pittock SJ, Weinshenker BG. The spectrum of neuromyelitis optica. Lancet Neurol. 2007;6:805–15.
38. Wingerchuk DM, Banwell B, Bennett J, Cabre P, Carroll W, Chitnis T, et al. Revised Diagnostic Criteria for Neuromyelitis Optica Spectrum Disorders. Abstract S 63.001, American Academy of Neurology 2014 Meeting.
39. Collongues N, Marignier R, Zéphir H, Blanc F, Vukusic S, Outteryck O, et al. High-risk syndrome for neuromyelitis optica: a descriptive and comparative study. Mult Scler. 2011;17:720–4.
40. Petzold A, Plant GT. Chronic relapsing inflammatory optic neuropathy: a systematic review of 122 cases reported. J Neurol. 2014;261:17–26.

41. Borchers AT, Gershwin ME. Transverse myelitis. Autoimmun Rev. 2012;11:231–48.
42. Krupp LB, Tardieu M, Amato MP, Banwell B, Chitnis T, Dale RC, et al. International Pediatric Multiple Sclerosis Study Group criteria for pediatric multiple sclerosis and immune-mediated central nervous system demyelinating disorders: revisions to the 2007 definitions. Mult Scler. 2013;19:1261–7.
43. de Seze J, Debouverie M, Zephir H, Lebrun C, Blanc F, Bourg V, et al. Acute fulminant demyelinating disease: a descriptive study of 60 patients. Arch Neurol. 2007;64:1426–32.
44. Ketelslegers IA, Neuteboom RF, Boon M, Catsman-Berrevoets CE, Hintzen RQ, Dutch Pediatric MS Study Group. A comparison of MRI criteria for diagnosing pediatric ADEM and MS. Neurology. 2010;74:1412–5.
45. Lassmann H. Acute disseminated encephalomyelitis and multiple sclerosis. Brain. 2010;133:317–9.
46. Hurst EW. Acute haemorrhagic leucoencephalitis: a previously undefined entity. Med J Aust. 1941;2:1–6.
47. Taieb G, Duflos C, Renard D, Audoin B, Kaphan E, Pelletier J, et al. Long-term outcomes of CLIPPERS (chronic lymphocytic inflammation with pontine perivascular enhancement responsive to steroids) in a consecutive series of 12 patients. Arch Neurol. 2012;69:847–55.

Chapter 2
Depression and Multiple Sclerosis: Clinical Aspects, Epidemiology, and Management

Marie Théaudin and Anthony Feinstein

Abstract Multiple sclerosis (MS) is associated with a broad array of neuropsychiatric problems of which depression is the commonest. Defining depression can present a potential problem because certain symptoms that underpin the diagnosis of depression may also be caused by multiple sclerosis. Certain self-report scales that take this symptom overlap into account have been validated for MS patients (Beck Fast Screen for Medical Patients and the Hospital Anxiety and Depression Scale). MS-related major depression has a lifetime prevalence of 25–50 %, well above the rate in the general population. Depression is linked to a poor quality of life, potentially greater cognitive impairment, an increase in suicidal ideation, and less compliance with disease-modifying drugs. Notwithstanding the high prevalence of depression in MS and its multiple adverse effects on the MS population, there are only two randomized trials of antidepressant medication (paroxetine and desipramine). Results are modest and side effects can be troubling. Treatment of choice is therefore cognitive behavioral therapy. Mindfulness-based therapy and exercise may also offer benefits to the depressed MS patient.

Keywords Depression • Anxiety • Beck Fast Screen for Medical Patients • Hospital anxiety and depression scale • Epidemiology • Suicide • Quality of life • Cognition • Antidepressant drug • Behavioral cognitive therapy • Mindfulness-based therapy • Exercise

M. Théaudin, MD, Ph.D.
Neurology Department, CHU Bicêtre, 78 rue du Général Leclerc, 94275 Le Kremlin Bicêtre, France
e-mail: marie.theaudin@bct.aphp.fr

A. Feinstein, MD, M.Phil., Ph.D., FRCP (✉)
Department of Psychiatry, University of Toronto and Sunnybrook Health Sciences Centre, 2075 Bayview Avenue, Toronto, ON M4N 3M5, Canada
e-mail: ant.feinstein@utoronto.ca

B. Brochet (ed.), *Neuropsychiatric Symptoms of Inflammatory Demyelinating Diseases*, Neuropsychiatric Symptoms of Neurological Disease,
DOI 10.1007/978-3-319-18464-7_2

Introduction

Multiple sclerosis is associated with a broad array of neuropsychiatric problems of which depression is the commonest. This chapter will describe the clinical features and epidemiology of depression before concluding with a section devoted to treatment. In a disease without cure, symptom management takes on an even greater weight, and this is particularly true for depression for as the chapter will make clear, the effects of low mood can be pervasive and severely debilitating.

Clinical Aspects

Depression is a broad term encompassing, on the one hand, the symptom of sadness and, on the other, the full syndrome diagnosis of major depression. The latter has been defined by the American Psychiatric Association as a collection of nine signs and symptoms of which five or more have to be present for at least a 2-week period in order to achieve the diagnosis. The symptoms include depressed mood for most of the day, a loss of interest or pleasure in activities that were formerly enjoyable, changes in appetite linked to weight loss or weight gain, insomnia or hypersomnia, psychomotor agitation or retardation, fatigue, feelings of worthlessness or inappropriate and excessive guilt, a reduction in concentration, and recurring thoughts of death. For clinicians working with MS patients, this definition can present a potential problem because certain of the symptoms that underpin the diagnosis of depression may also be caused by multiple sclerosis itself. Here, the most frequent overlapping symptoms are those of fatigue, reduced concentration, and difficulties with sleep.

Researchers have attempted to address this symptom overlap in a number of ways. Psychometric, self-report scales have been developed specifically for use in medically unwell patients with the aim of removing the somatic confounders. The best examples of these are the Beck Fast Screen for Medical Patients (BFS) [1] and the Hospital Anxiety and Depression Scale (HADS) [2]. Both of these self-report measures have been validated for use in patients with multiple sclerosis [3, 4]. The advantage of the HADS is that the scale contains an index for anxiety too. Both measures are easy and quick to complete and can be introduced into routine clinical practice without difficulty. They have also been translated into different languages. In addition to these two indices, Mohr and colleagues [5] recommend the two-question approach, i.e., asking patients about pervasively low mood and the inability to enjoy activities, as before. An even briefer approach is the Yale Single Question screen for depression [6]. The brevity of such an approach is attractive, but the sensitivity is understandably on the lower side. A recent critical review from the American Academy of Neurology (AAN) endorses the Beck Depression Inventory (Revised Edition) and the two-question approach mentioned above [7]. Clinicians therefore have considerable choice in selecting a measure, but in doing so should remember that only two scales, namely, the BFS and HADS, have been specifically validated for use in an MS population.

Anxiety

As occurs in the general population, depression in MS patients is often associated with anxiety. The [8, 9] clinical importance of this kind of morbidity should not be overlooked because MS patients who have both anxiety and depression are more likely to have increased thoughts of self-harm, greater somatic complaints, and more extensive social dysfunction than MS patients with depression or anxiety alone [10, 11]. Anxiety as a symptom occurs more frequently than depression as a symptom [10, 12], and rates of generalized anxiety, panic disorder, obsessive-compulsive disorder, and social phobia are all increased significantly in MS patients relative to the general population [13].

Epidemiology

It is estimated that major depression has a lifetime prevalence of 25–50 % [8, 9, 14–16]. These figures generally come from tertiary referral clinics and as such the data may be slightly skewed. That said, community-based data support the elevated prevalence. In a study of 115,071 adult Canadians, the 12-month prevalence of depression in MS subjects was elevated relative both to healthy individuals and those with long-term medical difficulties. The highest rate of depression was found in individuals aged 18–45 years where the 1-year prevalence rate for depression approached 25 % [9]. Additional evidence supporting the frequency of MS-related depression comes from administrative data bases which have the advantage of robust sample sizes [17–19]. Here, the frequency of depression is comparable to that reported in the community and tertiary clinics reported above. It is also important to note that data do not support an increase prevalence of depression prior to the onset of multiple sclerosis [20]. This observation is important for it points to a closer link between neuropathological changes and/or psychosocial factors and a disturbance in mood.

Disease Duration, Disability, and Depression

There is no clear association between the presence of depression and disease-related variables. The relationship with physical disability is equivocal [21, 22]. The same situation pertains to disease duration [8]. The reasons for these mixed findings could be due to the diversity of the disease itself. For example, patients with the same disease duration may have a markedly differently relapse rate or disease course. Moreover, the degree of physical disability may be determined by a combination of cerebral and spinal involvement, each having a potentially different effect on mood. Therefore, the important determinant for mood may be less closely related to the

Expanded Disability Status Scale (EDSS) than how the individual adjusts to adversity and the adaptive strategies he or she uses.

The Clinical Significance of Depression

Depression is linked to a poor quality of life, greater cognitive impairment, an increase in suicidal ideation, and less compliance with disease-modifying drugs (DMDs). Each of these will be discussed in turn.

Quality of Life Depression in MS is associated with decreased quality of life whatever the neurological or functional impairments associated with MS [23]. Carta et al. [24] revealed that patients with MS and a co-morbid lifetime diagnosis of a mood disorder had significantly lower scores on the SF-12 (a measure of quality of life) than MS patients with no history of mood disturbance. Significantly, MS patients with a past history of depression, whatever their mood status at the time of that evaluation, endorsed significantly lower MSQOL-54 scores relating to energy, mental health, cognitive function, general quality of life, and sexual function [25].

Depression and Cognition Approximately 40–70 % of MS patients will have cognitive dysfunction depending on the disease type. Evidence now suggests that clinically significant depression may lead to a further deterioration in a patient's cognitive abilities. Works from Arnett and colleagues have shown that depression can impair working memory, in particular the executive component of this [26–28]. These findings raise the intriguing possibility that successfully treating depression could, in theory, lead to a concomitant improvement in an MS patient's cognitive ability. To date, no specific study has explored this possibility.

Adherence to Disease-Modifying Drugs (DMDs) A number of studies have connected depression with poor compliance with respect to DMDs [29–31]. Bruce et al. [29] showed that MS patients with a current mood or anxiety disorder are almost five times less likely than MS patients with no psychiatric diagnosis to adhere to disease-modifying therapy. Significantly, treating depressed MS patients for at least 6 months with antidepressant medication has been associated with better compliance with DMDs.

Suicidal Risks One in three MS patients will entertain thoughts of suicide [32]. The predictors here are the presence of a major depression, the severity of the depression, social isolation, and concomitant alcohol abuse [33]. Suicidal intent is also a risk factor for a suicide attempt. Epidemiological data from Scandinavia reveal that MS patients are twice as likely to commit suicide as individuals in the general population [34–36]. The figure from British Columbia in Canada is significantly higher than this with a 7.5 increase documented [37]. The data also suggest that males within the first 5 years of diagnosis may be at a particularly high risk for suicide [36, 38].

Management of Depression

It comes as a surprise to find that notwithstanding the high prevalence of depression in MS and the multiple adverse effects of depression on the MS population, there are only two controlled treatment studies of an antidepressant medication that are considered methodically robust by the Cochrane review committee [39]. The first of these studies involves the old tricycle drug desipramine which was found to be partly effectively in treating depression. However, treatment was linked to troubling anticholinergic side effects for patients, such as dry mouth, sedation, and constipation among others, thereby, in some cases, limiting the attainment of a therapeutic dose [40]. While not mentioned in the report, one of the disadvantages of a drug like desipramine is that it is potentially fatal in overdose. This point is important given the frequency with which MS patients think of suicide and the high completed suicide rate of this population. This concern does not pertain to the second drug that is mentioned in the Cochrane review, namely, paroxetine, a selective serotonin reuptake inhibitor (SSRI) [41]. Paroxetine, like desipramine, was found to be modestly effective in helping mood, but the drug is also with side effects, most notably sexual dysfunction. This has introduced challenges with treatment compliance. Should sexual side effects lead to treatment discontinuation, clinicians may wish to consider two other antidepressant medications, namely, bupropion and mirtazapine, both of which spare sexual function. Neither have, however, been assessed in an MS-related RCT. Here it is germane to note that the recent AAN critical review article concluded that there are no sufficient data at present to endorse the use of antidepressant medication for MS-related depression [7]. While once cannot refute the AAN's rigorous review process in arriving at this conclusion, it is important for clinicians not to lose sight to the fact that antidepressant medication can bring about some symptom relief. Should there be no recourse to psychotherapies, and this is a reality faced by many practitioners in a smaller health care setting, rather than admit to therapeutic defeat, a trial of an SSRI is warranted. Here, the old neuropsychiatric dictum of start low and go slow with dosing applies. Drug management is presented in more detail in Chap. 8.

More promising data have been reported with certain psychotherapies. Cognitive behavioral therapy (CBT) has emerged as a treatment of choice for MS-related depression [42–44] and is endorsed by both the Cochrane review committee [45] and the AAN [7]. Moreover, CBT may be effectively given over the telephone to MS patients [46], an important observation given that mobility issues can make it difficult for some patients to attend clinic. CBT of course does not come with the troubling side effects of sexual dysfunction, dry mouth, and weight gain that can bedevil the use of an SSRI, but in many centers CBT might not be available. A recommendation from the Goldman consensus panel [47] was that in a situation such as this, a neurologist should treat the depressed MS patient with medication. No specific drug was endorsed.

Other treatments reportedly effective in helping MS-related depression are mindfulness-based therapy [48, 49] and exercise [50]. In relation to the latter, a number of studies have been undertaken with exercise as a secondary outcome variable. The definitive study is therefore awaited, but preliminary evidence suggests that exercise may not only elevate depressed mood; it may also lead to improvement in certain cognitive difficulties as well [51]. Finally, an intriguing observation has recently emerged with respect to stress management therapy (SMT). A randomized controlled treatment study over 24 weeks revealed that SMT was effective in reducing cumulative new T2 and contrast-enhancing lesion burdens relative to MS patients who had not received the therapy [52]. Unfortunately, these improvements in brain MRI metrics were not accompanied by benefits with respect to the patients' mood, this point underscoring the complex relationship between brain MRI changes and depression.

Finally, any treatment recommendation would be incomplete without brief mention of electroconvulsive therapy, reserved for severe depression often medication refractory or associated with intense suicidal intent where time is of the essence in providing symptom relief. The treatment is generally well tolerated in MS patients although the literature here is small [53].

Conclusion

There is inconvertible evidence linking clinically significant depression in people with MS to a multiplicity of negative effects with respect to activities of daily living. It is therefore imperative that clinicians from many disciplines who treat MS patients not miss the diagnosis. This point is further underlined by studies that demonstrate the effectiveness of treatments for depression in this population. Not only will successful treatment reduce the morbidity associated with MS; it holds out the promise of also lessening suicide-related mortality.

References

1. Beck AT, Steer RA, Brown GK. Manual for the beck depression inventory-II. San Antonio: Psychological Corporation; 1996.
2. Zigmond AS, Snaith RP. The hospital anxiety and depression scale. Acta Psychiatr Scand. 1983;67(6):361–70.
3. Benedict RHB, Fishman I, McClellan MM, Bakshi R, Weinstock-Guttman B. Validity of the Beck Depression Inventory-Fast Screen in multiple sclerosis. Mult Scler Houndmills Basingstoke Engl. 2003;9(4):393–6.
4. Honarmand K, Feinstein A. Validation of the Hospital Anxiety and Depression Scale for use with multiple sclerosis patients. Mult Scler Houndmills Basingstoke Engl. 2009;15(12):1518–24.
5. Mohr DC, Hart SL, Julian L, Tasch ES. Screening for depression among patients with multiple sclerosis: two questions may be enough. Mult Scler Houndmills Basingstoke Engl. 2007;13(2):215–9.

6. Avasarala JR, Cross AH, Trinkaus K. Comparative assessment of Yale Single Question and Beck Depression Inventory Scale in screening for depression in multiple sclerosis. Mult Scler Houndmills Basingstoke Engl. 2003;9(3):307–10.
7. Minden SL, Feinstein A, Kalb RC, Miller D, Mohr DC, Patten SB, et al. Evidence-based guideline: assessment and management of psychiatric disorders in individuals with MS: report of the Guideline Development Subcommittee of the American Academy of Neurology. Neurology. 2014;82(2):174–81.
8. Chwastiak L, Ehde DM, Gibbons LE, Sullivan M, Bowen JD, Kraft GH. Depressive symptoms and severity of illness in multiple sclerosis: epidemiologic study of a large community sample. Am J Psychiatry. 2002;159(11):1862–8.
9. Patten SB, Beck CA, Williams JVA, Barbui C, Metz LM. Major depression in multiple sclerosis: a population-based perspective. Neurology. 2003;61(11):1524–7.
10. Feinstein A, O'Connor P, Gray T, Feinstein K. The effects of anxiety on psychiatric morbidity in patients with multiple sclerosis. Mult Scler Houndmills Basingstoke Engl. 1999;5(5):323–6.
11. José SM. Psychological aspects of multiple sclerosis. Clin Neurol Neurosurg. 2008;110(9):868–77.
12. Dahl O-P, Stordal E, Lydersen S, Midgard R. Anxiety and depression in multiple sclerosis. A comparative population-based study in Nord-Trøndelag County, Norway. Mult Scler Houndmills Basingstoke Engl. 2009;15(12):1495–501.
13. Korostil M, Feinstein A. Anxiety disorders and their clinical correlates in multiple sclerosis patients. Mult Scler Houndmills Basingstoke Engl. 2007;13(1):67–72.
14. Sadovnick AD, Remick RA, Allen J, Swartz E, Yee IM, Eisen K, et al. Depression and multiple sclerosis. Neurology. 1996;46(3):628–32.
15. Siegert RJ, Abernethy DA. Depression in multiple sclerosis: a review. J Neurol Neurosurg Psychiatry. 2005;76(4):469–75.
16. Feinstein A. Multiple sclerosis and depression. Mult Scler Houndmills Basingstoke Engl. 2011;17(11):1276–81.
17. Jones KH, Ford DV, Jones PA, John A, Middleton RM, Lockhart-Jones H, et al. A large-scale study of anxiety and depression in people with multiple sclerosis: a survey via the web portal of the UK MS register. PLoS One. 2012;7(7), e41910.
18. Marrie RA, Horwitz R, Cutter G, Tyry T, Campagnolo D, Vollmer T. The burden of mental comorbidity in multiple sclerosis: frequent, underdiagnosed, and undertreated. Mult Scler Houndmills Basingstoke Engl. 2009;15(3):385–92.
19. Marrie RA, Fisk JD, Yu BN, Leung S, Elliott L, Caetano P, et al. Mental comorbidity and multiple sclerosis: validating administrative data to support population-based surveillance. BMC Neurol. 2013;13:16.
20. Minden SL, Orav J, Reich P. Depression in multiple sclerosis. Gen Hosp Psychiatry. 1987;9(6):426–34.
21. Patten SB, Lavorato DH, Metz LM. Clinical correlates of CES-D depressive symptom ratings in an MS population. Gen Hosp Psychiatry. 2005;27(6):439–45.
22. Ron MA, Logsdail SJ. Psychiatric morbidity in multiple sclerosis: a clinical and MRI study. Psychol Med. 1989;19(4):887–95.
23. D'Alisa S, Miscio G, Baudo S, Simone A, Tesio L, Mauro A. Depression is the main determinant of quality of life in multiple sclerosis: a classification-regression (CART) study. Disabil Rehabil. 2006;28(5):307–14.
24. Carta MG, Moro MF, Lorefice L, Picardi A, Trincas G, Fenu G, et al. Multiple sclerosis and bipolar disorders: the burden of comorbidity and its consequences on quality of life. J Affect Disord. 2014;167:192–7.
25. Wang JL, Reimer MA, Metz LM, Patten SB. Major depression and quality of life in individuals with multiple sclerosis. Int J Psychiatry Med. 2000;30(4):309–17.
26. Arnett PA, Higginson CI, Voss WD, Wright B, Bender WI, Wurst JM, et al. Depressed mood in multiple sclerosis: relationship to capacity-demanding memory and attentional functioning. Neuropsychology. 1999;13(3):434–46.

27. Arnett PA, Higginson CI, Randolph JJ. Depression in multiple sclerosis: relationship to planning ability. J Int Neuropsychol Soc. 2001;7(6):665–74.
28. Arnett PA, Higginson CI, Voss WD, Bender WI, Wurst JM, Tippin JM. Depression in multiple sclerosis: relationship to working memory capacity. Neuropsychology. 1999;13(4):546–56.
29. Bruce JM, Hancock LM, Arnett P, Lynch S. Treatment adherence in multiple sclerosis: association with emotional status, personality, and cognition. J Behav Med. 2010;33(3):219–27.
30. Mohr DC, Goodkin DE, Likosky W, Gatto N, Baumann KA, Rudick RA. Treatment of depression improves adherence to interferon beta-1b therapy for multiple sclerosis. Arch Neurol. 1997;54(5):531–3.
31. Tarrants M, Oleen-Burkey M, Castelli-Haley J, Lage MJ. The impact of comorbid depression on adherence to therapy for multiple sclerosis. Mult Scler Int. 2011;2011:271321.
32. Pompili M, Forte A, Palermo M, Stefani H, Lamis DA, Serafini G, et al. Suicide risk in multiple sclerosis: a systematic review of current literature. J Psychosom Res. 2012;73(6):411–7.
33. Feinstein A. An examination of suicidal intent in patients with multiple sclerosis. Neurology. 2002;59(5):674–8.
34. Brønnum-Hansen H, Stenager E, Nylev Stenager E, Koch-Henriksen N. Suicide among Danes with multiple sclerosis. J Neurol Neurosurg Psychiatry. 2005;76(10):1457–9.
35. Brønnum-Hansen H, Koch-Henriksen N, Stenager E. Trends in survival and cause of death in Danish patients with multiple sclerosis. Brain J Neurol. 2004;127(Pt 4):844–50.
36. Fredrikson S, Cheng Q, Jiang G-X, Wasserman D. Elevated suicide risk among patients with multiple sclerosis in Sweden. Neuroepidemiology. 2003;22(2):146–52.
37. Sadovnick AD, Eisen K, Ebers GC, Paty DW. Cause of death in patients attending multiple sclerosis clinics. Neurology. 1991;41(8):1193–6.
38. Stenager EN, Stenager E, Koch-Henriksen N, Brønnum-Hansen H, Hyllested K, Jensen K, et al. Suicide and multiple sclerosis: an epidemiological investigation. J Neurol Neurosurg Psychiatry. 1992;55(7):542–5.
39. Koch MW, Glazenborg A, Uyttenboogaart M, Mostert J, De Keyser J. Pharmacologic treatment of depression in multiple sclerosis. Cochrane Database Syst Rev. 2011;2, CD007295.
40. Schiffer RB, Wineman NM. Antidepressant pharmacotherapy of depression associated with multiple sclerosis. Am J Psychiatry. 1990;147(11):1493–7.
41. Ehde DM, Kraft GH, Chwastiak L, Sullivan MD, Gibbons LE, Bombardier CH, et al. Efficacy of paroxetine in treating major depressive disorder in persons with multiple sclerosis. Gen Hosp Psychiatry. 2008;30(1):40–8.
42. Firth N. Effectiveness of psychologically focused group interventions for multiple sclerosis: a review of the experimental literature. J Health Psychol. 2013;19(6):789–801.
43. Foley FW, Bedell JR, LaRocca NG, Scheinberg LC, Reznikoff M. Efficacy of stress-inoculation training in coping with multiple sclerosis. J Consult Clin Psychol. 1987;55(6):919–22.
44. Mohr DC, Boudewyn AC, Goodkin DE, Bostrom A, Epstein L. Comparative outcomes for individual cognitive-behavior therapy, supportive-expressive group psychotherapy, and sertraline for the treatment of depression in multiple sclerosis. J Consult Clin Psychol. 2001;69(6):942–9.
45. Thomas PW, Thomas S, Hillier C, Galvin K, Baker R. Psychological interventions for multiple sclerosis. Cochrane Database Syst Rev. 2006;1, CD004431.
46. Mohr DC, Likosky W, Bertagnolli A, Goodkin DE, Van Der Wende J, Dwyer P, et al. Telephone-administered cognitive-behavioral therapy for the treatment of depressive symptoms in multiple sclerosis. J Consult Clin Psychol. 2000;68(2):356–61.
47. Goldman Consensus Group. The Goldman Consensus statement on depression in multiple sclerosis. Mult Scler Houndmills Basingstoke Engl. 2005;11(3):328–37.
48. Burschka JM, Keune PM, Oy UH, Oschmann P, Kuhn P. Mindfulness-based interventions in multiple sclerosis: beneficial effects of Tai Chi on balance, coordination, fatigue and depression. BMC Neurol. 2014;14:165.
49. Grossman P, Kappos L, Gensicke H, D'Souza M, Mohr DC, Penner IK, et al. MS quality of life, depression, and fatigue improve after mindfulness training: a randomized trial. Neurology. 2010;75(13):1141–9.

50. Ensari I, Motl RW, Pilutti LA. Exercise training improves depressive symptoms in people with multiple sclerosis: results of a meta-analysis. J Psychosom Res. 2014;76(6):465–71.
51. Beier M, Bombardier CH, Hartoonian N, Motl RW, Kraft GH. Improved physical fitness correlates with improved cognition in multiple sclerosis. Arch Phys Med Rehabil. 2014;95(7):1328–34.
52. Burns MN, Nawacki E, Kwasny MJ, Pelletier D, Mohr DC. Do positive or negative stressful events predict the development of new brain lesions in people with multiple sclerosis? Psychol Med. 2014;44(2):349–59.
53. Rasmussen KG, Keegan BM. Electroconvulsive therapy in patients with multiple sclerosis. J ECT. 2007;23(3):179–80.

Chapter 3
Depression and Multiple Sclerosis: Imaging, Mechanisms

Kristoffer Romero and Anthony Feinstein

Abstract Advances in neuroimaging over the past three decades have substantially improved our understanding of multiple sclerosis (MS) and its effects on the brain. Lesions in white matter and gray matter, as well as atrophy of normal appearing brain tissue, are frequent and varied across MS patients and correlate with neuropsychological performance and various measures of disability. Studies of depression in MS suggest prefrontal gray and white matter atrophy is correlated with depressive symptoms, although which structural and/or functional metrics are most sensitive to depression, and how these regions theoretically contribute to the development or maintenance of disturbed mood, is not yet clear. In this chapter, we review the major structural and functional changes associated with major depressive disorder in the general population and the existing work on the neural correlates of depression in MS. We also review major themes regarding neurobiological theories of depression and whether they apply to MS, particularly in terms of decreased executive control of affect due to degradation of prefrontal regions.

Keywords Depression • Multiple sclerosis • Magnetic resonance imaging • Diffusion tensor imaging • Functional magnetic resonance imaging • Resting state • Hippocampus • Prefrontal cortex • Executive control • Default-mode network

Depression and MS: Imaging, Mechanisms

In terms of clinical presentation, depression is the most common psychiatric complaint in patients with multiple sclerosis (MS), affecting between 25 and 50 % of the patient population over the course of the illness, which is between two and five times higher than the prevalence rate in the general population. The etiology of depression in MS patients is only now becoming clearer with pathophysiological changes in the brain, and psychosocial variables likely play a role. Despite an

K. Romero, Ph.D. • A. Feinstein, MD, M.Phil., Ph.D., FRCP (✉)
Department of Psychiatry, University of Toronto and Sunnybrook Health Sciences Centre, Room FG52, 2075 Bayview Ave, Toronto, ON M4N 3M5, Canada
e-mail: kris.romero@utoronto.ca; ant.feinstein@utoronto.ca

B. Brochet (ed.), *Neuropsychiatric Symptoms of Inflammatory Demyelinating Diseases*, Neuropsychiatric Symptoms of Neurological Disease,
DOI 10.1007/978-3-319-18464-7_3

extensive literature on brain changes in MS, the majority of extant literature has focused on the association between various metrics and cognitive impairment. However, determining the relation between depressive symptoms and structural and functional brain changes, as well as disentangling the association between depression and other factors such as cognitive deficits and fatigue, will be crucial in understanding the nature of depression in MS. In this chapter, we first briefly review structural and functional abnormalities associated with MS. We then briefly outline the general brain abnormalities associated with depression in non-neurological populations. Finally, we review the burgeoning neuroimaging literature on depression in MS and the applicability of theoretically derived models of depression to MS patients.

Neuroimaging Findings in MS

The pattern of MS-related pathology can be quite variable across patients and has been extensively measured with the advent of neuroimaging techniques. Common findings include the presence of white matter (WM) lesions that appear hyperintense on T2-weighted or Fluid-attenuated inversion recovery (FLAIR) MRI sequences as well as hypointense lesions on T1 scans (i.e., black holes) which can be found throughout the parenchyma and are likely indicative of neuronal loss when they are persistent. Atrophy of normal appearing gray matter (GM) and WM can be seen on T1-weighted images and more accurately measured using automated quantification techniques such as voxel-based morphometry. Although the location of atrophy and lesions can be quite varied, GM volume reduction is common in deep gray matter structures including the thalamus, caudate, putamen, and hippocampus, as well as the cortex [1, 2]. In terms of WM integrity, degraded WM can be detected via diffusion tensor imaging (DTI) in many of the major fiber bundles, including the corpus callosum, superior and inferior longitudinal fasciculi, cingulum bundle, uncinate fasciculus, as well as corticospinal tracts [3]. Associations between different metrics and neuropsychiatric symptoms have focused mainly on cognitive impairment and functional disability. Importantly, although lesion load is an important factor, atrophy of normal appearing GM tends to be more robustly associated with cognitive impairment [4, 5]. However, fewer studies have examined the association between cerebral abnormalities in MS and other clinical aspects such as depression. In order to discuss the putative effects of depression on the brain in MS, we briefly review the neuroimaging findings of major depressive disorder (MDD).

Neuroimaging Studies of Depression

There is a considerable neuroimaging literature on major depressive disorder across various neurological populations. Yet, despite the identification of consistent effects in terms of structural and functional abnormalities, a comprehensive model of the neural substrates underlying depression remains elusive. In terms of structural changes in patients with major depressive disorder (MDD), Kempton et al. conducted one of the largest meta-analyses of changes in gray matter, aggregating results from 143 studies [6]. Overall, compared to controls, MDD patients showed decreased volume in the hippocampus, thalamus, basal ganglia (caudate, putamen, globus pallidus), and medial prefrontal regions (orbitofrontal cortex, gyrus rectus). Because of the size of their sample, Kempton and colleagues were also able to compare the effects of clinical variables on hippocampal volumes. Neither the use of antidepressants, number of depressive episodes, nor age accounted for decreased hippocampal volume. However, patients currently in a depressive episode had smaller hippocampi compared to healthy controls, whereas patients with remitted MDD did not.

These effects may also be observed in first-episode depressed patients. Han et al. examined patients who had been depressed for less than 6 months and found GM reductions in orbitofrontal cortex, anterior cingulate, and middle frontal gyrus, although the authors also found increased hippocampal volume relative to controls [7]. A more recent meta-analysis of structural changes in treatment-naïve MDD patients showed GM atrophy in bilateral hippocampi and parahippocampal gyri as well as reduced volumes in inferior, middle, and superior frontal gyri [8]. Similarly, reduced volume in medial prefrontal structures may even be found in individuals with subclinical levels of depressive symptoms, suggesting they are among the first regions to be affected [9].

More recent work has shown that hippocampal and amygdala volumes are negatively correlated with the number of depressive episodes, suggesting they are particularly sensitive to duration of illness [10]. It should be noted that some regions tend to show increased GM volume in MDD patients, particularly when examining those who are drug naïve. Specifically, evidence suggests that thalamic volumes are actually increased in first-episode MDD patients [8].

With respect to WM integrity, results using DTI have been equivocal. A few investigations found significant declines in fractional anisotropy (FA), a measure of white matter integrity, whereas others have failed to find such a difference. Choi et al. recently compared FA values between134 MDD patients and 54 matched controls, using whole brain analyses and strict statistical thresholds [11]. The authors found no significant difference in FA values in any brain region: however, this does not rule out significant WM reductions in specific subgroups of depressed patients. Indeed, a recent study reported significant FA reductions in several limbic and subcortical WM tracts, but only when comparing melancholic MDD patients to controls [12]. In a similar vein, a recent investigation of depressive symptoms in over 810 community-dwelling adults found an association between depressive symptomatology and decreased FA, but only in women [13]. Regions identified in the analysis included several frontal areas such as the anterior cingulum bundle, uncinate fasciculus, as well as subcortical tracts such as the fornix and external capsule. Thus, the effects of depression on WM are more nuanced and likely interact with other factors such as disorder subtype and age and other clinical variables.

In terms of functional activation, results are divided into studies of resting-state activity and studies examining neural activation in response to various tasks.

With respect to resting-state activity, patients with MDD consistently show both increased and decreased activation at rest, compared to healthy controls. In a meta-analysis using activation likelihood estimation, Fitzgerald et al. (2008) reported that across 25 studies of PET and SPECT, MDD patients showed increased activity in inferior, medial, and superior frontal gyri as well as increased activity in the hippocampus, amygdala, and thalamus [14]. More recently, Hamilton et al. confirmed that the thalamus, in particular the pulvinar nucleus, was consistently hyperactive in depressed patients. Studies employing resting-state fMRI note converging evidence with studies of cerebral perfusion, showing increased activity in regions comprising the "default-mode" network [15], including ventromedial prefrontal cortex and subgenual anterior cingulate cortex (see [16] for a review). However, depressed patients also tend to show decreased resting activity in lateral prefrontal regions such as the dorsolateral prefrontal cortex, which is highly implicated in tasks with an executive control component, such as working memory, and is considered part of an "executive control" network.

The results of functional activation studies tend to show MDD patients exhibit hyperactivation and/or hypoactivation depending on the experimental task conditions. A well-studied paradigm requires subjects to view emotional stimuli (e.g., affectively valenced words or pictures), in order to measure differences in the neural response between patients and controls. For example, Surguladze et al. 2005 showed happy and sad faces to patients with MDD ($n=16$) and healthy controls ($n=14$) and found that depressed patients showed increased activation in response to sad faces in left parahippocampal gyrus, left amygdala, left putamen, and right fusiform gyrus [17]. Conversely, healthy controls showed increased activation to happy faces in the right putamen and bilateral fusiform gyri. Averaging across many of such studies, a meta-analysis by Diener et al. found that MDD patients overall showed decreased activity to negatively valenced stimuli in the rostral anterior cingulate cortex, medial frontal gyrus, anterior insula, inferior parietal lobes, and caudate. In addition, these patients showed hyperactivity in response to negative stimuli in the thalamus, medial temporal regions including the hippocampus, parahippocampal gyrus/amygdala, and to some extent the subgenual anterior cingulate cortex [18]. A more recent meta-analysis of 44 neuroimaging studies also found that within MDD patients, neural activation within many of the same subcortical regions was modulated by emotional valence: specifically, patients showed increased activation to negative stimuli in the parahippocampal gyrus, amygdala, putamen, and anterior cingulate. However, in response to positive stimuli, MDD patients showed decreased activation in the insula, striatum, amygdala, parahippocampal gyrus, and hippocampus [19].

In sum, the emerging picture is that depression is generally associated with GM atrophy in medial prefrontal regions, including the orbitofrontal cortex, anterior cingulate cortex, and ventromedial prefrontal cortex. The hippocampus and other subcortical regions are also atrophied in depression, with the hippocampus particularly sensitive to the number of depressive episodes. Effects of depression on WM are less consistent but may include prefrontal and subcortical fiber tracks. Furthermore, the regions tending to show structural abnormalities in MDD are also those showing aberrant functional activation. Overall, patients with depression show hyperactivity of subcortical limbic structures and prefrontal regions (i.e., anterior cingulate/medial prefrontal cortices), both in terms of activity at rest and in terms of reactivity to negative stimuli. Depressed patients also show decreased activity in the dorsolateral prefrontal cortex and adjacent lateral prefrontal regions implicated in executive control.

Based on these findings, several neurobiological models of depression have been posited, a full discussion of which is beyond the scope of this chapter. However, several key themes emerge that are useful when considering depression in MS. One is that the hyperactive limbic, subcortical, and prefrontal regions create a bias toward processing of negative emotional stimuli in depression. In addition, hypoactive lateral prefrontal cortices, which are heavily implicated in executive control, may suggest that depressed patients may have difficulty with cognitive reappraisal or exerting top-down control on affective states [20–23]. Another relevant notion is that many of the regions showing increased activity at rest comprise the default-mode network, a network implicated in self-referential processing, autobiographical memory, and even simulating nonpersonal future events [24–26]. Consequently, the increased in default-mode network activity may reflect the ruminative aspects of depression [16, 27].

MS, Neuroimaging, and Depression

We turn now to a review of the existing work on neuroimaging of depression in MS. In the first study relating depressive symptoms to cerebral dysfunction, Pujol et al. found that hyperintense lesions in the left arcuate fasciculus were correlated with BDI scores [28]. Subsequent work also found that lesion load correlated with depressive symptoms but hinted that GM atrophy may be a stronger predictor of depression than lesion load [29]. Feinstein et al. (2004) compared structural MRI MS of patients with a DSM-IV diagnosis of major depressive disorder to a group on nondepressed MS patients, who were matched in terms of age, disease duration and course, overall disability, and cognitive functioning. Crucially, the study only included patients who were diagnosed with MDD after having a definite diagnosis of MS, thus ruling out other co-morbid factors as the source of the depression. Using semiautomated tissue segmentation algorithms, the authors found that depressed MS patients showed more hyperintense and hypointense lesions in left medial inferior frontal regions as well as decreased GM volume in the left anterior temporal lobe. Moreover, the inclusion of these regions into a logistic regression analysis showed that these two factors predicted 42 % of the variance in the likelihood of being diagnosed with depression [30]. This finding is confirmed by recent work using more refined measures of cortical volume. Specifically, Gobbi and colleagues examined 123 MS patients, splitting them into depressed or nondepressed based on a cutoff score of 9 on the Montgomery-Asberg Depression Rating Scale (MADRS) and into fatigued or non-fatigued based on one question on the scale that measured "lassitude" as a proxy of fatigue [31]. Using voxel-based morphometry, the authors found that depressed patients showed decreased GM in the right inferior frontal gyrus and left middle frontal gyrus, but only when these comparisons were masked with the contrast comparing fatigued and non-fatigued patients.

There are hints of similar effects emerging even in those patients in the earliest stages of the disease course. Nygaard et al. compared patients with early relapsing-remitting MS (mean disease duration = 26 months) to healthy controls in terms of GM integrity, while also obtaining measures of cognition, depression, and fatigue. Compared to controls, patients with early relapsing-remitting MS showed the expected pattern of decreased cortical thickness in several areas throughout the brain and increased white matter lesion load. Notably, however, within MS patients, depression scores emerged as the only measure significantly associated with GM integrity. Specifically, depression scores were negatively correlated with overall cortical thickness in medial and superior prefrontal cortices, medial temporal regions, and inferior parietal lobes. Furthermore, this relation remained even after covarying out age, gender, and disease duration [32].

Finally, in line with the effects of depression on the hippocampus in non-neurological populations, Gold et al. also found that depressed MS patients showed decreased hippocampal volumes. Crucially, the authors acquired high-resolution T2 scans of the medial temporal lobes, allowing them to isolate atrophy into specific subregions. Compared to healthy controls, MS patients showed decreased CA1 subfield volumes: however, when comparing depressed MS patients (scores of 13 or higher on the Beck Depression Inventory) to nondepressed patients, CA2/CA3/dentate gyrus subfield volumes were significantly lower in the depressed group. Moreover, CA2/CA3 volume was negatively correlated with consistently elevated cortisol levels, suggesting a role for hypothalamic-pituitary axis (HPA) dysregulation as a potential mechanism for hippocampal atrophy [33].

Despite the prominent role of WM in the pathology of MS, there are only a handful of studies that have examined WM integrity as it relates to depression in MS. Feinstein et al. (2010) obtained DTI metrics within the major WM tracts of depressed and nondepressed MS patients and found that depressed patients showed decreased average integrity (decreased fractional anisotropy and increased mean diffusivity values) in the left anterior temporal lobe. In addition, depressed patients showed decreased normal appearing white matter volume in the left inferior prefrontal region [34]. Gobbi et al. acquired DTI scans in 147 MS patients, characterized as depressed ($n=92$) or nondepressed ($n=55$) based on a cutoff of 9 on the MADRS. No differences in FA were found when comparing patient groups on a voxel-wise basis, but when FA values were averaged within the major WM fiber bundles, depressed patients showed decreased WM integrity in the forceps minor [35]. Although these findings are promising, additional studies are crucial to determine which WM regions are consistently affected in depression.

To date, there are no studies comparing functional activity in depressed MS patients. However, one study did use PET to compare serotonin transporter (SERT) availability in MS patients, compared to healthy controls. MS patients showed decreased SERT uptake in the thalamus, medial temporal lobes, insula, and cingulate gyrus but increased SERT availability in the orbitofrontal cortex. Interestingly, SERT binding in the insula was positively correlated to BDI scores, suggesting that poor serotonergic regulation may also be present in depressed MS patients [36].

In sum, an emerging picture seems to be that there may be consistent effects of depression on brain tissue in MS, which may be dissociable from MS-related atrophy. At this time, the most consistent evidence points to GM volume reduction in depressed MS patients, particularly in terms of hippocampal subregions, anterior temporal cortex, and lateral prefrontal cortices.

Neuroimaging, Depression, and MS: A Synthesis

Given the consistent structural and functional abnormalities in non-neurological depressed patients and their associated models of depression, the question arises to what extent imaging results from MDD patients inform our understanding of depression in MS. One common theme of the extant work on depression is hyperactivity of limbic-prefrontal circuits that biases attention toward negative stimuli, which when coupled with dysfunctional prefrontal regions involved in executive control, may result in a bias to negative emotions or stimuli in the environment without the necessary means to regulate or reappraise the situation. Such a notion is in line with cognitive models of depression stressing the role of dysfunctional cognitive schemas (see [37] for a review). Given the amount of atrophy that can occur in MS, it is quite possible that atrophy of prefrontal GM and WM could contribute to the maintenance of depression by impairing emotional regulation. Indeed, the few studies that have examined depression in MS using neuroimaging have found atrophy primarily in prefrontal GM and WM, which overlaps partially with those areas found to be atrophied and/or show decreased activation to emotional stimuli. Moreover, there is ample evidence to suggest that MS patients show dysfunctional prefrontal activation in response to cognitive control tasks, such as working memory tasks [38]. Thus, it is quite plausible that prefrontal volume reductions have a detrimental effect on affect regulation and the use of cognitive reappraisals in MS patients.

Another relevant notion for MS patients is models of depression that stress ruminative tendencies and default-mode network activity. That is, given that default-mode activity is associated with autobiographical memory, self-referential processing, and simulating future events, the increase in network activity at rest may reflect the rumination that often occurs in depression. Anecdotally, patients with MS often report ruminative tendencies [39, 40], particularly regarding general concerns about the future. In terms of resting-state activation and rumination, it is interesting to note that MS patients also show altered resting-state activity in the default-mode network, which is predictive of poorer memory performance [41]. Whether aberrant resting-state activity in MS is also related to depression or other psychiatric symptomatology is an intriguing avenue of future investigation.

However, some models of depression based on neuroimaging findings in MDD patients do not map well to MS patients. Notably, one hypothesis suggests hyperactivity of the thalamus and subcortical regions reflects their involvement in detecting emotional stimuli. In particular, the amygdala and more recently the pulvinar nucleus of the thalamus both show increased activity at rest, which some suggest make patients more biased toward negative stimuli [42]. However, in MS the thalamus and basal ganglia are among the most commonly affected GM regions, and so it is difficult to reconcile this aspect of depression (i.e., bias to negative emotional stimuli) with the patterns of atrophy seen in MS.

Finally, it is worth noting that the presence of depression may potentially have a negative impact on memory performance in MS patients. That is, given that depression is associated with hippocampal atrophy due to dysregulation of the HPA axis [43] and that hippocampal atrophy is also found in MS [44], it is possible that prolonged depression may have sizable consequences on brain tissue and cognition, particularly in these patients whose brains are already vulnerable [45–47], though the reverse causal association is also possible [48].

In terms of treatment, preliminary findings of disrupted serotonergic neurotransmission in MS would also suggest the use of SSRIs in treatment and management of depressive symptoms [36]. It is interesting to note that treatment of MDD in nonneurological populations can return patterns of aberrant functional activation to those seen in nondepressed controls. For example, several studies have found that MDD patients who respond to treatment with SSRIs show increased activation of prefrontal regions and decreased activation of medial temporal and medial prefrontal regions [14]. More recently, McGrath et al. demonstrated in a randomized treatment study that MDD patients who achieved full remission (either by treatment with escitalopram or Cognitive Behavioral Therapy (CBT)) showed significant changes in resting insular glucose metabolism, suggesting the region may serve as a candidate biomarker for treatment selection [49]. Whether similar pharmacological or psychotherapeutic effects can be found in MS is not known [47].

Conclusions

It is now evident that the functional networks associated with depression in the general psychiatric population overlap partially with current neuroimaging findings in MS. Future studies providing more precise links between depressive symptoms and specific structural abnormalities are essential, particularly given the heterogeneity of atrophy and lesions in MS. Finally, tracking the deterioration of these networks with disease progression as well as mapping brain changes to the different depressive symptoms (i.e., reactivity to situational factors, rumination, poor emotional regulation) will also be necessary to untangling this facet of a disease with a complex clinical presentation.

References

1. Messina S, Patti F. Gray matters in multiple sclerosis: cognitive impairment and structural MRI. Mult Scler Int [Internet]. Hindawi Publishing Corporation; 2014 [cited 2014 Aug 19];2014:609694. Available from: http://www.pubmedcentral.nih.gov/articlerender.fcgi?artid=3920616&tool=pmcentrez&rendertype=abstract.
2. Lansley J, Mataix-Cols D, Grau M, Radua J, Sastre-Garriga J. Localized grey matter atrophy in multiple sclerosis: a meta-analysis of voxel-based morphometry studies and associations with functional disability. Neurosci Biobehav Rev [Internet]. Elsevier Ltd; 2013 [cited 2014 Aug 4];37(5):819–30. Available from: http://www.ncbi.nlm.nih.gov/pubmed/23518268.
3. Hulst HE, Steenwijk MD, Versteeg A, Pouwels PJW, Vrenken H, Uitdehaag BMJ, et al. Cognitive impairment in MS: impact of white matter integrity, gray matter volume, and lesions. Neurology [Internet]. 2013;80(11):1025–32. Available from: http://www.ncbi.nlm.nih.gov/pubmed/23468546.
4. Filippi M, Rocca MA, Benedict RHB, DeLuca J, Geurts JJG, Rombouts SARB, et al. The contribution of MRI in assessing cognitive impairment in multiple sclerosis. Neurology [Internet]. 2010;75(23):2121–8. Available from: http://www.pubmedcentral.nih.gov/articlerender.fcgi?artid=3385423&tool=pmcentrez&rendertype=abstract.
5. Benedict RHB, Zivadinov R. Risk factors for and management of cognitive dysfunction in multiple sclerosis. Nat Rev Neurol [Internet]. Nature Publishing Group; 2011 [cited 2014 Aug 5];7(6):332–42. Available from: http://www.ncbi.nlm.nih.gov/pubmed/21556031.
6. Kempton MJ, Salvador Z, Munafo MR, Geddes JR, Simmons A, Frangou S, et al. Structural neuroimaging studies in major depressive disorder: meta-analysis and comparison with bipolar disorder. Arch Gen Psychiatry. 2011;68(7):675–90.
7. Han K-M, Choi S, Jung J, Na K-S, Yoon H-K, Lee M-S, et al. Cortical thickness, cortical and subcortical volume, and white matter integrity in patients with their first episode of major depression. J Affect Disord. Elsevier; 2014;155:42–8
8. Zhao Y-J, Du M-Y, Huang X-Q, Lui S, Chen Z-Q, Liu J, et al. Brain grey matter abnormalities in medication-free patients with major depressive disorder: a meta-analysis. Psychol Med [Internet]. 2014 [cited 2014 Sept 12];44(14):2927–37. Available from: http://www.ncbi.nlm.nih.gov/pubmed/25065859.
9. Webb CA, Weber M, Mundy EA, Killgore WDS. Reduced gray matter volume in the anterior cingulate, orbitofrontal cortex and thalamus as a function of mild depressive symptoms: a voxel-based morphometric analysis. Psychol Med [Internet]. 2014 [cited 2014 Sept 12];44(13):2833–43. Available from: http://www.ncbi.nlm.nih.gov/pubmed/25066703.
10. Stratmann M, Konrad C, Kugel H, Krug A, Schöning S, Ohrmann P, et al. Insular and hippocampal gray matter volume reductions in patients with major depressive disorder. PLoS One [Internet]. 2014 [cited 2014 Sept 12];9(7):e102692. Available from: http://www.pubmedcentral.nih.gov/articlerender.fcgi?artid=4106847&tool=pmcentrez&rendertype=abstract.
11. Choi KS, Holtzheimer PE, Franco AR, Kelley ME, Dunlop BW, Hu XP, et al. Reconciling variable findings of white matter integrity in major depressive disorder. Neuropsychopharmacology [Internet]. Nature Publishing Group; 2014 [cited 2014 Sept 12];39(6):1332–9. Available from: http://www.ncbi.nlm.nih.gov/pubmed/24352368
12. Korgaonkar MS, Grieve SM, Koslow SH, Gabrieli JDE, Gordon E, Williams LM. Loss of white matter integrity in major depressive disorder: evidence using tract-based spatial statistical analysis of diffusion tensor imaging. Hum Brain Mapp [Internet]. 2011 [cited 2014 Sept 12];32(12):2161–71. Available from: http://www.ncbi.nlm.nih.gov/pubmed/21170955.
13. Hayakawa YK, Sasaki H, Takao H, Hayashi N, Kunimatsu A, Ohtomo K, et al. Depressive symptoms and neuroanatomical structures in community-dwelling women: A combined voxel-based morphometry and diffusion tensor imaging study with tract-based spatial statistics. NeuroImage Clin [Internet]. Elsevier Ltd.; 2014 [cited 2014 Sept 12];4:481–7. Available from: http://www.pubmedcentral.nih.gov/articlerender.fcgi?artid=3984445&tool=pmcentrez&rendertype=abstract.

14. Fitzgerald PB, Laird AR, Maller J, Daskalakis ZJ. A meta-analytic study of changes in brain activation in depression. Hum Brain Mapp [Internet]. 2008 [cited 2014 Jan 9];29(6):683–95. Available from: http://www.pubmedcentral.nih.gov/articlerender.fcgi?artid=2873772&tool=pmcentrez&rendertype=abstract.
15. Buckner RL, Andrews-Hanna JR, Schacter DL. The brain's default network: anatomy, function, and relevance to disease. Ann N Y Acad Sci. 2008;1124:1–38.
16. Northoff G, Wiebking C, Feinberg T, Panksepp J. The "resting-state hypothesis" of major depressive disorder-A translational subcortical-cortical framework for a system disorder. Neurosci Biobehav Rev. 2011;35:1929–45.
17. Surguladze S, Brammer MJ, Keedwell P, Giampietro V, Young AW, Travis MJ, et al. A differential pattern of neural response toward sad versus happy facial expressions in major depressive disorder. Biol Psychiatry [Internet]. 2005 [cited 2014 July 26];57(3):201–9. Available from: http://www.ncbi.nlm.nih.gov/pubmed/15691520.
18. Diener C, Kuehner C, Brusniak W, Ubl B, Wessa M, Flor H. A meta-analysis of neurofunctional imaging studies of emotion and cognition in major depression. Neuroimage [Internet]. Elsevier Inc.; 2012 [cited 2014 July 15];61(3):677–85. Available from: http://www.ncbi.nlm.nih.gov/pubmed/22521254.
19. Groenewold NA, Opmeer EM, de Jonge P, Aleman A, Costafreda SG. Emotional valence modulates brain functional abnormalities in depression: evidence from a meta-analysis of fMRI studies. Neurosci Biobehav Rev [Internet]. Elsevier Ltd; 2013 [cited 2014 July 18];37(2):152–63. Available from: http://www.ncbi.nlm.nih.gov/pubmed/23206667.
20. Drevets WC. Prefrontal cortical-amygdalar metabolism in major depression. Ann N Y Acad Sci. 1999;877:614–37.
21. Hamilton JP, Etkin A, Furman DJ, Lemus MG, Johnson RF, Gotlib IH. Functional neuroimaging of major depressive disorder: a meta-analysis and new integration of baseline activation and neural response data. Am J Psychiatry. 2012;169:693–703.
22. Hamilton JP, Furman DJ, Chang C, Thomason ME, Dennis E, Gotlib IH. Default-mode and task-positive network activity in major depressive disorder: implications for adaptive and maladaptive rumination. Biol Psychiatry. Elsevier Inc.; 2011;70:327–33.
23. Mayberg HS. Modulating dysfunctional limbic-cortical circuits in depression: towards development of brain-based algorithms for diagnosis and. Br Med Bull. 2003;65:193–207.
24. Andrews-Hanna JR. The brain's default network and its adaptive role in internal mentation. Neuroscience. 2012;18(3):251–70.
25. Schacter DL, Addis DR, Hassabis D, Martin VC, Spreng RN, Szpunar KK. The future of memory: remembering, imagining, and the brain. Neuron [Internet]. Elsevier Inc.; 2012 [cited 2013 May 21];76(4):677–94. Available from: http://www.ncbi.nlm.nih.gov/pubmed/23177955.
26. Romero K, Moscovitch M. Episodic memory and event construction in aging and amnesia. J Mem Lang [Internet]. Elsevier Inc.; 2012;67(2):270–84. Available from: http://dx.doi.org/10.1016/j.jml.2012.05.002.
27. Nejad AB, Fossati P, Lemogne C. Self-referential processing, rumination, and cortical midline structures in major depression. Front Hum Neurosci. 2013;7:1–9.
28. Pujol J, Bello J, Deus J, Marti-Vilata JL, Capdevila A. Lesions in the left arcuate fasciculus region and depressive symptoms in multiple sclerosis. Neurology. 1997;49:1105–10.
29. Bakshi R, Czarnecki D, Shaikh ZA, Priore RL, Janardhan V, Kaliszky Z, et al. Brain MRI lesions and atrophy are related to depression in multiple sclerosis. Neuroreport. 2000;11(6):1153–8.
30. Feinstein A, Roy P, Lobaugh N, Feinstein K, O'Connor P, Black S. Structural brain abnormalities in multiple sclerosis patients with major depression. Neurology [Internet]. 2004 [cited 2014 Sept 9];62(4):586–90. Available from: http://www.neurology.org/cgi/doi/10.1212/01.WNL.0000110316.12086.0C.
31. Gobbi C, Rocca M a, Riccitelli G, Pagani E, Messina R, Preziosa P, et al. Influence of the topography of brain damage on depression and fatigue in patients with multiple sclerosis. Mult Scler [Internet]. 2014 [cited 2014 Sept 9];20(2):192–201. Available from: http://www.ncbi.nlm.nih.gov/pubmed/23812284

32. Nygaard GO, Walhovd KB, Sowa P, Chepkoech J-L, Bjornerud A, Due-Tonnessen P, et al. Cortical thickness and surface area relate to specific symptoms in early relapsing-remitting multiple sclerosis. Mult Scler J [Internet]. 2014 [cited 2014 Aug 20]; Available from: http://msj.sagepub.com/cgi/doi/10.1177/1352458514543811.
33. Gold SM, O'Connor M-F, Gill R, Kern KC, Shi Y, Henry RG, et al. Detection of altered hippocampal morphology in multiple sclerosis-associated depression using automated surface mesh modeling. Hum Brain Mapp [Internet]. 2014 [cited 2014 Sept 9];35(1):30–7. Available from: http://www.ncbi.nlm.nih.gov/pubmed/22847919.
34. Feinstein A, O'Connor P, Akbar N, Moradzadeh L, Scott CJM, Lobaugh NJ. Diffusion tensor imaging abnormalities in depressed multiple sclerosis patients. Mult Scler [Internet]. 2010 [cited 2014 Sept 9];16(2):189–96. Available from: http://www.ncbi.nlm.nih.gov/pubmed/20007425.
35. Gobbi C, Rocca M, Pagani E, Riccitelli G, Pravatà E, Radaelli M, et al. Forceps minor damage and co-occurrence of depression and fatigue in multiple sclerosis. Mult Scler [Internet]. 2014 [cited 2014 Sept 9]; Available from: http://www.ncbi.nlm.nih.gov/pubmed/24740370.
36. Hesse S, Moeller F, Petroff D, Lobsien D, Luthardt J, Regenthal R, et al. Altered serotonin transporter availability in patients with multiple sclerosis. Eur J Nucl Med Mol Imaging [Internet]. 2014 [cited 2014 Sept 9];41(5):827–35. Available from: http://www.ncbi.nlm.nih.gov/pubmed/24562640.
37. Disner SG, Beevers CG, Haigh EAP, Beck AT. Neural mechanisms of the cognitive model of depression. Nat Rev Neurosci [Internet]. 2011 [cited 2014 July 9];12(8):467–77. Available from: http://www.ncbi.nlm.nih.gov/pubmed/21731066.
38. Kollndorfer K, Krajnik J, Woitek R, Freiherr J, Prayer D, Schöpf V. Altered likelihood of brain activation in attention and working memory networks in patients with multiple sclerosis: an ALE meta-analysis. Neurosci Biobehav Rev [Internet]. Elsevier Ltd; 2013 [cited 2014 Sept 8];37(10 Pt 2):2699–708. Available from: http://www.pubmedcentral.nih.gov/articlerender.fcgi?artid=3878376&tool=pmcentrez&rendertype=abstract.
39. Bruce JM, Arnett P. Clinical correlates of generalized worry in multiple sclerosis clinical correlates of generalized worry in multiple sclerosis. J Clin Exp Neuropsychol. 2009;31(6):698–705.
40. Thornton EW, Tedman S, Rigby S, Bashforth H, Young C. Worries and concerns of patients with multiple sclerosis: development of an assessment scale. Mult Scler J. 2006;12:196–203.
41. Sumowski JF, Wylie GR, Leavitt VM, Chiaravalloti ND, DeLuca J. Default network activity is a sensitive and specific biomarker of memory in multiple sclerosis. Mult Scler [Internet]. 2013 [cited 2014 July 23];19(2):199–208. Available from: http://www.ncbi.nlm.nih.gov/pubmed/22685065.
42. Hamilton JP, Chen MC, Gotlib IH. Neural systems approaches to understanding major depressive disorder: an intrinsic functional organization perspective. Neurobiol Dis. Elsevier Inc.; 2013;52:4–11.
43. Sapolsky RM. Depression, antidepressants, and the shrinking hippocampus. Proc Natl Acad Sci U S A [Internet]. 2001;98(22):12320–2. Available from: http://www.pubmedcentral.nih.gov/articlerender.fcgi?artid=60045&tool=pmcentrez&rendertype=abstract.
44. Pardini M, Bergamino M, Bommarito G, Bonzano L, Luigi Mancardi G, Roccatagliata L. Structural correlates of subjective and objective memory performance in multiple sclerosis. Hippocampus [Internet]. 2014 [cited 2014 Aug 19];24(4):436–45. Available from: http://www.ncbi.nlm.nih.gov/pubmed/24375730.
45. Julian LJ, MOhr DC. Cognitive predictors of response to treatment for depression in multiple sclerosis. J Neuropsychiatry Clin Neurosci. 2006;18:356–63.
46. Julian L, Merluzzi NM, Mohr DC. The relationship among depression, subjective cognitive impairment, and neuropsychological performance in multiple sclerosis. Mult Scler J. 2007;13:81–6.
47. Mohr DC, Lovera J, Brown T, Cohen B, Neylan T, Henry R, et al. A randomized trial of stress management for the prevention of new brain lesions in MS. Neurology. 2012;79:412–9.

48. Barwick FH, Arnett PA. Relationship between global cognitive decline and depressive symptoms in multiple sclerosis. Clin Neuropsychol. 2011;25(2):193–209.
49. McGrath CL, Kelley ME, Holtzheimer PE, Dunlop BW, Craighead WE, Franco AR, et al. Toward a neuroimaging treatment selection biomarker for major depressive disorder. JAMA Psychiatry [Internet]. 2013 [cited 2014 Aug 8];70(8):821–9. Available from: http://www.ncbi.nlm.nih.gov/pubmed/23760393.

Chapter 4
Anxiety and Multiple Sclerosis

Giuseppe Magistrale and Ugo Nocentini

Abstract Normal worry and fear are largely adaptive, pervading anxiety is considered maladaptive and, in its various forms, it can characterize clinical disorders. Anxiety levels and anxiety disorders are more frequent in MS patients than in general population; nevertheless, they are overlooked and undertreated. The increase of anxiety in MS patients, more frequent in women than in men, seems related mainly to the diagnostic work-up period and prognostic uncertainties; the groundless fear of becoming rapidly wheelchair dependent plays some role. Exacerbations are also related to anxiety increase, while anxiety decreases as time elapses after the diagnosis and in the remitting phase. Anxiety levels have been studied in relationship with many other variables of interest in MS patients. The outcome of anxiety, the role of coping strategies, and other aspects are of support to the reactive nature of pathological anxiety. Few studies have investigated possible organic contributors, mainly with negative results. Several studies show how high levels of anxiety are associated with low health-related quality of life and decreased performance of some cognitive functions. Unfortunately, some of the available studies have important limitations (e.g., small samples, retrospective collection of data, exclusion of some subjects). In the matter of treatment, randomized controlled trials assessing pharmacologic and non-pharmacologic therapies for anxiety in MS showed inconsistent results. An important effort in covering the existing gaps, particularly about assessment and treatment of anxiety in MS patients, seems timely and relevant.

Keywords Anxiety • Anxiety disorders • Multiple sclerosis • Assessment • Clinical features • Treatment • Diagnosis • Pharmacologic therapy • Psychotherapy

---

G. Magistrale, MSc
Neurology and NeuroRehabilitation Unit, I.R.C.C.S. "Santa Lucia" Foundation,
Via Ardeatina 306, Rome 00179, Italy
e-mail: g.magistrale@hsantalucia.it

U. Nocentini, MD (✉)
Neurology and NeuroRehabilitation Unit, I.R.C.C.S. "Santa Lucia" Foundation,
Via Ardeatina 306, Rome 00179, Italy

Department of Systems Medicine, University of Rome "Tor Vergata",
Via Ardeatina 306, Rome 00179, Italy
e-mail: u.nocentini@hsantalucia.it

B. Brochet (ed.), *Neuropsychiatric Symptoms of Inflammatory Demyelinating Diseases*, Neuropsychiatric Symptoms of Neurological Disease,
DOI 10.1007/978-3-319-18464-7_4

Definition of Anxiety

According to the definition given by the American Psychological Association, anxiety is "an emotion characterized by feelings of tension, worried thoughts, and physical changes like increased blood pressure" which emerges from an expectation of future threat or a motivational conflict [1]. It is an emotional state defined by aversive cognitive (thoughts of apprehensive expectations), physiological (hyperarousal and somatic activation), and behavioral (i.e., avoidance, paralysis) components [2]. Anxiety can be partially distinguished from fear, as the latter is the emotional response to immediate threat, although fear and anxiety are strictly related [3]. Moreover, fear and anxiety can be distinguished on the basis of duration, temporal focus, threat specificity, and motivated direction: while fear is immediate, focused on the present, and targeted to a specific threat in order to avoid it, anxiety has a longer duration without a specific threat. Following danger, fear and anxiety activate a sequence of adaptive behaviors aiming to reduce the unpleasant physiological response and to escape the environmental threat or resolve the underlying motivational conflict.

As the Roman philosopher Lucius Annaeus Seneca once wrote, "There is nothing so wretched or foolish as to anticipate misfortunes. What madness it is in your expecting evil before it arrives!" [4]. Normal anxiety and fear are largely adaptive, as they mobilize one's resources in order to cope with an environmental challenge. Nevertheless, pervading anxiety is considered maladaptive, and in its various forms it can characterize clinical disorders, in so far as it compromises the normal functioning and quality of life of the individual. Moreover, abnormal anxiety also occurs as a symptom in other psychiatric disorders, such as clinical depression.

Anxiety disorders appear to have a number of biological and environmental contributing factors [3]. For example, dysfunctional anxiety can be "learned" from the social environment (i.e., the family) in the presence of a biological predisposition [3], or it can be the result of negative life events such as accumulated trauma [5].

Anxiety Disorders According to the Diagnostic and Statistical Manual of Mental Disorders *Fifth Edition*

According to the *Diagnostic and Statistical Manual of Mental Disorders* fifth edition [6], anxiety disorders differ from normal fear or anxiety in as much as they are persistent (generally lasting 6 months or more) and characterized by overestimation of the danger represented by the situation that is feared or avoided.

Anxiety disorders can be distinguished on the basis of the underlying cognitive ideation and the eliciting contexts that induce anxious and fearful behaviors. The various forms of phobias and anxiety disorders in the DSM-V include separation anxiety disorder, selective mutism, specific phobia, social anxiety disorder, panic disorder/panic attack, agoraphobia, generalized anxiety disorder (GAD), substance-/

medication-induced anxiety disorder, anxiety disorder due to another medical condition, and other specified anxiety disorders and unspecified anxiety disorders. Selective mutism (code 312.23) and separation anxiety (309.21) are typically considered developmental disorders and have been recently reclassified and moved into the broader category of the anxiety disorders.

Specific phobias (300.29) affect individuals who fear and avoid particular things or situations in a way clearly exaggerated in respect to the real risk posed by the situation, and the emotional reaction arises immediately and lastingly in response to the feared situation (i.e., animals, injections, a specific place, etc.). Social anxiety (300.23) is diagnosed when there is a persistent fear of being awkward, humiliated, or rejected. These cognitive ideations force individuals suffering from social anxiety to avoid social situations.

Panic disorder (300.01) is characterized by frequent panic attacks, which are abrupt episodes of intense fear and apprehension of variable duration (from minutes to hours) that can be expected or unexpected. Since individuals suffering from this disorder constantly fear panic attacks, they often change their habits in a dysfunctional way in order to avoid their insurgence.

Agoraphobia (300.22) is the fear of being in open spaces or uncontrollable social situations such as public transportations, malls, or crowds. Fear of these situations is accompanied and elicited by the cognitive ideation of being unable to escape or get help once the anxiety symptomatology has arisen.

Individuals suffering from generalized anxiety disorder (300.02) are excessively and chronically anxious about many things and situations (i.e., health, money, and family). The persistence of generalized anxiety results in anguish and several functional impairments due to the physical symptoms experienced by the individual, such as restlessness, irritability, lack of concentration, and muscle tension.

In the substance-/medication-induced anxiety disorder, the prominent anxiety symptoms are due to the effects of a psychoactive substance, while anxiety disorder due to another medical condition (293.84) is clinically significant anxiety that can be attributed to the secondary effects of the different medical condition.

Anxiety Disorders in Multiple Sclerosis

Although anxiety disorders are often diagnosed in patients with MS, they are often overlooked and have been investigated less deeply than other neuropsychiatric disorders, such as depression [7]. Nevertheless, the prevalence of anxiety disorders in MS is significantly higher if compared to the general population [7]. According to Korostil and Feinstein [7], lifetime rates of anxiety in MS are higher than in other chronic medical illnesses such as diabetes [8], chronic obstructive airway disease [9], and rheumatoid arthritis [10]. Nevertheless, these disorders are underdiagnosed and undertreated [7]. Therefore, a deeper understanding of anxiety disorders in this pathology is necessary.

Epidemiology

Although many studies assessed anxiety in MS, much of the literature on the subject is affected by several limitations, such as the absence of clinical interviews, which are necessary in order to formulate a clinical diagnosis. Indeed, although informative, the majority of the studies regarding emotional disturbances in MS uniquely relied on self-report instruments (such as the Hospital Anxiety and Depression Scale [HADS]) for the assessment of clinically significant anxiety (see Table 4.1), with few exceptions. One study conducted in Italy [11] recorded the presence of anxiety disorders in 36 % of a sample of 50 outpatients with definite MS using the Structured Clinical Interview for DSM-IV disorders (SCID-IV). Moreover, MS patients were more likely to meet diagnostic criteria for obsessive-compulsive disorder, a result replicated by another study conducted on Iranian patients with MS [12]. A large Canadian study conducted on 140 participants with a definite MS diagnosis using both the SCID-IV and the HADS reported a lifetime prevalence of anxiety disorders of 35.7 % [7]. In particular, the authors reported a lifetime prevalence of 7.8 % for social phobia, 8.6 % for obsessive-compulsive disorder, 10 % for panic disorder, 10.8 % for specific phobias, and 18.6 % for GAD. In this study, risk factors for developing an anxiety disorder included being female, comorbid depression, and lack of social support. The authors also reported that anxiety disorders were largely underdiagnosed in their sample, thus preventing the possibility of a necessary targeted treatment. Differences in the occurrence of specific anxiety disorders in MS patients between the study by Korostil and Feinstein and the other two studies could be attributed to the difference in sample size and sample selection. In another study conducted in Mexico on 37 consecutive MS patients and 37 healthy controls using the SCID-IV and the Hamilton Anxiety Rating Scale (HARS), 21.6 % of the people with MS had clinically relevant anxiety [13]. Notably, a Dutch study showed that 8 months after the diagnosis of MS, high self-reported anxiety was present in 34 % of patients [14], while after 2 years, 69 % still showed significant anxiety levels [15]. Similar results were reported in several other studies, which employed self-reported anxiety questionnaires (see Table 4.1). Moreover, a recent study by Poder and co-workers [16] conducted on a cohort of 251 patients found that 30.6 % had clinically significant social anxiety symptoms. A recent population-based study conducted in Canada detected the presence of anxiety disorders in 35.6 % of the MS population ($n = 4{,}192$) using administrative data from sanitary records (i.e., ICD-9/10 codes). In summary, notwithstanding a shortage of robust epidemiologic studies regarding anxiety disorders, the current body of evidence shows that a large proportion of patients diagnosed with MS suffer from abnormal anxiety [17].

Table 4.1 Studies which assessed the presence of clinically significant anxiety in multiple sclerosis [7, 11, 13–28]

Study	Country	Population	Measures	% of patients with anxiety
Joffe et al. (1987) [18]	Canada	Clinic outpatients with MS ($n=100$)	SADS-L, RDC diagnosis	11 %
Minden et al. (1987) [19]	United Kingdom	50 MS patients	SADS	4 % prior to MS
				12 % since MS
				16 % lifetime (generalized anxiety)
Arias Bal et al. (1991) [20]	Spain	50 Patients with MS	CIS	12 %
Stenager et al. (1994) [21]	Denmark	94 MS outpatients	STAI	20.2 % high state score
				24.5 % high trait score
Diaz-Olavarrieta (1999) [13]	Mexico	MS outpatients ($n=44$), control subjects ($n=25$)	NPI, indirect evaluation	37 %
Smith and Young (2000) [22]	United Kingdom	88 patients with definite MS	HADS	34 %
Nicholl et al. (2001) [23]	United Kingdom	MS patients in contact with a rehabilitation consultant ($n=96$)	HAD anxiety	39 %
Mendes et al. (2003) [24]	Brazil	84 patients with relapsing-remitting MS	HADS (cutoff, 8)	34.5 %
Jannsens et al. (2003, 2006) [14, 15]	Netherlands	101 MS outpatients	HADS	34 % (8 months after diagnosis) of which 69 % had high anxiety after 2 years
Galeazzi et al. (2005) [11]	Italy	50 outpatients with definite MS diagnosis	SCID-I	36 %
Figved et al. (2005) [25]	Norway	Patients with MS (86), compared with 49 SLE controls	NPI	19.8 %
Korostil and Feinstein (2007) [7]	Canada	140 consecutive clinic attendees	SCID-I	35.7 %
Beiske et al. (2008) [26]	Norway	MS population-based study ($n=140$)	HSC-25	19.3 %

(continued)

Table 4.1 (continued)

Study	Country	Population	Measures	% of patients with anxiety
Dahl et al. (2009) [27]	Norway	172 MS patients	HADS	30.2 %
Poder et al. (2009) [16]	Canada	251 patients with MS	SPI	30.6 % (social anxiety)
Espinola-Nadurille et al. (2010) [28]	Mexico	37 outpatients with MS and 37 healthy controls	SCID-I	21.6 %
Marrie et al. (2013) [17]	Canada	MS population-based study (n=4192)	Case definition based on ICD-9/10 codes and physician-assigned diagnosis	35.6 %

SADS Schedule for Affective Disorders and Schizophrenia; *NPI* Neuropsychiatric Inventory; *SCID-I* Structural Clinical Interview for DSM Axis I Disorders; *CIS* Clinical Interview Scale; *HADS* Hospital Anxiety and Depression Scale; *ICD* International Classification of Diseases; *STAI* State-Trait Anxiety Inventory; *HSC* Hopkins Symptom Checklist; *SPI* Social Phobia Inventory

Clinical Presentation

Several aspects should be taken into account when considering the clinical features of anxiety in MS. In this regard, a common issue in the clinical assessment of psychiatric disorders such as anxiety and depression is the symptom overlap with somatic features of MS. Indeed, some of the somatic symptoms of anxiety such as unsteadiness, dizziness, fainting, and leg wobbliness can be often found among the somatic manifestations of MS. While this issue has been explored with regard to depression [29], the literature studying symptom overlap between anxiety and MS is scarce. Donnchadha et al. [30] indirectly explored this issue when validating the Beck Anxiety Inventory (BAI) in a group of patients with MS. With the use of hierarchical cluster analysis, they found three distinct symptom clusters in the BAI. Since all items of cluster one and some items of the second are also common somatic complaints in MS patients, they proposed to consider the development of a "trunk and branch" model for anxiety, a model originally conceptualized by Strober and Arnett for depression [29]: while "trunk" symptoms are shared between anxiety and MS, "branch" symptoms are specific for anxiety. However, as the authors point out, the evidence base for the development of a specific model for anxiety is insufficient. The symptom overlap between anxiety and MS has been also highlighted in a retrospective study by Brousseau et al. [31]. In their study, the authors sought to identify psychiatric diagnoses among 63 MS patients whose first clinical assessment suggested a primary psychiatric etiology for their symptoms. 92 % of patients in the Brousseau et al. study met diagnostic criteria for one or more psychiatric disorders including mood, somatoform, and anxiety disorders. In conclusion, clinicians should be cautious in the identification of psychiatric conditions producing pseudo-neurological and nonspecific symptoms (such as anxiety), and they should pay special attention when assessing anxiety exclusively with screening measures, as this symptom overlap could inflate scale scores and require further assessment with a clinical interview.

The largest study that assessed the prevalence of specific anxiety disorders with a clinical interview found that GAD is the most common among MS patients [7]. GAD is characterized by uncontrollable worry accompanied by several physical symptoms such as headaches, nausea, muscle tension, and swallowing difficulty. The presence of generalized worry and health anxiety in MS patients has been underlined in several studies [32–34], and it is no surprise given the unpredictable nature of MS. As distinctly described by Bruce and Arnett [33], "Some MS patients must awaken each morning not knowing whether a restaurant will be wheelchair accessible, whether an exacerbation will prevent a vacation, or whether sudden onset visual disturbances will make a trip to the grocery store nearly impossible." Patients with MS have several reasons to be constantly worried and anxious about their health complications. In this connection, Janssens and colleagues [35] found an association between the perception of prognostic risk and anxiety in MS patients. Patients who thought that they would become dependent from a wheelchair within 2 years had high levels of anxiety and depression. Notably, in the same study, patients were inclined to overestimate their short-term risk of wheelchair dependence. Another study conducted by Jopson et al. [36] showed an association between illness identity and anxiety, explained by the authors with the fact that the tendency to attribute unpredictable symptoms (such as headache and sore throat) to MS could make patients anxious if they interpret that as a signal of disease progression. Excessive health anxiety results in greater medical care [37] and increased physical disability [38]. Kehler and Hadjistavropoulos [34] found that MS patients with elevated health anxiety are less likely to use problem-focused coping, preferring emotional preoccupation and social support as main coping strategies. They also showed how MS individuals with high levels of health anxiety experience greater disability and GAD. These results are in line with a study conducted by Feinstein et al. [39], which showed that comorbid anxiety and depression in MS patients result in increased somatic preoccupations and social dysfunction. The authors also found that suicidal thoughts in MS patients are the result of comorbid anxiety and depression and not depression alone.

As highlighted by these studies, the clinical manifestations of anxiety in MS patients could be interpreted as a reaction to the disease. While this view is sustained by a neuroimaging study that found no evidence of a cerebral correlate of anxiety in MS [40], it should be noted that two recent studies have found an association between inflammatory processes in the central nervous system and anxiety in animal models of MS [41, 42] and one MRI study found an association between gray matter atrophy in the superior and middle gyri of the right frontal lobe and anxiety scores [43]. Further investigations are needed in order to understand whether elevated anxiety in MS could be linked to specific features of the disease.

Relationships of Anxiety with Other Aspects of MS

As previously reported, a consistent percentage of MS patients show anxiety symptoms. However, it is undeniable that not all MS patients are affected by clinically relevant anxiety. This means that one or more factors can facilitate or protect a MS patient by developing an anxious state.

Many of the possible candidate factors have been examined, but a fundamental question remains unanswered about the direction of the causality between anxiety and these other aspects.

About the relationships between anxiety and other clinical aspects of MS, in some cases, the reasoning that has informed the research is that anxiety can influence another clinical feature of the disease; in other cases, a reversed direction has been hypothesized. We shall examine the details about these points case by case.

Anxiety and MS Phases, Course, Relapses, and Induced Disability

Anxiety and MS Diagnosis

A first aspect to be examined is the relationship between anxiety and the phases of the disease. Notwithstanding the possible interest of the topic, few studies have specifically addressed the issue. The diffusion of the intuitive belief that anxiety is given by a reaction to the disease and by its course over time is a possible explanation of this scarcity.

Already in 1994, it was found that the level of anxiety was influenced by the uncertainty of the diagnosis following the appearance of neurological disturbances resembling the picture of MS; after MS was confirmed or disconfirmed, anxiety more likely decreased even in the subjects which received a diagnosis of MS; subjects with no definite diagnosis tend to be more anxious [44, 45]. To be noted, the above reported studies were conducted when immune-modulating drugs were not available.

Di Legge et al. [46] reported that subjects with a clinically isolated syndrome that can be considered as the first manifestation of a possible MS showed trait anxiety scores higher than controls at the baseline evaluation; at follow-up performed on average 33 ± 6 months later, no more difference appeared. This result is at odd with what could be expected, as trait anxiety should be stable over time.

A recent (0–24 months) diagnosis of MS has a significant impact in terms of anxiety on both patients and their partners: 34 % of MS patients and 40 % of partners showed significantly higher levels of anxiety than those observed in healthy individuals from a population sample; patients (36 %) were more frequently distressed than partners. A higher EDSS ($\geq$3) score corresponded to higher levels of anxiety in patients, but disability levels did not influence anxiety in partners [35].

Anxiety seems to decrease as the time since diagnosis elapses as demonstrated by a 1-month [47], 6-month [48], 24-month [49], and 30-month [50] follow-up. In the Bianchi et al. study, anxiety scores were related to "accepting responsibility" and "seeking social support" coping, and at 24-month follow-up a reduction in "seeking social support" coping and an increase in "planful problem solving" was detected. The changes in anxiety were strongly related to those in depression at follow-up.

An increase of anxiety is among the changes perceived by close relatives in MS patients in relationship with MS appearance [51]. In this study, the behavioral changes perceived in MS patients were similar to those found in subjects with other inflammatory diseases not involving the CNS: this does not support an MS-specific behavioral profile or the connection with the damage caused by MS pathology, even if behavioral changes were associated with dysexecutive and cognitive dysfunctions in MS patients.

Anxiety and MS Exacerbations

The influence of exacerbations on the mood state has been explored by Warren et al. [52], McCabe [53], and Burns et al. [54]. Warren et al. [52] found that the experience of exacerbation increases the level of emotional disturbance in comparison with the remission phase. In the McCabe [53] study, MS patients who experienced an exacerbation in 6 months before the start of the study had anxiety levels higher than both MS patients without exacerbation and control subjects; the anxiety level registered at baseline remained stable over the 18-month observation period.

Burns et al. [54] have prospectively examined the relations of anxiety and depression to exacerbations and pseudo-exacerbations: increase in anxiety symptoms relative to baseline has predictive value for subsequent pseudo-exacerbation, while increased somatic depressive symptoms predicted confirmed exacerbation.

Warren et al. [52] and McCabe [53] have obtained different results about the use of coping strategies by MS patients in relation to exacerbations, but the differences in timing relative to exacerbation (Warren et al.'s patients were having an exacerbation when tested, while McCabe's patients experienced the exacerbation in the previous 6 months) can explain the discrepant results [52].

The relationship between anxiety and first stages of MS or relapses can be examined from a reverse point of view, that is, the possibility that anxiety, as well as other emotional disturbances, could precipitate MS onset and increase the relapse occurrence risk.

Actually, MS patients report frequently that in the period of time preceding the onset of symptoms, later diagnosed as the debut of MS, or before relapses, important stressful events happened. Some MS patients report also the experience of unusual and incomprehensible feelings preceding relapse symptoms.

Systematic observations [55–59] have shown that stressful events are associated with an increase occurrence of exacerbation, independently from infections, and psychosocial factors (negative familiar and social events related to anxiety) are associated with MS onset. Trait and state anxiety per se were not correlated with the occurrence of relapses in the Brown et al. [57] study.

Anxiety and Disability

The association between anxiety and disability levels is an issue difficult to examine. So many variables (e.g., gender, education, time since diagnosis, disease course and exacerbation number, concomitant medications, measurement of both disability and anxiety) can play a role in the relationship between the measured or perceived disability and anxiety that a single study cannot consider all of them. The cross-sectional nature of many studies makes it difficult to derive firm conclusions.

Colombo et al. [60] were the first to investigate the relationship between anxiety (using the Symptom Rating Scale), disease severity, disease duration, and age and did not find any significant association.

The studies reporting data on the association between the objective measure of disability represented by the EDSS and anxiety can be summarized with the conclusions of the Tsivgoulis et al. [61] study that "disability status is an independent but moderate determinant of depression and anxiety in MS patients."

Anxiety does not seem to influence the perception of disability in MS patients [22].

A partial conclusion that could be derived by the data on the impact of discovering of being affected by MS or of experiencing a disease exacerbation is that anxiety levels increase in a substantial percentage of these patients and that this increase seems due to reactive psychological mechanisms. However, these studies suggest something that can be underlined for many of the studies we have considered: it is very difficult to establish the direction of the causality when exploring the relations between anxiety and other aspects, e.g., is the coping strategy influencing the increase of anxiety? Or is anxiety increasing the use of a certain coping strategy?

The cross-sectional nature of some of the above reported studies does not help to derive a firm conclusion; even an observation period of 18 months seems not sufficient to clarify the point, due to the apparent stability of anxiety levels.

We shall see that data derived by different approaches concur on the idea that anxiety increase is related to reactive psychological mechanisms.

Anxiety and Worry

Worries and concerns are two topics strongly connected with anxiety: following Bruce and Arnett [33], excessive, uncontrollable worry is the hallmark of GAD, and GAD has resulted in the most common anxiety-related disorders in MS patients [7].

The study performed by Bruce and Arnett [33] on the relationships between worry and anxiety in MS patients has confirmed that, notwithstanding the strong relation with anxiety, worry can be considered as a separable and unitary construct. The relevance of worry and concerns for MS patients and their peculiarities in these patients have prompted the development of an assessment scale suitable for testing MS patients [32].

Anxiety and Cognitive Functioning

The possible correlation of anxiety, as well as of other affective symptoms, with cognitive dysfunctions has attracted the interest of various researchers. Impairment of some cognitive function affects a high percentage of MS patients (see the specific chapter in this volume), and it is therefore relevant to identify all the possible determinants.

Following a study of Simioni et al. [62], anxiety typically found at an early stage of MS seems also related to the presence of cognitive impairment; however, after adjustment for QoL levels, the relation between cognitive deficits and mood state was no longer significant; this could mean that QoL summarizes the effects of other factors influencing the same QoL.

Stenager et al. [21] have reported that the only cognitive test showing significant correlation with both trait and state anxiety was the Trail Making Test.

Summers et al. [63] found that high anxiety levels were associated with poor performance in the working memory, information processing speed, attention, and memory scores in a sample of MS patients evaluated 7 years after a clinically isolated syndrome onset.

By means of regression analyses, Julian and Arnett [64] evidenced that state, but not trait, anxiety contributed, independently from depression, to the variance of an executive function index.

On the other hand, state anxiety seems a predictor of cognitive changes over a 1-year observation period together with other negative affects [65].

The relationships between anxiety/depression and the objective performances in cognitive tests of executive functions were confirmed even in a study by Bol et al. [66]; this study has also registered that anxiety (and depression) was a significant contributor to the levels of cognitive complaints by MS patients. Unfortunately, the authors have taken into account the cumulative score of the HADS and have not separated anxiety and depression scores.

The significant relationship between anxiety and perceived cognitive functioning has been confirmed by a subsequent research by Middleton et al. [67]. However, the authors did not evaluate the influence of anxiety on objective cognitive functioning.

On the contrary, in the Karadayi et al. [68] study, retrieval from long-term memory and psychomotor speed was not related to anxiety or depression, but to other clinical variables.

Goretti et al. [69] showed that state anxiety was related to a worse performance in the SDMT; the relation with the performances in the PASAT-3 and with the presence of cognitive impairment was almost significant.

Bruce et al. [70] reported that trait anxiety was associated with self-reported memory problems. This study suggests also that normative dissociation (i.e., the disruption of an individual's usually integrated cognitive processes, such as consciousness, memory, identity, or perception) partially mediated the relationship between emotional problems and perceived memory difficulties.

Other studies support the role of anxiety in the perception or self-evaluation of cognitive functioning in MS patients: anxiety, together with other variables, was a significant predictor of the scores in the patient report of the Multiple Sclerosis Neuropsychological Questionnaire [71]; the perception of cognitive slowing was related to trait anxiety, as well as to motor speed, impulsivity, and increased introversion, more than to real performances of processing speed [72]; in the van der Hiele et al. [73] study, MS patients underestimating their executive performances showed higher levels of anxiety, as well as of depression and psychosocial stress, and used a different coping style with respect to accurate estimators and overestimators; the underestimators' awareness of some objective cognitive impairment in processing speed and cognitive flexibility might lead to psychological distress and negative report bias.

Lastly, in the study of Lester et al. [74], the relationship between anxiety and self-reported cognitive impairment was evaluated by a reverse angle, that is, the possibility that the estimate of cognitive functioning level could influence anxiety levels. Perceived cognitive impairment accounted for a 17 % of the variance in anxiety, added to the 21 % accounted for by the MS physical impact subscale. MS physical impact and perceived cognitive impairment produce, in the authors' opinion, a sense of helplessness or a feeling of lack of control.

Anxiety and Other Psychiatric Disturbances

The relationship between anxiety and other psychiatric disturbances in MS patients is obviously of interest both from a clinical and a theoretical perspective, but even this topic has not received great attention by researchers.

First, we are going to examine the connection between anxiety and depression: this association is statistically significant in all the studies that have examined both matters. But, as far as it has been reported in the existing literature, not all MS patients showing clinically significant anxiety have been classified as depressed and vice versa [7, 39, 48, 75].

A longitudinal assessment of anxiety, together with depression and fatigue, over a 2-year period was performed by Brown et al. [76]. Even if also unhealthy behaviors and psychological factors predicted psychological distress and immunotherapy status predicted state anxiety, depression at baseline was the stronger predictor of anxiety and fatigue, and anxiety and fatigue at baseline were the stronger predictors of depression. In the authors' opinion, co-morbidity of anxiety, depression, and fatigue and the overlap of their symptoms are the most suitable explanation of their results. The anxiety-lowering effect exerted by being on immunotherapy can be interpreted as a reaction of feeling safe from the disease pathological process.

Gay et al. [75] applied statistics based on causal path analysis models looking for predictors of depression in MS patients; functional status (measured by EDSS), trait anxiety, alexithymia, and social support satisfaction were the predicting factors of depression. Trait anxiety and functional status were independent and simultaneous predictors of depression; trait anxiety played a predominant role, and alexithymia and social support play as mediators for trait anxiety.

The influence of anxiety on self-harm, the degree of somatic complaints, and the functioning at social level have been examined by the above reported study of Feinstein et al. [39]: the association of depression and anxiety represents a more relevant risk than anxiety or depression alone.

Bruce and Lynch [77] have explored the relationship between personality traits and mood and anxiety disorders: anxious MS patients showed more neuroticism and were less extroverted, open, agreeable, and conscientious than both healthy controls and MS patients without an anxiety disorder. The authors suggest that MS patients with Axis I mood or anxiety disturbances are likely to experience concomitant personality changes and that suffering from MS does not mean experiencing a personality change.

No study has specifically explored the relationship between fundamental emotions and anxiety: indirect information can be derived by the lack of a significant correlation between both state and trait anxiety and the type of anger expression reported in a study devoted to anger phenomenology in a cohort of MS patients [78].

Relationship Between Anxiety and Health-Related Quality of Life

There is extensive evidence showing how anxiety is associated with low health-related quality of life (HRQoL) in MS. The first study to explore this relationship, conducted by Fruehwald et al. [79], found highly significant correlations between the majority of the scales of the Functional Status Questionnaire (FSQ) and anxiety levels measured with the Zung Anxiety Rating Scale. Benito-Leòn et al. [80] obtained very similar results using the Functional Assessment of Multiple Sclerosis (FAMS) and the Hamilton Rating Scale for Anxiety (HRSA). It has to be noted that these first studies only used bivariate correlations in order to assess the association between anxiety and HRQoL and did not control demographic data or other confounding variables such as depression or EDSS.

Spain and co-workers [81] studied the relationship between HRQoL and illness perception using multivariate models in order to predict the Short Form Health Survey (SF-36) scales. They found that anxiety was a significant predictor of all the SF-36 scores except the Physical Function Scale after controlling for age, disease duration, processing speed, fatigue, pain, and depression. Goretti and co-workers [69] obtained similar results. In their study, they tried to predict Multiple Sclerosis Quality of Life-54 (MSQoL-54) Mental and Physical Health domains accounting for anxiety (STAI-Y), mood (measured with the BDI), disability (EDSS), personality (Eysenck Personality Questionnaire, EPQ), coping (Coping Orientation for Problem Experiences, COPE), and fatigue (Fatigue Severity Scale) using multivariate regression analysis and found a significant association between lower anxiety and the mental health summary score of the MSQoL-54. Similarly, Dubayova and colleagues [82] found an association between anxiety and lower scores in the mental health composite scores of the MsQoL-54 after controlling for demographical variables and disability measured with EDSS.

Interestingly, the relationship between quality of life and clinical variables in MS patients has also been explored using path analysis in two studies [83, 84]. Salehpoor et al. [83] report an indirect relationship between anxiety and fatigue, mediated by the physical components of quality of life. Using their model obtained through path analysis, they hypothesize that an increase of fatigue levels could be a consequence of the heightened stress and tension originated from physical impairments. Using structural equation modeling, Kikuchi et al. [84] found a twofold influence of anxiety and depression on FAMS thinking and fatigue scores if compared to the EDSS.

Lifestyle and Anxiety

The interest for lifestyle in MS patients has more and more increased in the last years as a consequence of the possible causal relationships between some habits and disease course or progression. Unfortunately, as for the actual topic, the interest of researchers up to now has been limited to the relationships of mood disorders with alcohol or drug abuse. The few published studies have obtained conflicting results, with Bombardier et al. [85], Quesnel and Feinstein [86], and Korostil and Feinstein [7] reporting an increased prevalence of anxiety and depression in excessive drinkers and Turner et al. [87] and Beier et al. [88] finding no significant association. The different results can be explained by differences in applied methodologies, e.g., use of clinical interviews vs. standardized scales.

In the Quesnel and Feinstein [86] study, high anxiety levels and a family history of mental illness represent warning signals for suspecting the drinking problem which appears to be also associated with suicidal ideation and abuse of other substances.

In the Korostil and Feinstein [7] study, the association disappears when a different level of statistical significance is applied.

Beier et al. [88] also reported that an increase in drug use was associated with lower self-reported anxiety, but with greater disability and depression.

Other Aspects Related to Anxiety in MS

Anxiety seems to be related to other relevant aspects of MS or MS patients' behavior, but as these relations emerge by sparse evidences, further confirmations should be welcomed.

Bruce et al. [89] have reported that in MS patients, problems adhering to disease-modifying schedules are connected to the presence of anxiety disorders.

Anxiety has been found to be related with fatigue, mainly with mental fatigue, but even in this case the literature shows scarce and conflicting results [66].

The relationship between anxiety and employment has been examined by Krokavcova et al. [90] and by Glanz et al. [91]. Krokavcova et al. reported that MS patients without anxiety had a 2.64 greater chance of being employed, while in the study by Glanz et al., exploring work productivity in a sample of MS patients, various parameters of work and daily activities have been taken into account; among these parameters, they have also examined presenteeism (impairment while working) that was the main cause of work productivity losses and was related to some other factors, anxiety included. Even overall work productivity and activity impairment (working plus not working subjects) were related to the same factors.

As already reported in a previous section, anxiety can be related to performances on neuropsychological tests. A particular aspect of mental activity, that is, social cognition, does not seem related to anxiety levels [92].

Trait anxiety is also associated with the disability level due to comorbid migraine in MS patients [93]. In MS female patients, chronic pain was significantly related to anxiety (and depression) [94].

Anxiety and Neuroimaging

Actually, only few studies explored the possible associations between neuroimaging data and anxiety levels: Zorzon et al. [40] have reported that MRI parameters (brain volume, regional and total lesion loads) did not correlate with anxiety, while some interesting data emerged about the relationship of depression and right frontal lesion load and right temporal volume; in this study, anxiety did not correlate with any other clinical parameter, and MS patients were not more anxious than patients with rheumatoid diseases. Considering the overall obtained results, the authors conclude that "anxiety is a reactive response to the psychosocial pressure put on the patients."

Diaz-Olavarrieta et al. [13] explored the prevalence of neuropsychiatric symptoms in a sample of MS patients and their relationship with MRI results, and once more, they could not find any significant association for anxiety, measured by the Neuropsychiatric Inventory.

Also Di Legge et al. [46] did not find any correlation between state and trait anxiety and any MRI parameters, regional lesion load included.

On the contrary, an association between some MRI data and anxiety levels has been found by a study of Fassbender et al. [95], aimed at studying the relationship between mood disorders and dysfunction of the hypothalamic-pituitary-adrenal axis in MS. The authors report that 8 out of 23 enrolled MS patients that showed active lesions had significantly higher levels of depression and anxiety. The relationship of anxiety levels with inflammatory phenomena was confirmed also by the correlation with higher cell counts in cerebrospinal fluid. Furthermore, the increase of cortisol production after corticotropin stimulation correlated with anxiety scales scores (the corticotropin effect was maintained elevated in MS patients even after the suppression of HPA axis by dexamethasone).

A more recent study [43] has applied the voxel-based morphometry in studying the relationship of cognitive and mood disorders with gray matter atrophy and has found that atrophy in the gray matter of superior and middle gyri of the right frontal lobe correlated with the scores in the Hamilton Anxiety Rating Scale.

Therapeutic Interventions

As underlined thus far, although anxiety represents a common psychiatric comorbidity in MS patients, researchers in the psychiatric field have overlooked it. Not surprisingly, the literature that explores the efficacy of pharmacologic and non-pharmacologic treatments for anxiety disorders is scarce and produced inconsistent results. In their comprehensive guideline, Minden et al. [96] reviewed several studies, which concerned the assessment and management of psychiatric disorders in MS patients. Despite pharmacologic and non-pharmacologic therapies are often used for the treatment of anxiety in MS, the authors found little evidence supporting their efficacy. Indeed, few specific studies assessed the efficacy of these treatments for anxiety disorders in MS.

Non-pharmacologic Treatments for Anxiety Disorders

The majority of the research regarding psychological therapies for anxiety in MS is based on manualized cognitive behavioral interventions, and the results supporting their efficacy are conflicting [96]. A randomized controlled trial (RCT) comparing the efficacy of a short-term protocol mixing cognitive behavioral therapy (CBT) and relaxation training with two sessions of supportive psychotherapy (namely, the "stress inoculation training" or SIT) found a reduction of the STAI scores after treatment for the SIT group [97]. Another RCT found that a treatment based on relaxation and imagery is more effective than no treatment in reducing high anxiety levels measured with the STAI [98]. However, in the same study, anxiety symptoms measured at baseline with the Profile of Mood Scales (POMS) were low, thus questioning the generalizability of the results. Another study assessing the efficacy of CBT-based group therapy on a group of 20 patients with MS showing elevated anxiety and depression found no significant difference in anxiety levels between the pre- and post treatment conditions [99]. More recently, a RCT comparing a CBT self-management program (MS Invigor8) with standard care found significant improvements in anxiety and depression in the treatment group [100].

Given the inconsistency in the literature, it has to be underlined that when evaluating non-pharmacologic and psychotherapeutic interventions for psychological disorders in MS, clinicians should be well aware that despite a growing body of evidence supporting the efficacy of "empirically supported treatments" for specific disorders, there is just as much evidence showing that the majority of the variance in the outcome of a therapeutic intervention is explained by nonspecific factors [101, 102]. As already established by solid research, the most important predictor of the success of psychotherapy is the working alliance between the therapist and the patient [103, 104]. In other words, the quality of the relationship between a healthcare professional and the patient (e.g., the agreement of both on therapeutic tasks and goals) is crucial for the success of the intervention, regardless of the diagnosis. For this reason, clinicians cannot disregard that the singularity of the patient is far more important than the specificity of the treatment to the diagnosis. As stated by the conclusions and recommendations of the interdivisional task force on evidence-based therapy relationships of the American Psychological Association, "efforts to promulgate best practices or evidence-based practices (EBPs) without including the relationship are seriously incomplete and potentially misleading" [101]. Therefore, further qualitative and quantitative studies are needed in order to widen the range of therapeutic interventions (i.e., including interpersonal, psychodynamic, and humanistic interventions) together with special focus on effective ways in which clinicians could tailor their work on the specificity of the problems faced by patients with MS.

Pharmacologic Therapy

There is no trial having considered a pharmacologic therapy of anxiety in MS patients. The absence of such trials in the literature is confirmed by a very recent and already cited report [96]. Therefore, we cannot advance any suggestion based on controlled data. The experience of the single specialist remains valid, based on the efficacy of various drugs in anxious patients in the general population, with some important warnings: benzodiazepines that could be used for acute anxious symptoms can cause excessive somnolence, mental slowing, and diffuse muscle relaxation, which could be problematic for MS patients; drugs in the categories of selective serotonin reuptake inhibitors or selective norepinephrine reuptake inhibitors are suitable for long-term treatment of some chronic anxiety disorders: they are more manageable than older drugs, like tricyclic antidepressants, but they can nevertheless have important side effects (e.g., sexual dysfunctions, weight increase, sedation, feelings of fatigue). Even a more recently introduced drug, mirtazapine, which does not impair sexual function, causes sedation and weight gain. These drugs are frequently abandoned by MS patients under treatment for depression, and therefore, they could appear even less acceptable to patients for treating isolated anxiety disorders. Patients affected by GAD from the general population have taken advantage by treatments based on pregabalin, a calcium channel modulator, or quetiapine, an atypical neuroleptic, both drugs at medium-high doses; this second drug can be charged by relevant side effects. The only suggestion that can be advanced, hoping in some well-conducted drug trial, is a case-by-case evaluation of a pharmacologic therapy for an anxiety disorder and to rely on the practice with non-MS patients.

Conclusion

Anxiety levels and anxiety disorders are more frequent in MS patients than in general population; pathological increase of anxiety levels is reported in the literature as more frequent than depression, when the two disorders have been examined in the same MS patient sample. While normal anxiety and fear are largely adaptive, pervading anxiety is considered maladaptive, and in its various forms, it can characterize clinical disorders. However, anxiety in MS is overlooked and undertreated.

The increase of anxiety seems mainly related to the diagnostic work-up period (in this case also connected to diagnostic uncertainties), after the diagnosis has been advanced and after an exacerbation. Anxiety tends to decrease as time elapses after diagnosis and exacerbations; however, the "longitudinal" studies that have been conducted till now have covered a maximum of 30-month period and have regarded the RR phase of the disease. Information is not available about patients in the SP phase or with PP course. Anxiety seems to increase more frequently in women than in men. While stressors resulted to increase the risk of relapses and have been also connected to MS onset, anxiety per se does not seem to influence those risks.

Assessment of anxiety has been mainly performed by standardized self-reported scales, being the Hospital Anxiety and Depression Scale the most extensively used. Rarely, a clinical interview (e.g., SCID) suitable for diagnosing also the presence of a specific anxiety disorder has been used. A moderate interest has been as well devoted to the identification of trait or state anxiety.

When specific anxiety disorders have been searched for, GAD, panic disorder, and obsessive-compulsive disorder showed to be the most frequently present. To be noted, however, obsessive-compulsive disorder is no longer considered in the classification of anxiety disorders in the DSM-V and described in a dedicated chapter with other obsessive-compulsive-related disorders, as substantial evidence suggests that obsessive-compulsive spectrum disorders are distinct from anxiety disorders both in their behavioral and phenomenological appearance [105].

Anxiety levels have been studied in connection with many other variables of interest in MS patients. Apart from the abovementioned relation with disease fundamental clinical features, the most frequently studied relationships were those with cognitive functioning and health-related quality of life (HRQoL): about cognitive functioning, even taking into account conflicting results, anxiety seems to influence complex attention and executive functioning; by a reverse angle, perceived cognitive dysfunctions seem to increase anxiety. HRQoL is clearly related to anxiety levels or severity.

Apart from anxiety and depression that have been evaluated in almost every study aimed at exploring MS patients' psycho-emotional state, the relationship between anxiety and other psychiatric dysfunctions, personality traits, or emotion expression has been largely ignored.

The relative neglect for anxiety and its relationships has probably its motivation in the diffuse belief, among clinicians, that increase of anxiety has a purely reactive nature and that an anxious reaction is unavoidable, especially in some occurrences like diagnosis disclosure or relapses.

The available data seem to confirm this position, as the few studies devoted to search for a connection between the various aspects of brain damage (lesion load and/or atrophy, both global and/or regional) and anxiety parameters have achieved mainly negative results. Together with the clinical data, these results support the "reactive" interpretation of anxiety.

Therefore, anxiety could be explained by the ineffectiveness of "buffer" mechanisms or the exhaustion of energies in response to an intense distress caused by the perspective connected to a serious and unpredictable disease. Anxiety in MS patients appears to be related to the beliefs concerning the pathology, such as the prognostic risk and the likelihood of being wheelchair dependent.

The role of coping strategies in moderating or favoring anxious reactions and the relationship between normative dissociation (defined as the disruption of usually integrated cognitive processes) and anxiety are of support to the reactive nature of pathological anxiety.

After this brief summary of the most interesting results, it's time to spend some words on the weaknesses and limitations of the available studies.

The cross-sectional nature of some studies and the relatively short length of the longitudinal ones do not allow inferring the real meaning of the reported relationships. Furthermore, it is not easy to establish the direction of the relationship between anxiety and other aspects, from both the conceptual and observational point of view. The hypothesis that anxiety and the other aspects are both a consequence of a third factor is worth being explored in many cases.

Some of the available studies have important limitations (e.g., small samples, retrospective collection of data, lack of important data, exclusion of some subjects). One of the most important points, the "reactive" vs. "organic" nature of anxiety disorders, not considering the complexity and uncertainty of the theme in general, has never been explored through functional neuroimaging studies.

The second most disappointing aspect appears from a very recent statement made in the framework of the report on evidence-based guidelines for the assessment and management of psychiatric disorders in individuals with MS [96]: not a single instrument among those used in the published studies for evaluating anxiety has obtained even the lower level of recommendation. We would like to point out that, even if the instruments that have been taken into account in our treatise are not completely satisfactory, the obtained data have some reliability, as in most cases they have been validated in MS samples. Otherwise, all results and considerations reported in this chapter have scarce meaning.

The main disappointment arises from the awareness that anxiety disorders are under-identified and, consequently, undertreated: these aspects are in reciprocal relation with the lack or paucity of clinical trials on the treatment of anxiety, both pharmacologic and non-pharmacologic.

A possible list of recommendations for future research in this field can be found in the abovementioned report [96]. For clinical practice, waiting for further information by the research side, we credit MS specialists for their capacity to suggest anxiety treatments when needed, relying on the knowledge deriving from their experiences about the same disturbances in the general population, with all the cautions requested by the peculiarities of MS patients. As a first step, we wish to recommend that a good-quality communication between healthcare providers and patients is crucial in order to prevent excessive anxiety.

References

1. Kazdin AE. Encyclopedia of psychology, Vol. 2 [Internet]. 2000. doi:10.1037/10517.
2. Craighead WE, Nemeroff CB. The Corsini encyclopedia of psychology and behavioral science, vol. 1–4. 3rd ed. New York: Wiley; 2001.
3. Barlow DH. Unraveling the mysteries of anxiety and its disorders from the perspective of emotion theory. Am Psychol. 2000;55:1247–63. doi:10.1037/0003-066X.55.11.1247.
4. Seneca LA. “Letter LXXXVI” Letters from a stoic. Trans. Robin Campbell. London: Penguin; 1975.
5. Dulin PL, Passmore T. Avoidance of potentially traumatic stimuli mediates the relationship between accumulated lifetime trauma and late-life depression and anxiety. J Trauma Stress. 2010;23:296–9. doi:10.1002/jts.20512.
6. American Psychiatric Association. Diagnostic and statistical manual of mental disorders. 5th ed. Washington, DC: American Psychiatric Association; 2013.
7. Korostil M, Feinstein A. Anxiety disorders and their clinical correlates in multiple sclerosis patients. Mult Scler. 2007;13:67–72. doi:10.1177/1352458506071161.
8. Grigsby AB, Anderson RJ, Freedland KE, Clouse RE, Lustman PJ. Prevalence of anxiety in adults with diabetes a systematic review. J Psychosom Res. 2002;53:1053–60. doi:10.1016/S0022-3999(02)00417-8.
9. Karajgi B, Rifkin A, Doddi S, Kolli R. The prevalence of anxiety disorders in patients with chronic obstructive pulmonary disease. Am J Psychiatry. 1990;147:200–201. doi:10.1176/ajp.1472.200.
10. Kirmayer LJ, Robbins JM, Dworkind M, Yaffe MJ. Somatization and the recognition of depression and anxiety in primary care. Am J Psychiatry. 1993;150:734–41.
11. Galeazzi GM, Ferrari S, Giaroli G, Mackinnon A, Merelli E, Motti L, et al. Psychiatric disorders and depression in multiple sclerosis outpatients: impact of disability and interferon beta therapy. Neurol Sci. 2005;26:255–62. doi:10.1007/s10072-005-0468-8.
12. Shabani A, Moghadam J, Panaghi L, Seddigh A. Anxiety disorders in multiple sclerosis: significance of obsessive-compulsive disorder comorbidity. J Res Med Sci. 2007;12(4):172–7.
13. Diaz-Olavarrieta C, Cummings JL, Velazquez J, Garcia de la Cadena C. Neuropsychiatric manifestations of multiple sclerosis. J Neuropsychiatry Clin Neurosci. 1999;11:51–57. http://neuro.psychiatryonline.org/article.aspx?articleid=100012#tab1.
14. Janssens ACJW, van Doorn PA, de Boer JB, Kalkers NF, van der Meche FGA, Passchier J, et al. Anxiety and depression influence the relation between disability status and quality of life in multiple sclerosis. Mult Scler. 2003;9:397–403. doi:10.1191/1352458503ms930oa.
15. Janssens ACJW, Buljevac D, van Doorn PA, van der Meché FGA, Polman CH, Passchier J, et al. Prediction of anxiety and distress following diagnosis of multiple sclerosis: a two-year longitudinal study. Mult Scler. 2006;12:794–801. doi:10.1177/1352458506070935.
16. Poder K, Ghatavi K, Fisk JD, Campbell TL, Kisely S, Sarty I, et al. Social anxiety in a multiple sclerosis clinic population. Mult Scler. 2009;15:393–8. doi:10.1177/1352458508099143.
17. Marrie RA, Fisk JD, Yu BN, Leung S, Elliott L, Caetano P, et al. Mental comorbidity and multiple sclerosis: validating administrative data to support population-based surveillance. BMC Neurol. 2013;13:16. doi:10.1186/1471-2377-13-16.
18. Joffe RT, Lippert GP, Gray TA, Sawa G, Horvath Z. Mood disorder and multiple sclerosis. Arch Neurol. 1987;44:376–8. doi:10.1001/archneur.1987.00520160018007.

19. Minden SL, Orav J, Reich P. Depression in multiple sclerosis. Gen Hosp Psychiatry. 1987;9(6):426–34. doi:Depression in multiple sclerosis. Gen Hosp Psychiatry.
20. Arias Bal MA, Vázquez-Barquero JL, Peña C, Miro J, Berciano JA. Psychiatric aspects of multiple sclerosis. Acta Psychiatr Scand. 1991;83:292–6. http://www.ncbi.nlm.nih.gov/pubmed/2028805.
21. Stenager E, Knudsen L, Jensen K. Multiple sclerosis: correlation of anxiety, physical impairment and cognitive dysfunction. Ital J Neurol Sci. 1994;15:97–101. doi:10.1007/BF02340120.
22. Smith SJ, Young CA. The role of affect on the perception of disability in multiple sclerosis. Clin Rehabil. 2000;14:50–4. doi:10.1191/026921500676724210.
23. Nicholl CR, Lincoln NB, Francis VM, Stephan TF. Assessment of emotional problems in people with multiple sclerosis. Clin Rehabil. 2001;15:657–68. doi:10.1191/0269215501cr427oa.
24. Mendes MF, Tilbery CP, Balsimelli S, Moreira MA, Barao-Cruz AM. Depression in relapsing-remitting multiple sclerosis. Arq Neuropsiquiatr. 2003;61(3A):591–5. doi:10.1590/S0004-282X2003000400012.
25. Figved N, Benedict R, Klevan G, Myhr KM, Nyland HI, Landrø NI, et al. Relationship of cognitive impairment to psychiatric symptoms in multiple sclerosis. Mult Scler. 2008;14:1084–90. doi:10.1177/1352458508092262.
26. Beiske AG, Svensson E, Sandanger I, Czujko B, Pedersen ED, Aarseth JH, et al. Depression and anxiety amongst multiple sclerosis patients. Eur J Neurol. 2008;15:239–45. doi:10.1111/j.1468-1331.2007.02041.x.
27. Dahl O-P, Stordal E, Lydersen S, Midgard R. Anxiety and depression in multiple sclerosis. A comparative population-based study in Nord-Trøndelag County, Norway. Mult Scler. 2009;15:1495–501. doi:10.1177/1352458509351542.
28. Espinnola-Nadurille M, Colin-Piana R, Ramirez-Bermudez J, Lopez-Gomez M, Flores J, Arrambide G, et al. Mental disorders in Mexican patients with multiple sclerosis. J Neuropsychiatry Clin Neurosci 2010;63–69. doi:10.1176/appi.neuropsych.22.1.63.
29. Strober LB, Arnett PA. Assessment of depression in multiple sclerosis: development of a "trunk and branch" model. Clin Neuropsychol. 2010;24:1146–66. doi:10.1080/13854046.2010.514863.
30. Donnchadha S, Burke T, Bramham J, O'Brien MC, Whelan R, Reilly R, et al. Symptom overlap in anxiety and multiple sclerosis. Mult Scler. 2013;19:1349–54. doi:10.1080/13854046.2010.514863.
31. Brousseau KM, Arciniegas DB, Carmosino MJ, Corboy JR. The differential diagnosis of Axis I psychopathology presenting to a university-based multiple sclerosis clinic. Mult Scler. 2007;13:749–53. doi:10.1177/1352458506075032.
32. Thornton EW, Tedman S, Rigby S, Bashforth H, Young C. Worries and concerns of patients with multiple sclerosis: development of an assessment scale. Mult Scler. 2006;12:196–203. doi:10.1191/135248506ms1273oa.
33. Bruce JM, Arnett P. Clinical correlates of generalized worry in multiple sclerosis. J Clin Exp Neuropsychol. 2009;31:698–705. doi:10.1080/13803390802484789.
34. Kehler MD, Hadjistavropoulos HD. Is health anxiety a significant problem for individuals with multiple sclerosis? J Behav Med. 2009;32:150–61. doi:10.1007/s10865-008-9186-z.
35. Janssens ACJW, van Doorn PA, de Boer JB, van der Meché FGA, Passchier J, Hintzen RQ. Impact of recently diagnosed multiple sclerosis on quality of life, anxiety, depression and distress of patients and partners. Acta Neurol Scand. 2003;108:389–95. doi:10.1034/j.1600-0404.2003.00166.x.
36. Jopson NM, Moss-Morris R. The role of illness severity and illness representations in adjusting to multiple sclerosis. J Psychosom Res. 2003;54:503–11. doi:10.1016/S0022-3999(02)00455-5.
37. Barsky AJ, Ettner SL, Horsky J, Bates DW. Resource utilization of patients with hypochondriacal health anxiety and somatization. Med Care. 2001;39:705–15. doi:10.1097/00005650-200107000-00007.

38. Gureje O, Simon GE, Ustun TB, Goldberg DP. Somatization in cross-cultural perspective: a World Health Organization study in primary care. Am J Psychiatry. 1997;154:989–95. doi:10.1017/S003329 1797005345.
39. Feinstein A, O'Connor P, Gray T, Feinstein K. The effects of anxiety on psychiatric morbidity in patients with multiple sclerosis. Mult Scler. 1999;5:323–6. doi:10.1177/135245859900500504.
40. Zorzon M, de Masi R, Nasuelli D, Ukmar M, Mucelli RP, Cazzato G, et al. Depression and anxiety in multiple sclerosis. A clinical and MRI study in 95 subjects. J Neurol. 2001;248:416–21. doi:10.1007/s004150170184.
41. Peruga I, Hartwig S, Thöne J, Hovemann B, Gold R, Juckel G, et al. Inflammation modulates anxiety in an animal model of multiple sclerosis. Behav Brain Res. 2011;220:20–9. doi:10.1016/j.bbr.2011.01.018.
42. Haji N, Mandolesi G, Gentile A, Sacchetti L, Fresegna D, Rossi S, et al. TNF-α-mediated anxiety in a mouse model of multiple sclerosis. Exp Neurol. 2012;237:296–303. doi:10.1016/j.expneurol.2012.07.010.
43. Lin A, Chen F, Liu F, Li Z, Liu Y, Lin S, et al. Regional gray matter atrophy and neuropsychologcal problems in relapsing-remitting multiple sclerosis. Neural Regen Res. 2013;8(21):1958–65. doi:10.3969/j.issn.1673-5374.2013.21.004.
44. O'Connor P, Detsky AS, Tansey C, Kucharczyk W. Effect of diagnostic testing for multiple sclerosis on patient health perceptions. Arch Neurol. 1994;51:46–51. doi:10.1001/archneur.1994.00540130072013.
45. Mushlin AI, Mooney C, Grow V, Phelps CE. The value of diagnostic information to patients with suspected multiple sclerosis. Rochester-Toronto MRI Study Group. Arch Neurol. 1994;51:67–72. doi:10.1001/archneur.1994.00540130093017.
46. Di Legge S, Piattella MC, Pozzilli C, Pantano P, Caramia F, Pestalozza IF, et al. Longitudinal evaluation of depression and anxiety in patients with clinically isolated syndrome at high risk of developing early multiple sclerosis. Mult Scler. 2003;9:302–6. doi:10.1191/1352458503ms921oa.
47. Mattarozzi K, Vignatelli L, Baldin E, Lugaresi A, Pietrolongo E, et al. G.E. Ro.N.I. Mu.S. Effect of the disclosure of MS diagnosis on anxiety, mood and quality of life of patients: a prospective study. Int J Clin Pract. 2012;66(5):504–14. doi:10.1111/j.1742-1241.2012.02912.x.
48. Giordano A, Granella F, Lugaresi A, Martinelli V, Trojano M, Confalonieri P, et al. Anxiety and depression in multiple sclerosis patients around diagnosis. J Neurol Sci. 2011;307:86–91. doi:10.1016/j.jns.2011.05.008.
49. Bianchi V, De Giglio L, Prosperini L, Mancinelli C, De Angelis F, Barletta V, et al. Mood and coping in clinically isolated syndrome and multiple sclerosis. Acta Neurol Scand. 2013. doi:10.1111/ane.12194.
50. Wood B, van der Mei IAF, Ponsonby A-L, Pittas F, Quinn S, Dwyer T, et al. Prevalence and concurrence of anxiety, depression and fatigue over time in multiple sclerosis. Mult Scler. 2012;19(2):217–24. doi:10.1177/1352458512450351.
51. Lima FS, Simioni S, Bruggimann L, Ruffieux C, Dudler J, Felley C, et al. Perceived behavioral changes in early multiple sclerosis. Behav Neurol. 2007;18:81–90. doi:10.1155/2007/674075.
52. Warren S, Warren KG, Cockerill R. Emotional stress and coping in multiple sclerosis (MS) exacerbations. J Psychosom Res. 1991;35(1):37–47. doi:10.1016/0022-3999(91)90005-9.
53. McCabe MP. Mood and self-esteem of persons with multiple sclerosis following an exacerbation. J Psychosom Res. 2005;59:161–6. doi:10.1016/j.jpsychores.2005.04.010.
54. Burns MN, Nawacki E, Siddique J, Pelletier D, Mohr DC. Prospective examination of anxiety and depression before and during confirmed and pseudoexacerbations in patients with multiple sclerosis. Psychosom Med. 2013;75:76–82. doi:10.1097/PSY.0b013e3182757b2b.
55. Buljevac D, Hop WCJ, Reedeker W, Janssens ACJW, van der Meché FGA, van Doorn PA, et al. Self reported stressful life events and exacerbations in multiple sclerosis: prospective study. BMJ. 2003;327:646. doi:10.1136/bmj.327.7416.646.

56. Brown RF, Tennant CC, Sharrock M, Hodgkinson S, Dunn SM, Pollard JD. Relationship between stress and relapse in multiple sclerosis: part I. Important features. Mult Scler. 2006;12:453–64. doi:10.1191/1352458506ms1295oa.
57. Brown RF, Tennant CC, Sharrock M, Hodgkinson S, Dunn SM, Pollard JD. Relationship between stress and relapse in multiple sclerosis: part II. Direct and indirect relationships. Mult Scler. 2006;12:465–75. doi:10.1191/1352458506ms1296oa.
58. Potagas C, Mitsonis C, Watier L, Dellatolas G, Retziou A, Mitropoulos P, et al. Influence of anxiety and reported stressful life events on relapses in multiple sclerosis: a prospective study. Mult Scler. 2008;14:1262–8. doi:10.1177/1352458508095331.
59. Liu XJ, Ye HX, Li WP, Dai R, Chen D, Jin M. Relationship between psychosocial factors and onset of multiple sclerosis. Eur Neurol. 2009;62:130–6. doi:10.1159/000226428.
60. Colombo G, Armani M, Ferruzza E, Zuliani C. Depression and neuroticism in multiple sclerosis. Ital J Neurol Sci. 1988;9(6):551–7. doi:10.1007/BF02337008.
61. Tsivgoulis G, Triantafyllou N, Papageorgiou C, Evangelopoulos ME, Kararizou E, Sfagos C, et al. Associations of the Expanded Disability Status Scale with anxiety and depression in multiple sclerosis outpatients. Acta Neurol Scand. 2007;115:67–72. doi:10.1111/j.1600-0404.2006.00736.x.
62. Simioni S, Ruffieux C, Bruggimann L, Annoni J-M, Schluep M. Cognition, mood and fatigue in patients in the early stage of multiple sclerosis. Swiss Med Wkly. 2007;137:496–501. http://www.smw.ch/for-readers/archive/backlinks/?url=/docs/archive200x/2007/35/smw-11874.html.
63. Summers M, Swanton J, Fernando K, Dalton C, Miller DH, Cipolotti L, et al. Cognitive impairment in multiple sclerosis can be predicted by imaging early in the disease. J Neurol Neurosurg Psychiatry. 2008;79:955–8. doi:10.1136/jnnp.2007.138685.
64. Julian LJ, Arnett PA. Relationships among anxiety, depression, and executive functioning in multiple sclerosis. Clin Neuropsychol. 2009;23:794–804. doi:10.1080/13854040802665808.
65. Christodoulou C, Melville P, Scherl WF, Macallister WS, Abensur RL, Troxell RM, et al. Negative affect predicts subsequent cognitive change in multiple sclerosis. J Int Neuropsychol Soc. 2009;15:53–61. doi:10.1017/S135561770809005X.
66. Bol Y, Duits AA, Hupperts RMM, Vlaeyen JWS, Verhey FRJ. The psychology of fatigue in patients with multiple sclerosis: A review. J Psychosom Res. 2009;66:3–11. doi:10.1016/j.jpsychores.2008.05.003.
67. Middleton L, Denney D, Lynch S, Parmenter B. The relationship between perceived and objective functioning in multiple sclerosis. Arch Clin Neuropsychol. 2006;21:487–94. doi:10.1016/j.acn.2006.06.008.
68. Karadayi H, Arisoy O, Altunrende B, Boztas MH, Sercan M. The relationship of cognitive impairment with neurological and psychiatric variables in multiple sclerosis patients. Int J Psychiatry Clin Pract. 2014;18(1):45–51. doi:10.3109/13651501.2013.845221.
69. Goretti B, Portaccio E, Zipoli V, Hakiki B, Siracusa G, Sorbi S, et al. Coping strategies, psychological variables and their relationship with quality of life in multiple sclerosis. Neurol Sci. 2009;30:15–20. doi:10.1007/s10072-008-0009-3. doi:10.1093/arclin/acp092.
70. Bruce JM, Bruce AS, Hancock L, Lynch S. Self-reported memory problems in multiple sclerosis: Influence of psychiatric status and normative dissociative experiences. Arch Clin Neuropsychol. 2010;25:39–48. doi:10.1093/arclin/acp092.
71. Akbar N, Honarmand K, Feinstein A. Self-assessment of cognition in Multiple Sclerosis: the role of personality and anxiety. Cogn Behav Neurol [Internet]. 2011;24:115–21. doi:10.1097/WNN.0b013e31822a20ae.
72. Roberg BL, Bruce JM, Lovelace CT, Lynch S. How patients with multiple sclerosis perceive cognitive slowing. Clin Neuropsychol. 2012;26:1278–95. doi:10.1080/13854046.2012.733413.
73. van der Hiele K, Spliethoff-Kamminga NGA, Ruimschotel RP, Middelkoop HAM, Visser LH. The relationship between self-reported executive performance and psychological characteristics in multiple sclerosis. Eur J Neurol. 2012;19:562–9. doi:10.1111/j.1468-1331.2011.03538.x.

74. Lester K, Stepleman L, Hughes M. The association of illness severity, self-reported cognitive impairment, and perceived illness management with depression and anxiety in a multiple sclerosis clinic population. J Behav Med. 2007;30(2):177–86. doi:10.1007/s10865-007-9095-6.
75. Gay MC, Vrignaud P, Garitte C, Meunier C. Predictors of depression in multiple sclerosis patients. Acta Neurol Scand. 2010;121:161–70. doi:10.1111/j.1600-0404.2009.01232.x.
76. Brown RF, Valpiani EM, Tennant CC, Dunn SM, Sharrock M, Hodgkinson S, et al. Longitudinal assessment of anxiety, depression, and fatigue in people with multiple sclerosis. Psychol Psychother. 2009;82:41–56. doi:10.1348/147608308X345614.
77. Bruce JM, Lynch SG. Personality traits in multiple sclerosis: association with mood and anxiety disorders. J Psychosom Res. 2011;70:479–85. doi:10.1016/j.jpsychores.2010.12.010.
78. Nocentini U, Tedeschi G, Migliaccio R, Dinacci D, Lavorgna L, Bonavita S, et al. An exploration of anger phenomenology in multiple sclerosis. Eur J Neurol. 2009;16:1312–7. doi:10.1111/j.1468-1331.2009.02727.x.
79. Fruehwald S, Loeffler-Stastka H, Eher R, Saletu B, Baumhackl U. Depression and quality of life in multiple sclerosis. Acta Neurol Scand. 2001;104:257–61. doi:10.1034/j.1600-0404.2001.00022.x.
80. Benito-León J, Morales JM, Rivera-Navarro J. Health-related quality of life and its relationship to cognitive and emotional functioning in multiple sclerosis patients. Eur J Neurol. 2002;9:497–502. doi:10.1046/j.1468-1331.2002.00450.x.
81. Spain LA, Tubridy N, Kilpatrick TJ, Adams SJ, Holmes ACN. Illness perception and health-related quality of life in multiple sclerosis. Acta Neurol Scand. 2007;116:293–299. doi:0.1111/j.1600-0404.2007.00895.x.
82. Dubayova T, Krokavcova M, Nagyova I, Rosenberger J, Gdovinova Z, Middel B, et al. Type D, anxiety and depression in association with quality of life in patients with Parkinson's disease and patients with multiple sclerosis. Qual Life Res. 2013;22:1353–60. doi:10.1007/s11136-012-0257-9.
83. Salehpoor G, Mozaffar H, Sajjad R. A preliminary path analysis: Effect of psychopathological symptoms, mental and physical dysfunctions related to quality of life and body mass index on fatigue severity of Iranian patients with multiple sclerosis. Iran J Neurol. 2012;11(3):96–105. http://www.ncbi.nlm.nih.gov/pmc/articles/PMC3829253/pdf/IJNL-11-096.pdf.
84. Kikuchi H, Mifune N, Niino M, Kira J-I, Kohriyama T, Ota K, et al. Structural equation modeling of factors contributing to quality of life in Japanese patients with multiple sclerosis. BMC Neurol. 2013;13:10. doi:10.1186/1471-2377-13-10.
85. Bombardier CH, Blake KD, Ehde DM, Gibbons LE, Moore D, Kraft GH. Alcohol and drug abuse among persons with multiple sclerosis. Mult Scler. 2004;10:35–40. doi:10.1191/1352458504ms989oa.
86. Quesnel S, Feinstein A. Multiple sclerosis and alcohol: a study of problem drinking. Mult Scler. 2004;10:197–201. doi:10.1191/1352458504ms992oa.
87. Turner AP, Hawkins EJ, Haselkorn JK, Kivlahan DR. Alcohol misuse and multiple sclerosis. Arch Phys Med Rehabil. 2009;90:842–8. doi:10.1016/j.apmr.2008.11.017.
88. Beier M, D'Orio V, Spat J, Shuman M, Foley FW. Alcohol and substance use in multiple sclerosis. J Neurol Sci. 2014;338:122–7. doi:10.1016/j.jns.2013.12.029.
89. Bruce JM, Hancock LM, Arnett P, Lynch S. Treatment adherence in multiple sclerosis: association with emotional status, personality, and cognition. J Behav Med. 2010;33:219–27. doi:10.1007/s10865-010-9247-y.
90. Krokavcova M, Nagyova I, Van Dijk JP, Rosenberger J, Gavelova M, Middel B, et al. Self-rated health and employment status in patients with multiple sclerosis. Disabil Rehabil. 2010;32:1742–8. doi:10.3109/09638281003734334.
91. Glanz BI, Dégano IR, Rintell DJ, Chitnis T, Weiner HL, Healy BC. Work productivity in relapsing multiple sclerosis: associations with disability, depression, fatigue, anxiety, cognition, and health-related quality of life. Value Health. 2012;15:1029–35. doi:10.1016/j.jval.2012.07.010.

92. Pöttgen J, Dziobek I, Reh S, Heesen C, Gold SM. Impaired social cognition in multiple sclerosis. J Neurol Neurosurg Psychiatry. 2013;84:523–8. doi:10.1136/jnnp-2012-304157.
93. Villani V, De Giglio L, Sette G, Pozzilli C, Salvetti M, Prosperini L. Determinants of the severity of comorbid migraine in multiple sclerosis. Neurol Sci. 2012;33:1345–53. doi:10.1007/s10072-012-1119-5.
94. Kalia LV, O'Connor PW. Severity of chronic pain and its relationship to quality of life in multiple sclerosis. Mult Scler. 2005;11:322–7. doi:10.1191/1352458505ms1168oa.
95. Fassbender K, Schmidt R, Mössner R, Kischka U, Kühnen J, Schwartz A, et al. Mood disorders and dysfunction of the hypothalamic-pituitary-adrenal axis in multiple sclerosis: association with cerebral inflammation. Arch Neurol. 1998;55:66–72. doi:10.1001/archneur.55.1.66.
96. Minden SL, Feinstein A, Kalb RC, Miller D, Mohr DC, Patten SB, et al. Evidence-based guideline: assessment and management of psychiatric disorders in individuals with MS Report of the Guideline Development Subcommittee of the American Academy of Neurology. Neurology. 2014;82:174–81. doi:10.1212/WNL.0000000000000013.
97. Foley FW, Bedell JR, LaRocca NG, Scheinberg LC, Reznikoff M. Efficacy of stress-inoculation training in coping with multiple sclerosis. J Consult Clin Psychol. 1987;55:919–22. doi:10.1037/0022-006X.55.6.919.
98. Maguire BL. The effects of imagery on attitudes and moods in multiple sclerosis patients. Altern Ther Health Med. 1996;2:75–9.
99. Forman AC, Lincoln NB. Evaluation of an adjustment group for people with multiple sclerosis: a pilot randomized controlled trial. Clin Rehabil. 2010;24:211–21. doi:10.1177/0269215509343492.
100. Moss-Morris R, McCrone P, Yardley L, van Kessel K, Wills G, Dennison L. A pilot randomised controlled trial of an Internet-based cognitive behavioural therapy self-management programme (MS Invigor8) for multiple sclerosis fatigue. Behav Res Ther. 2012;50:415–21. doi:10.1016/j.brat.2012.03.001.
101. Norcross JC, Wampold BE. Evidence-based therapy relationships: research conclusions and clinical practices. Psychotherapy (Chic). 2011;48:98–102. doi:10.1037/a0022161.
102. Wampold BE, Goodheart CD, Levant RF. Clarification and elaboration on evidence-based practice in psychology. Am Psychol. 2007;62:616–8. doi:10.1037/0003-066X62.6.616.
103. Wampold BE. The great psychotherapy debate: models, methods, and findings [Internet]. 2001. Available from: http://www.amazon.com/dp/0805832025.
104. Ahn H, Wampold BE. Where oh where are the specific ingredients? A meta-analysis of component studies in counseling and psychotherapy. J Couns Psychol. 2001;48:251–7. doi:10.1037/0022-0167.48.3.251.
105. Van Ameringen M, Patterson B, Simpson W. DSM-5 obsessive-compulsive and related disorders: clinical implications of new criteria. Depress Anxiety. 2014;31:487–93. doi:10.1002/da.22259.

Chapter 5
Multiple Sclerosis and Bipolar Disorders

Mauro Giovanni Carta, Maria Francesca Moro, Giuseppina Trincas, Lorena Lorefice, Eleonora Cocco, and Maria Giovanna Marrosu

Abstract Multiple sclerosis (MS) is a neurological disorder with high prevalence among young adults and a heavy impact on quality of life. Co-morbidity of MS and mood disorders has relevance because these disorders may interact with MS, thus increasing the burden of disability and worsening the course of illness and quality of life.

This chapter briefly summarizes recent works that have well established that bipolar disorder (BD) is quite common in MS and that the co-morbidity between MS and BP seriously compromises the quality of life of the patient.

The major pathogenic theories that have tried to explain the association between MS and BP are also taken into account, with particular attention to assumptions about a possible common pathogenesis in the alterations of the oxidation mechanisms in the brain.

From a clinical perspective, BP may be well treated if correctly diagnosed in MS, but the risk of an underdiagnosis of BP, and of type II in particular, in MS suggests caution in prescribing antidepressants to people with depressive episodes in MS without prior excluding BD diagnosis.

Keywords Multiple sclerosis • Bipolar disorders • Diagnosis • Epidemiology • Treatment • Oxidative damage hypothesis

M.G. Carta, MD (✉) • M.F. Moro, MD • G. Trincas, MD
Department of Public Health and Clinical and Molecular Medicine,
University of Cagliari, Cagliari, Italy

Consultation Liaison Psychiatric Unit at the University Hospital of Cagliari,
University of Cagliari and AOU Cagliari, Cagliari, Italy
e-mail: mgcarta@tiscali.it

L. Lorefice, MD • E. Cocco, MD • M.G. Marrosu, MD
Department of Public Health and Clinical and Molecular Medicine,
University of Cagliari, Cagliari, Italy

Centro Sclerosi Multipla, Cagliari, Italy

B. Brochet (ed.), *Neuropsychiatric Symptoms of Inflammatory Demyelinating Diseases*, Neuropsychiatric Symptoms of Neurological Disease,
DOI 10.1007/978-3-319-18464-7_5

Introduction

Multiple sclerosis (MS) is a neurological disorder with high prevalence among young adults and a heavier impact on quality of life than other disabling conditions [1, 2]. These characteristics are relevant when assessing co-morbidity of MS with mood disorders, which, by themselves, may strongly disable the patient and may then interact with MS in increasing the burden of disability and worsening the course of illness and quality of life.

The co-morbidity of MS with depression has been well documented [3] showing that rates of major depressive disorders (MDD) have a twofold prevalence in MS patients than in other chronic illnesses or neurological disorders [4, 5]. MS patients with comorbid MDD have a lower quality of life and increased suicidal ideation risk [6]; they are also characterized by delays in MS diagnosis and poorer outcome [7] than those without MDD.

Neuroimaging has highlighted that depressive symptoms are related to structural and functional brain abnormalities, suggesting that the demyelination process may have a role in the pathogenesis of depressive symptoms [8, 9].

In the past few decades, the concept of "bipolar spectrum disorders" has evolved, so much so that psychiatrists now diagnose bipolar disorders (BD) more frequently than in the past [10]. In fact some mild forms of this disorder (with hypomania) were frequently diagnosed as major "unipolar" depression [11] in the past. The difficulties identifying bipolar depression are frequently underlined today, especially in detecting manic or hypomanic episodes from the medical history of depressed patients in both clinical and epidemiological settings. The patients could remember euphoric periods as phases of well-being or as depressive remissions, while they are less often able to recognize the psychopathology component of such episodes [12]. During these episodes, the patient does not see the physician, or, when forced to do so, they are not convinced about their psychiatric illness. The consequence is that the physician sees the depressive episodes only and not the hypomanic ones, because patients usually seek help from the doctor when they are depressed and not when they are hypomanic [12].

Another factor favoring misdiagnosis needs consideration. In people suffering from BD, and type II in particular, depressive symptoms have a considerably longer duration than hypomanic symptoms (37–50 % vs. 4–9 % of the total time) [13, 14]. Due to the higher recurrence of this kind of disorder, probably the life of these patients is burdened by depression for a longer time span than MDD subjects.

These new concepts have also cast doubts on the validity and reliability of the diagnosis of BD conducted through highly structured clinical interviews by lay interviewers and not by psychiatrists. As of now, most epidemiological and research studies used interviews carried out by lay interviewers which only reflect what the patient is aware of regarding his condition and his clinical history (the patient's point of view). As a consequence since the individual remembers euphoric periods as phases of well-being, the result of a series of closed questions (yes/no) is the underdiagnosis of BD [12].

The research into BD today is better geared to the use of semistructured clinical interviews conducted by clinicians or to the use of specific screening tools for bipolar spectrum disorders such as HCL-32 [15] and MDQ [15], although the latter showed severe limitations in case-finding studies [16].

Even the field of co-morbidity between mood disorders and chronic conditions is undergoing a review process in order to attach greater importance to BD than in the past [17–19].

In the light of these new perspectives, also the burden of the co-morbidity of BD in MS has been the subject of a review. In this field no research was made using clinical semi-standardized tools carried out by clinicians to study co-morbidity between mood disorders and MS. Most research works have adopted specific screening tools for depressive symptoms, particularly the CES Depression Scale (CES-D) [20] and the Hospital Anxiety and Depression Scale (HADS) [21].

Bipolar Disorders and Multiple Sclerosis: The Measure of Association and the Interaction in the Impairment of Quality of Life

Joffe and coworkers have conducted a systematic psychiatric evaluation on 100 consecutive patients attending an MS clinic in Toronto. Forty-two percent of the patients had a lifetime history of depression, and 13 % fulfilled the criteria for manic-depressive illness. Only 28 % of the patients had no psychiatric diagnosis [22].

A survey on administrative data from Canada found that the risk of mental comorbidities (including depression, anxiety, and BD but excluding schizophrenia) is increased in MS compared to the general population [23].

A case-control study on 201 consecutive patients with MS and 804 sex-and-age-matched controls without MS was recently carried out [24]. Psychiatric diagnoses according to DSM-IV were determined by physicians using semistructured interview tools (ANTAS-SCID); in addition, the frequency of wide bipolar spectrum in cases and controls and the odds ratio for cases were measured by means of the screening tool MDQ. The MS group showed a 24.7 % prevalence of DSM-IV MDD, 2.0 % of BP I, 11.9 % of BPII, and 3.0 % of cyclothymic disorder; the bipolar spectrum frequency detected by MDQ was found at 47.5 %. Higher lifetime prevalence was found in MS patients than in the control group for MDD (X2=93.4; OR=7.4; CI 95 % 4.7–11.7; $P<0.0001$), BD I (X2=3.98; OR and CI not calculable, $P=0.05$), BD II ($\chi 2=46.0$; OR=36.2; CI95 %; 8.20–159.9 $P<0.0001$), cyclothymic disorder ($\chi 2=5.57$; OR and CI not calculable; $P=0.0001$), and bipolar spectrum positivity detected by MDQ (X2=122.71; OR=16.5; P<0.0001; CI95 % 10.2–26.5).

The survey confirms that major depressive disorder is the most frequent psychiatric disorder associated with MS [25, 26], but it can be underlined that BD and BPII, in particular, are not uncommon, and BPII had the highest OR in comparison with the sample without MS.

This study found that MS diagnosis is associated with a worse quality of life when compared to psychiatric diagnosis of BD, MDD, or eating disorder or to another neurological condition such as Wilson's disease [27]. In patients with MS, the impairment in the quality of life attributable to comorbid bipolar II disorders was even higher than the one caused by major depressive disorders [2].

Etiopathogenic Hypothesis

The etiopathogenic mechanisms explaining the association between MS and BD have not yet been understood [28].

BD and, more generally, mood disorders were interpreted as an early manifestation of MS [29], as the presenting of mood symptoms was described to begin even before the development of the neurological picture [30]. This perspective was supported by a brain imaging systematic survey on a large sample of 6-year consecutive psychiatric patients, referred from an inpatient ward of a psychiatric hospital. The survey found a prevalence of MS like brain white matter hyperintensities (WMH) at 0.83 % without MS diagnosis: this figure was almost 15 times the prevalence of MS in the general population in the United States. According to the authors, these results supported the hypothesis of pure "psychiatric fits" in MS [31, 32]. Patients with WMH have longer hospitalization episodes, more psychiatric admissions, a higher risk of brain atrophy, and more neurological symptoms than those without WMH [33].

A Danish study examined the records of 9,478 autopsies and confirmed the high frequency of MS in psychiatric patients, but did not find silent and, perhaps, pure mental forms of MS without neurological signs [34].

It is well known that adverse effects on mood are common during corticosteroid therapy [35]. Two meta-analyses found that a large amount of psychiatric disturbances can occur as a consequence of steroid use on mood, cognition, sleep, and behavior as well psychotic symptoms; largely, the most common side effect of short-term corticosteroid therapy was found to be hypomania. Dosage was shown to be directly related to the incidence of adverse effects, but it is not related to the timing, severity, or duration of these effects [36]. According to these evidences, some authors have argued that the frequent use of corticosteroids could be the cause of the excess of BD in MS. In fact 40 % of MS patients treated with corticosteroids or ACTH in a sample survey had depressive episodes, 31 % had hypomanic symptoms, 11 % had mixed state, and 16 % had psychotic symptoms [37]. Please note that the aforementioned case-control study had adopted the psychiatric DSM-IV diagnosis that, in the case of BD, excludes drug-induced cases; therefore, this study made it clear that the excess could not be due only to the concomitant use of corticosteroids, although the use of these medications should be better monitored in the light of recent evidence.

A common genetic susceptibility of both MS and BD is today under debate. The analysis of a small sample of patients with both MS and BD revealed a higher frequency of the HLA-DR2 and -DR3 haplotypes than expected and a decrease in the frequency of HLA-DR1 and -DR4 [38]. These data were partially confirmed by another study that analyzed five members of the same family over three generations to check for MS, BD, and the HLA class I and II specificities. The class II, HLA-DR2, DQ1 haplotype shared among the persons with BD, which is well known to be associated with MS in some Caucasians groups, "suggests a possible susceptibility locus for BD, mapped on chromosome 6, very close to the HLA region, underlying the clinical co-morbidity of the two disorders" [39].

The results of some studies, however, seem to disagree with the hypothesis of a common genetic base between MS and BD. A recent study evaluated pleiotropy in immune-related single-nucleotide polymorphisms associated with schizophrenia (SCZ), BD, and MS. The study found significant genetic overlap between SCZ and MS and identified 21 independent loci associated with SCZ, conditioned on association with MS, but found no genetic overlap between BD and MS [40].

A recent theory concerning the role of oxidative damage in the brain can offer some elements to explain the association between MS and BP.

Oxidative damage can affect the central nervous system particularly through polysaturated fatty acids, because of the absence of a valid antioxidant activity in the brain [41]. Lipid peroxidation induced by neuronal oxidative stress (OxS) can produce alterations on signal transduction and damage cellular plasticity and resilience [42].

Oxidative stress is known to play a role in the pathogenesis of MS. Recently techniques of metabolomic and lipidomic analyses were applied to compare cerebrospinal fluid (CSF) in samples of MS patients, with samples of non-MS subjects. The results supported the hypothesis that autoimmunity producing epitopes derived from lipid peroxidation can be a relevant pathogenic factor in MS [42, 43].

Changes in transcriptome patterns have been found in BD in different brain areas. In these disorders research has found downregulation of the genes related to the processes of energy metabolism, mitochondrial function and oligodendrocytes activity, and upregulation of the genes involved in immune response and inflammation [44].

Interesting and close links between energy metabolism, inflammation, and demyelination have been found in MS. Therefore, the consequences of oxidative stress in oligodendrocytes can range from downregulation of oligodendrocyte genes – as observed in psychiatric disorders and in BD in particular – to cell death and brain lesions typical of MS [44].

The vulnerability for BD could be due to the impairment of the brain circuits regulating emotions, motor behavior, and pleasure [45, 46].

A recent study has investigated the fatty acid composition of the postmortem superior temporal gyrus, a cortical region implicated in emotional processing in normal controls and in samples of patients with BP, MDD, schizophrenia, and MS. Patients with BD, but not affected by MDD or schizophrenia, exhibited abnormal elevations in the saturated fatty acids as palmitic, stearic, linoleic, arachidonic, and docosahexaenoic acid and reductions in oleic monounsaturated fatty acid. In MS patients, a pattern of fatty acid abnormalities similar to the one observed in BD patients was found [47].

The pathophysiology of response to the therapeutic agent of BD resulted to be related to the contrast of oxidative stress parameters, such as superoxide dismutase (SOD), thiobarbituric acid reactive substances (TBARS), and catalase (CAT) [48]. Lithium – the oldest and still the most advanced mood stabilizer agent used in the treatment of BD – was proved to limit enzyme activity, lower hydrogen peroxide, and formation of hydroxyl radicals. Lithium, as well the other stabilizer as valproate, reverses the alteration of oxidative stress parameters in BD [45, 49].

Clinical Implications

The treatment of BD in MS patients can be done with some particular attention and a strict collaboration between the neurologist and the psychiatrist.

The first element is early diagnosis. The neurologist must know that the depressive syndrome is very common in bipolar II disorder and that it is very difficult to identify the history of a previous hypomanic episode.

In this regard, utmost importance must be vested in inquiry about familiarity, the temperamental characteristics of the individual, and his/her medical history, along with hyperactivity and euphoria, the elements related to the dysregulation of biological rhythms (like sleep, eating, and so on), and the components of irritability.

The clinician should also be aware of the low accuracy of current screening questionnaires when they are used as case-finding tools. He must therefore keep in mind that the diagnosis of hypomania is eminently clinical. The risk of underdiagnosis of BD in MS, and of type II in particular, imposes caution when prescribing antidepressants (ADs) to people with depressive episodes in MS without prior excluding a diagnosis.

If the diagnosis of BP disorder is ascertained, a treatment with mood stabilizers (lithium, sodium valproate, carbamazepine, and lamotrigine) is generally effective – in patients without MS – to prevent recurrence during the manic (lithium, sodium valproate, carbamazepine) and the depressive phases (lamotrigine). The association of atypical antipsychotics during manic episodes as well as the use of quetiapine, lurasidone, or of an SSRI antidepressant in association with olanzapine in bipolar depressive episodes should be evaluated in every specific case. We must emphasize, however, that the data on the use of stabilizers in MS is mostly anecdotal and large RCT studies are lacking.

In patients with sphincter disorders, the tendency to reduce fluid intake may cause high serum levels of lithium; therefore, the close monitoring of serum levels should be considered, or the use of lithium should be discussed because of the risk of toxic doses [50].

The risk of the onset of manic episodes induced by corticosteroids or the exacerbation of psychiatric symptoms should not delay the use of corticosteroids if these compounds are needed [51]. The episodes can be prevented by a treatment with a stabilizer if required, or, in any case, the clinician can simply remember that the episodes can be treated very effectively with an early intervention. The clinician must therefore be careful in monitoring and weighing the risk-benefit ratio, and he should be aware of the possibility that steroid-related episodes (particularly depressive episode with mixed states) can emerge when the steroid treatment is discontinued, and the clinician should therefore be ready to treat it.

Interferon beta (IFN-β) treatment has been carried out in patients with BP disorders, with good tolerance [28].

Cases of suicide and risk of depression in patients treated with IFN-β have been reported, but the randomized controlled trials on the use of this compound on MS did not show any increase in depressive episodes despite standardized monitoring [52].

The presence of a mood disorder and of BD is not an absolute contraindication to treatment with IFN-β. The decision about the type of treatment can be taken after evaluating the risk-benefit ratio in each specific case and adopting a potentially preventive stabilizer therapy if required.

Conclusion

The association between BD and MS is probably more frequent than generally supposed in the past. The association between immune degenerative diseases (such as MS) and BD may be an interesting field to study the pathogenic hypothesis, particularly concerning a possible common role of oxidative processes in the brain. The risk of underdiagnosis of BD, and of type II in particular, in MS suggests caution when prescribing ADs to people with depressive episodes in MS, without prior excluding a BD diagnosis.

References

1. Patten SB, Williams JV, Lavorato DH, Berzins S, Metz LM, Bulloch AG. Health status, stress and life satisfaction in a community population with MS. Can J Neurol Sci. 2012;39(2):206–12.
2. Carta MG, Moro MF, Lorefice L, Picardi A, Trincas G, Fenu G, Cocco E, Floris F, Bessonov D, Akiskal HS, Marrosu MG. Multiple sclerosis and bipolar disorders: the burden of comorbidity and its consequences on quality of life. J Affect Disord. 2014;167:192–7.
3. Byatt N, Rothschild AJ, Riskind P, Ionete C, Hunt AT. Relationships between multiple sclerosis and depression. J Neuropsychiatry Clin Neurosci. 2011;23:198–200.
4. Sadovnick AD, Remick RA, Allen J. Depression in multiple sclerosis. Neurology. 1996;46:628–32.
5. Siegert RJ, Abernerhy DA. Depression in multiple sclerosis: a review. J Neurology Neurosurg Psychiatry. 2005;76:469–75.
6. Feinstein A. An examination of suicidal intent in patients with multiple sclerosis. Neurology. 2002;59:674–8.
7. Marrie R, Horwitz R, Cutter G, et al. The burden of mental comorbidity in multiple sclerosis: frequent, underdiagnosed, and undertreated. Mult Scler. 2009;15:385–92.

8. Bakshi R, Czarnecki D, Zubair S, et al. Brain MRI lesions are related to depression in multiple sclerosis. Brain Imaging. 2000;11:1153–8.
9. Feinstein A, Roy P, Lobaugh N, et al. Structural brain abnormalities in multiple sclerosis patients with major depression. Neurology. 2004;62:586–90.
10. Carta MG, Aguglia E, Bocchetta A, Balestrieri M, Caraci F, Casacchia M, Dell'Osso L, Di Sciascio G, Drago F, Faravelli C, Lecca ME, Moro MF, Morosini PL, Marcello N, Palumbo G, Hardoy MC. The use of antidepressant drugs and the lifetime prevalence of major depressive disorders in Italy. Clin Pract Epidemiol Ment Health. 2010;6:94–100.
11. Carta MG, Aguglia E, Balestrieri M, Calabrese JR, Caraci F, Dell'osso L, Di Sciascio G, Drago F, Faravelli C, Lecca ME, Moro MF, Nardini M, Palumbo G, Hardoy MC. The lifetime prevalence of bipolar disorders and the use of antidepressant drugs in bipolar depression in Italy. J Affect Disord. 2012;136(3):775–80.
12. Carta MG, Angst J. Epidemiological and clinical aspects of bipolar disorders: controversies or a common need to redefine the aims and methodological aspects of surveys. Clin Pract Epidemol Ment Health. 2005;1(1):4.
13. Judd LL, Akiskal HS, Schettler PJ, Coryell W, Endicott J, Maser JD, Solomon DA, Leon AC, Keller MB. A prospective investigation of the natural history of the long-term weekly symptomatic status of bipolar II disorder. Arch Gen Psychiatry. 2003;60:261–9.
14. Kupka RW, Altshuler LL, Nolen WA, Suppes T, Luckenbaugh DA, Leverich GS, Frye MA, Keck Jr PE, McElroy SL, Grunze H, Post RM. Three times more days depressed than manic or hypomanic in both bipolar I and bipolar II disorder. Bipolar Disord. 2007;9(5):531–5.
15. Angst J, Meyer TD, Adolfsson R, Skeppar P, Carta M, Benazzi F, Lu RB, Wu YH, Yang HC, Yuan CM, Morselli P, Brieger P, Katzmann J, Teixeira Leão IA, Del Porto JA, Hupfeld Moreno D, Moreno RA, Soares OT, Vieta E, Gamma A. Hypomania: a transcultural perspective. World Psychiatry. 2010;9(1):41–9.
16. Hirschfeld RM, Calabrese JR, Weissman MM, Reed M, Davies MA, Frye MA, Keck Jr PE, Lewis L, McElroy SL, McNulty JP, Wagner KD. Screening for bipolar disorder in the community. J Clin Psychiatry. 2003;64(1):53–9.
17. Carta MG, Sorbello O, Moro MF, Bhat KM, Demelia E, Serra A, Mura G, Sancassiani F, Piga M, Demelia L. Bipolar disorders and Wilson's disease. BMC Psychiatry. 2012;12(1):52.
18. Carta M, Ruggiero V, Sancassiani F, Cutrano F, Manca A, Peri M, Fais A, Cacace E. The use of antidepressants in the long-term treatment should not improve the impact of fibromyalgia on quality of life. Clin Pract Epidemiol Ment Health. 2013;9:120–4.
19. Mura G, Bhat KM, Pisano A, Licci G, Carta M. Psychiatric symptoms and quality of life in systemic sclerosis. Clin Pract Epidemiol Ment Health. 2012;8:30–5. doi:10.2174/1745017901208010030. Epub 2012 Apr 20.
20. Chwastiak L, Ehde DM, Gibbons LE, Sullivan M, Bowen JD, Kraft GH. Depressive symptoms and severity of illness in multiple sclerosis: epidemiologic study of a large community sample. Am J Psychiatry. 2002;159(11):1862–8.
21. Giordano A, Granella F, Lugaresi A, Martinelli V, Trojano M, Confalonieri P, Radice D, Solari A, SIMS-Trial Group. Anxiety and depression in multiple sclerosis patients around diagnosis. J Neurol Sci. 2011;307(1–2):86–91. Epub 2011 May 31.
22. Joffe RT, Lippert GP, Gray TA, Sawa G, Horvath Z. Mood disorder and multiple sclerosis. Arch Neurol. 1987;44(4):376–8.
23. Marrie R, Fisk JD, Yu BN, Leung S, et al. Mental comorbidity and multiple sclerosis: validating administrative data to support population-based surveillance. BMC Neurol. 2013;13:16. doi:10.1186/1471-2377-13-16.
24. Carta MG, Moro MF, Lorefice L, Trincas G, Cocco E, Giudice ED, Fenu G, Colom F, Marrosu MG. The risk of bipolar disorders in multiple sclerosis. J Affect Disord. 2014;155:255–60.
25. Sadovnick AD, Remick RA, Allen J, et al. Depression and multiple sclerosis. Neurology. 1996;46:628–32.
26. Simmons RD. Life issues in multiple sclerosis. Nat Rev Neurol. 2010;6:603–10.

27. Carta MG, Preti A, Moro MF, Aguglia E, Balestrieri M, Caraci F, Dell'Osso L, Di Sciascio G, Drago F, Faravelli C, Hardoy MC, D'Aloja E, Cossu G, Calò S, Palumbo G, Bhugra D. Eating disorders as a public health issue: prevalence and attributable impairment of quality of life in an Italian community sample. Int Rev Psychiatry. 2014;26(4):486–92.
28. Sidhom Y, Ben Djebara M, Hizem Y, Abdelkefi I, Kacem I, Gargouri A, Gouider R. Bipolar disorder and multiple sclerosis: a case series. Behav Neurol. 2014;2014:536503. doi:10.1155/2014/536503. Epub 2014 Mar 17.
29. Blanc F, Berna F, Fleury M, et al. Inaugural psychotic events in multiple sclerosis? Rev Neurol. 2010;166(1):39–48.
30. Jongen PJ. Psychiatric onset of multiple sclerosis. J Neurol Sci. 2006;245(2):59–62.
31. Skegg K, Corwin PA, Skegg DC. How often is multiple sclerosis mistaken for a psychiatric disorder? Psychol Med. 1988;18(3):733–6. [PubMed].
32. Asghar-Ali AA, Taber KH, Hurley RA, Hayman LA. Pure neuropsychiatric presentation of multiple sclerosis. Am J Psychiatry. 2004;161(2):226–31.
33. Lyoo IK, Seol HY, Byun HS, Renshaw PF. Unsuspected multiple sclerosis in patients with psychiatric disorders: a magnetic resonance imaging study. J Neuropsychiatry Clin Neurosci. 1996;8(1):54–9.
34. Johannsen LG, Stenager E, Jensen K. Clinically unexpected multiple sclerosis in patients with mental disorders. A series of 7301 psychiatric autopsies. Acta Neurol Belg. 1996;96(1):62–5.
35. Bolanos SH, Khan DA, Hanczyc M, Bauer MS, Dhanani N, Brown ES. Assessment of mood states in patients receiving long-term corticosteroid therapy and in controls with patient-rated and clinician-rated scales. Ann Allergy Asthma Immunol. 2004;92(5):500–5.
36. Warrington TP, Bostwick JM. Psychiatric adverse effects of corticosteroids. Mayo Clin Proc. 2006;81(10):1361–7.
37. Minden SL, Orav J, Schildkraut JJ. Hypomanic reactions to ACTH and prednisone treatment for multiple sclerosis. Neurology. 1988;38(10):1631–4.
38. Schiffer RB, Wineman NM, Weitkamp LR. Association between bipolar affective disorder and multiple sclerosis. Am J Psychiatry. 1986;143(1):94–5.
39. Bozikas VP, Anagnostouli MC, Petrikis P, et al. Familial bipolar disorder and multiple sclerosis: a three-generation HLA family study. Prog Neuropsychopharmacol Biol Psychiatry. 2003;27(5):835–9.
40. Andreassen OA, Harbo HF, Wang Y, et al. Genetic pleiotropy between multiple sclerosis and schizophrenia but not bipolar disorder: differential involvement of immune-related gene loci. Mol Psychiatry. 2014. doi:10.1038/mp.2013.195. [Epub ahead of print].
41. Andreazza AC, Kauer-Sant'Anna M, Frey BN, et al. Effects of mood stabilizers on DNA damage in an animal model of mania. J Psychiatry Neurosci. 2008;33:516–24. [PMC free article].
42. Khairova R, Pawar R, Salvadore G, De Sousa RT, Soeiro-De-Souza MG, Salvador M, Zarate CA, Gattaz WF, Machado-Vieira R. Effects lithium oxidative stress parameters healthy subjects. Mol Med Report. 2012;5(3):680–2. doi:10.3892/mmr.2011.732. Published online 2011 December 22.
43. Gonzalo H, Brieva L, Tatzber F, Jové M, Cacabelos D, Cassanyé A, Lanau-Angulo L, Boada J, Serrano JC, González C, Hernández L, Peralta S, Pamplona R, Portero-Otin M. Lipidome analysis in multiple sclerosis reveals protein lipoxidative damage as a potential pathogenic mechanism. J Neurochem. 2012. doi:10.1111/j.1471-4159.2012.07934.x. [Epub ahead of print].
44. Konradi C, Sillivan SE, Clay HB. Mitochondria, oligodendrocytes and inflammation in bipolar disorder: evidence from transcriptome studies points to intriguing parallels with multiple sclerosis. Neurobiol Dis. 2012;45(1):37–47. Epub 2011 Feb 17.
45. Machado-Vieira R, Andreazza AC, Viale CI, et al. Oxidative stress parameters in unmedicated and treated bipolar subjects during initial manic episode: a possible role for lithium antioxidant effects. Neurosci Lett. 2007;421:33–6.

46. Zarate CA, Singh Jr J, Manji HK. Cellular plasticity cascades: targets for the development of novel therapeutics for bipolar disorder. Biol Psychiatry. 2006;59:1006–20.
47. McNamara R, Rider T, Jandacek R, Tso P. Abnormal fatty acid pattern in the superior temporal gyrus distinguishes bipolar disorder from major depression and schizophrenia and resembles multiple sclerosis. Psychiatry Res. 2014;215(3):560–7. doi:10.1016/j.psychres.2013.12.022. Epub 2014 Jan 2.
48. Andreazza AC, Cassini C, Rosa AR, et al. Serum S100B and antioxidant enzymes in bipolar patients. J Psychiatr Res. 2007;41:523–9.
49. Shao L, Young LT, Wang JF. Chronic treatment with mood stabilizers lithium and valproate prevents excitotoxicity by inhibiting oxidative stress in rat cerebral cortical cells. Biol Psychiatry. 2005;58:879–84.
50. Paparrigopoulos T, Ferentinos P, Kouzoupis A, Koutsis G, Papadimitriou GN. The neuropsychiatry of multiple sclerosis: focus on disorders of mood, affect and behaviour. Int Rev Psychiatry. 2010;22(1):14–21.
51. Judd LL, Schettler PJ, Brown ES, et al. Adverse consequences of glucocorticoid medication: psychological, cognitive, and behavioral effects source of the document. Am J Psychiatry. 2014;171:1045–51.
52. Porcel J, Río J, Sánchez-Betancourt A, et al. Long-term emotional state of multiple sclerosis patients treated with interferon beta. Mult Scler. 2006;12(6):802–7.

Chapter 6
Psychiatric Comorbidity

Olivier Heinzlef

Abstract Multiple sclerosis (MS) is a chronic inflammatory disorder of the central nervous system. Psychiatric comorbidities, in particular depression and anxiety, are frequent in MS patients.

These disorders are underdiagnosed and undertreated although they have been associated with decreased adherence to treatment, functional status, and quality of life. Behavioral disorders are more common than severe psychiatric disorders and are probably secondary to cognitive impairment. Addictions may be underestimated. Although the high frequency of psychiatric co-morbidity might be due to psychosocial factors, the role of demyelinization and inflammation is possible. Psychiatric comorbidities in MS deserve clinical attention because they are associated with an increase risk of suicide.

Keywords Multiple sclerosis • Suicide • Depression • Bipolar disorders • Anxiety disorders • Psychosis • Schizophrenia • Personality disorders • Euphoria • Pseudobulbar affect • Somatoforms disorders • Substance abuse

Introduction

Patients with multiple sclerosis are more likely to have psychiatric symptoms or disorders than people without MS (Table 6.1).

In a study of co-morbidity at the diagnosis of multiple sclerosis, Fromont et al. found that 42.5 % of women and 35.6 % of men had a psychiatric disorder associated with the diagnosis of multiple sclerosis versus 29.9 % of women and 33.9 % of men who did not have multiple sclerosis [21]. These disorders are underdiagnosed and undertreated, and they have been associated with decreased adherence to treatment [22], functional status, and quality of life [23]. Additionally, the occurrence of a psychiatric disorder prior to the onset of MS delayed the diagnosis of MS

O. Heinzlef, MD (✉)
Department of Neurology, Poissy-Saint-Germain Hospital,
10 rue Champ Gaillard, Poissy 78000, France
e-mail: oheinzlef@chi-poissy-st-germain.fr

B. Brochet (ed.), *Neuropsychiatric Symptoms of Inflammatory Demyelinating Diseases*, Neuropsychiatric Symptoms of Neurological Disease,
DOI 10.1007/978-3-319-18464-7_6

Table 6.1 Increased prevalence of psychiatric disorders among persons with MS compared to the general population [1–20]

Disorder	Prevalence in MS	Prevalence in general population
Major depressive disorder, 12 months	15.7 % (33) [1]	7.4 % (33) [1]
Major depressive disorder, lifetime	22.8 % (100) [2]	16.2 % (101) [3]
Anxiety disorder, lifetime	36 % (14, 90) [4, 5]	25 % (102) [6]
Generalized anxiety disorder	18.6 % (14, 90) [4, 5]	3 % (103) [7]
Bipolar disorder lifetime	0.3 % (65) [8]	0.2 % (63) [9]
Schizophrenia	0.5 % (66) [10]	0.3–0.66 % (80) [11]
Brief psychotic disorders	3 % (84) [12]	0 % (84) [12]
Somatoform disorders	2 % (90) [5]	0 % (90) [5]
Paranoid disorders	25 % (85) [13]	3 % (85) [13]
Borderline disorders	25 % (85) [13]	0 % (85) [13]
Alcohol abuse, lifetime	13.6 % (92) [14]	7.4 % (99) [15]
Euphoria	15 % (72) [16]	0 % (72) [16]
Pseudobulbar affect	10 % (75) [17]	–
Substance misuse, past month	18.7 % (93) [18]	11.1 %

by 3.2 years [21]. This review aims to summarize the existing literature on the epidemiology, impact, and treatment of psychiatric disorders among persons with MS.

Suicide

The risk of suicide has been reported to be increased among patients with MS. In Sweden [24], the standardized mortality ratio (SMR) was significantly elevated: SMR=2.3 among MS patients compared with the general population. Suicide risk was particularly high in the first year after initial admission with an MS diagnosis and among younger male MS patients. The crude suicide rate among MS patients during the study period was 71 per 100,000 person-years. In a large community-based study in Denmark, the suicide risk among people with multiple sclerosis was more than twice that of the general population (SMR=2.12). The increased risk was similarly high during the first year after diagnosis (SMR=3.15) [25]. In London, Ontario, the proportion of suicides among MS deaths was 7.5 times that of the age-matched general population [26]. The prevalence rate of suicide in MS ranges from 2.5 to 28.6 % [26–28]. This wide range of estimation is probably due to cultural variation and methodological considerations. In men (but not in women), risk factors for suicide were psychiatric co-morbidity, major depression, past suicide attempt, moderate disability, or recent accentuation of disability [25].

The frequency of suicidal intent in MS patients was 28.6 % in a study of 140 patients. The main risk factors were social isolation, familial psychiatric co-morbidity, social stress, past history of major depression, anxiety, and alcohol abuse [29].

Anxiety

As defined in the DSM-IV, symptoms can include prominent generalized anxiety symptoms, panic attacks, obsessions, or compulsions [30]. Anxiety is widely distributed in multiple sclerosis patients: A recent study using responses gained by the web portal of the UK MS register included 4,178 respondents. Over half of the respondents (54.1 %) scored>8 for anxiety on the Hospital Anxiety and Depression Scale (HADS). Women were more frequently anxious than men: 56.4 % compared to 48.0 %. Anxiety was most frequent among people with relapsing remitting MS (RRMS) (56.5 %). Among the patients enrolled, 27.6 % had mild anxiety, 37.4 % moderate anxiety, and 10.6 % severe anxiety. Only 24.4 % did not have anxiety (score HADS<8) [31]. The prevalence rates observed and the mean anxiety scores were higher than those found for the general UK population [32]. These figures were higher than in most of the previously published studies. The literature about anxiety and MS is reviewed in Chap. 4.

Depression

The essential feature of a major depressive episode (MDD), as described in the DSM-IV, is a period of at least 2 weeks during which there is either depressed mood or the loss of interest or pleasure in nearly all activities. The individual must also experience at least four additional symptoms drawn from a list that includes changes in appetite or weight, sleep, and psychomotor activity; decreased energy; feelings of worthlessness or guilt; difficulty thinking, concentrating, or making decisions; or recurrent thoughts of death or suicidal ideation, plans, or attempts [30].

Bipolar Affective Disorders

Bipolar affective disorders refer to a group of affective disorders characterized by depressive and manic or hypomanic episodes. The DSM-IV contains four main types of bipolar disorders. Bipolar disorders type I (BD I) is defined by the occurrence of episodes of depression and at least one full-blown manic episode, bipolar type II (BD II) by several episodes of depression and at least one hypomanic episodes, cyclothymic disorders by many periods of hypomanic and depressive symptoms (not fulfilling criteria for depressive episodes), and bipolar disorders not otherwise specified [30].

Bipolar disorders are underdiagnosed in the general population because it is difficult to differentiate these disorders from unipolar depression (defined by recurrent episodes of depression) when hypomanic or manic episodes are not identified [33].

Bipolar disorders type I affects 0.2–4 % of the general population, bipolar II disorders 0.3–4.8 %, 0.5–6.3 % for cyclothymic disorders, and 5 % for bipolar spectrum [9].

Mania is characterized by euphoria or irritable mood, decreased need for sleep, talkativeness, racing thoughts, increased sexual activity and aggressive activity, increased motor activity or agitation, and poor judgment. It makes the severity of the disease as it interferes with patients' capability to work and familial functioning. Postpartum period is associated with an increased risk of exacerbations, and the prognosis of manic episodes in this period is associated with a severe prognosis. Atypic depression and mixed depression (defined by the combination of depression and non-euphoric subsyndromal manic or hypomanic symptoms) are more frequent than in unipolar depression [9].

Alcohol abuse and drug abuse are frequently associated with bipolar disorders and complicate the care of these patients.

A family history of bipolar disorders is found in 50 % of the patients. Studies of twins showed that the concordance for bipolar illness is between 40 % and 80 % in monozygotic twins and only 10–20 % in dizygotic twins suggesting a genetic component.

Acute mania is usually treated with antipsychotic drugs [34]. Classical and atypical neuroleptics are effective treatment. Lithium valproate and carbamazepine have established efficacy in the treatment of acute mania, but they work slowly. Bipolar depression responds to tricyclic antidepressants, selective serotonin reuptake inhibitors, and monoamine oxidase inhibitors. Lithium is effective in the treatment of acute bipolar depression and the prevention of recurrences of mania hypomania and depression. Other mood-stabilizing agents are used like valproic acid, carbamazepine, and lamotrigine or neuroleptics, but their use is limited by the risk of tardive dyskinesias. Cognitive-behavioral therapy is effective in nonpsychotic depressive disorders [34].

Most of the epidemiological studies showed an increased prevalence rate of bipolar disorders in MS patients [8]. The NARCOMS registry is a self-report registry for patients with MS in the USA. Among 8,828 responders (55.7 % response rate), the prevalence rate was 2.4 % for bipolar disorder in MS [35].

Using administrative data in Canada in 4,192 persons with MS and 20,940 matched persons, Marrie et al. found that the age-standardized prevalence of bipolar disorder in 2005 was 5.83 % (95 % CI: 5.01–6.65 %) in the MS population and 3.45 % (95 % CI: 3.17–3.73 %) in the general population (PR 1.70; 95 % CI: 1.55–1.87) [10]. In another study with a general population control group, hospitalized MS patients had bipolar affective disorder twice as often as hospitalized controls (1.97 % vs. 0.92 %) [36]. However, they found a slight to moderate agreement between the administrative case definitions and medical records, which illustrates the difficulty to accurately diagnose bipolar affective disorders.

A recent case-control study including 201 consecutive MS patients and 804 sex- and age-matched persons without MS used structured interview tools to perform psychiatric diagnoses according to DSM-IV. Compared to controls, MS patients had a higher lifetime prevalence of DSM-IV major depressive disorders (MDD; $P<0.0001$), BD I ($P=0.05$), BD II ($P<0.0001$), and cyclothymia ($P=0.0001$) [37].

The relationship between bipolar disorders and MS is not well understood but is regarded as being multifactorial. It is attributed to medications, demyelinating brain lesions, genetics, psychological reactions, and adjustment difficulties. Several treatments used in MS could induce hypomanic or manic episodes as corticosteroids, baclofen, dantrolene, tizanidine, and illicit drugs [38]. Although cases of depressive episodes have been reported with interferon beta, sometimes accompanied by psychotic or manic behavior, there is no clear evidence that the administration of interferon to patients with MS increases the risk of depressive disorders [39].

Concerning the treatment, there are no controlled trials of mood stabilizers use in treating affective bipolar disorders in MS in particular.

Euphoria

Euphoria is defined as a stable elation of humor, an unsuitable cheerfulness, or a lack of concern to the consequences of the disease. It is secondary to personality disorders and it is not considered as a mood disorder. Euphoria is distinguished from mania. It is associated with childishness, disinhibition, impulsivity, emotional lability, anger outbursts, and lack of empathy. In modern studies the estimated prevalence rate of euphoria in MS patients is around 15 %. Two studies using the neuropsychiatric inventory (NPI), which covers ten domains including delusions, hallucinations, agitation, dysphoria, anxiety, euphoria, apathy, disinhibition, irritability, and aberrant motor activity, have evaluated the prevalence of euphoria in MS. In a study including 44 MS patients and 25 controls without MS, Diaz-Olavarietta et al. identified euphoria in 13 % of patients as compared to 0 % in the control group [40]. In 75 patients enrolled in a MS clinic, Fishman et al. found a prevalence rate of 7 % or 9 % as compared to 0 % in the control group and was associated with secondary progressive course, low agreeableness on personality testing, poor insight, and impaired cognition [16]. Euphoria is associated with the severity of the total T2 lesions load and the atrophy of gray and white matter [41].

Pseudobulbar Affect

Pseudobulbar affect (PBA) is defined as episodes of involuntary crying, laughing, or both that are inconsistent with the patient's underlying mood. The clinical condition has been known by different names, but the most widely used terms are "pseudobulbar affect," "emotional lability," "emotional incontinence," and "pathological laughter and crying" or "pathological laughing and crying (PLC)" [42]. Uncontrollable crying seems to be more common than laughing. This emotional incontinence causes significant social embarrassment. Although it is not a mood disorder, it can be associated with depression. In a cohort of 152 consecutive MS patients, Feinstein et al. found a point prevalence of 10 % [17]. Patients were severely disabled with a mean EDSS score of 6.5 and had progressive course. Emotional expression in PBA is secondary to a disconnection from cortical voluntary control or cortico-pontine-cerebellar control responsible for appropriate emotional adjustments to social situations. It is thought that loss of voluntary control results in involuntary activation of laughing/crying centers. An MRI study has correlated the occurrence of PBA with lesions in the brainstem, the inferior parietal (bilateral) and medial inferior frontal (bilateral), and the right medial superior frontal region [43].

Agents that are effective for the treatment of mood disorders are also effective for the treatment of PLC. Most commonly, MS patients are treated with tricyclic antidepressants or selective serotonin reuptake inhibitors. Levodopa and amantadine have been proposed. More recently dextromethorphan/quinidine has been shown to be efficacious in a randomized control trial in MS and has been approved in the USA [11, 44–46].

Psychotic Disorders

In the DSM-IV, psychotic disorders are defined by the presence of prominent hallucinations or delusions and other positive symptoms of schizophrenia like disorganized speech, disorganized speech or behavior. Psychotic disorders include schizophrenia, schizophreniform disorder, schizoaffective disorder, brief psychotic disorders, shared psychotic disorders, and psychotic disorders due to general medical condition or substance-induced psychotic disorders.

Schizophrenia is applied to a syndrome characterized by long-duration, bizarre delusions, negative symptoms, and few affective symptoms (non-affective psychosis) [30]. In the general population, the lifetime prevalence and incidence are 0.30–0.66 % and 10.2–22.0 per 100 000 person-years [11, 47]. The risk of schizophrenia and related categories increases with an urbanized environment during childhood and with the exposure to dronabinol, the main psychotropic component of cannabis [11].

Occasionally, acute bizarre behavioral symptoms leading to a diagnosis of psychosis could inaugurate the course of the disease or be associated with a relapse. Of the four patients described by Blanc et al., who developed psychotic symptoms that led to the diagnosis of multiple sclerosis, two developed persecutory delusions, one presented a manic episode and the fourth melancholia with catatonia [48].

In a population-based study in Canada, Patten et al., using administrative data, found that the prevalence of psychotic disorders was 1.3 % and that of organic psychotic disorders was 0.5 % in MS patient (N= 10,367). The prevalence of psychotic disorders was highest in the 15–24-year age group [49]. Although the prevalence of psychotic disorders increased with age in MS patients and controls, people with MS consistently had a higher prevalence of psychotic disorders than people without MS. A later study in the same population did not find an increase of the age-standardized prevalence of schizophrenia in MS as compared to the general population (0.93 % vs. 0.93 %) [10].

In a case-control study including 37 consecutive MS patients and 37 matched controls, a psychiatrist administered a structured clinical interview. Among the MS patients, 1 (2.7 %) had brief psychotic disorder vs. none in the control group [12].

Personality Disorders and Behavioral Symptoms

As defined in the DSM-IV, personality traits are enduring patterns of perceiving, relating to, and thinking about the environment and oneself that are exhibited in a wide range of social and personal contexts. Only when personality traits are inflexible and maladaptive and cause significant functional impairment or subjective distress do they constitute personality disorders [30]. Several patterns are described: paranoid, schizoid, schizotypal, antisocial, borderline, histrionic, narcissistic, avoidant, dependant, and obsessive-compulsive disorders. The diagnosis of personality disorders requires an evaluation of the individual's long-term patterns of functioning.

One case-control study specifically examined the frequency of personality disorders in multiple sclerosis in 20 MS patients and 35 healthy controls. Paranoid disorders and borderline disorders were observed more frequently in MS patients than in healthy control, respectively, 25 % vs. 3 % and 25 % vs. 0 %. There was no difference for the frequency of narcissistic, histrionic, avoidant passive aggressive, and dependant personality between MS patients and healthy control [13].

In their review and meta-analysis of 23 controlled studies, Rosti-Otajarvi and Hamalainen found that MS patients were more likely to manifest behavioral symptoms such as aggression (23 %), apathy (22 %), euphoria (12 %), and lack of insight (11 %) as well as impairments such as adjustment disorder (17 %) than healthy controls and other patients with SLE, chronic fatigue syndrome, and muscular dystrophy [20].

These disorders might be due to a psychological reaction to a chronic disease, but the association with neuropsychological deficits suggests that it could be secondary to the cerebral lesions. In a study in 34 MS patients and 14 healthy controls, Benedikt et al. found that MS patients with cognitive impairment were more neurotic and less empathic, agreeable, and conscientious as compared with normal control subjects [50].

Unusual episodes of hypersexuality have been described in MS. A patient with recurrent periods of heightened sexual desire has been reported. These episodes were thought to be due to strategically located plaques in the frontal lobes [51].

Somatoform Disorders

It is well known that patients, without any evidence of neurological disease, presenting features or symptoms that suggest the diagnosis of MS can be misdiagnosis. Allanson et al. reported 25 patients with severe functional assessment, without pathology to explain their neurological disability. The most common putative diagnosis was multiple sclerosis. Height of the patients had a final diagnosis of somatoform disorders and 13 of motor conversion disorder [52]. A more difficult situation could occur when hysterical symptoms add to MS symptoms during the course of the disease as the four cases reported by Caplan and Nadelson in 1980 [53].

Also well-recognized few studies have evaluated the frequency of this association. In one case-control study in 50 MS patients and 50 healthy control patients, Galeazzi et al. have found somatization disorder in 2 % of patients as compared to 0 % of the HC [5].

Substance Abuse

Alcohol and illicit drug abuse represent a growing challenge for the health of general populations. In their study on the global burden of disease attributable to mental and substance use disorders, Whiteford et al. found that illicit drug use disorders accounted for 10.9 % and alcohol disorders for 4.2 % of disability-adjusted life years [54]. In multiple sclerosis, substance abuse may be associated with mental disorders, may worsen neurological deficits, interact with MS treatment, and be associated with poor adherence. Several studies have found an increase in the prevalence of alcohol abuse in MS patients. In a study assessing drinking patterns in 140 MS patients, Quesnel and Feinstein found that 13.6 % have alcohol abuse. Patients with alcohol abuse were more likely to have suicidal ideation or other substances abuse [14].

In a large community-based study including 739 MS patients, Bombardier et al. found that 19 % of patients had alcohol or illicit drugs misuse. Alcohol abuse or dependence was detected in 14 % of the patients and illicit drugs in 7.4 % [18].

Cannabis and cannabinoids are used by MS patients to alleviate MS-related symptoms like pain, spasticity, tremor, and bladder dysfunction. Recently the nabiximols (Sativex) that contain two principal cannabinoids—delta-9-tetrahydrocannbinol (THC) and cannabidiol (CBD)—was approved for spasticity in MS [55]. The rate of cannabis misuse is high in MS patients, and in a survey in 220 patients, 36 % reported ever having used cannabis for any purpose. Use of cannabis was reported for symptom treatment by 14 % to relieve stress, sleep, mood, stiffness/spasm, and pain [56]. Although the majority of MS patients claimed that they use cannabis for medical purpose [57], patients and clinicians should be aware of the negative side effects of this use. Among patients treated with cannabinoids, the following side effects have been reported: nausea, increased weakness, behavioral or mood changes (or both), suicidal ideation or hallucinations (or both), dizziness or vasovagal symptoms (or both), fatigue, and feelings of intoxication. Psychosis, dysphoria, and anxiety are associated with high concentrations of THC.

Cognitive impairment in MS patients is also a matter of concern. Patients with MS who used cannabis performed significantly more poorly than nonusers on measures of cognitive working memory, information processing speed, executive functions, and visuospatial perception [58, 59]. They are twice as likely to be classified as globally cognitively impaired as those who did not use cannabis [59].

Alcohol and substance disorders complicate assessment and treatment of MS and psychiatric problems. Clinicians should routinely screen for alcohol and illicit drug abuse. Treatment can include motivational interviewing, interventions to facilitate more healthy behaviors, detoxification to address withdrawal symptoms, cognitive-behavioral therapies to avoid relapses, and the use of drugs to diminish cravings or discourage relapses [15].

Conclusion

Psychiatric comorbidities are common in multiple sclerosis. The risk of suicide is particularly serious in the first years of the disease. Anxiety is frequent in the disease and is the most powerful predictor of depression. Major depressive disorder is underdiagnosed and undertreated. Behavioral disorders are more common than severe psychiatric disorders and are probably secondary to cognitive impairment. Addictions may be undervalued.

References

1. Patten SB, Beck CA, Williams JVA, Barbui C, Metz LM. Major depression in multiple sclerosis. A population based perspective. Neurology. 2003;61:1524–7.
2. Patten SB, Metz LM, Reimer MA. Biopsychosocial correlates of lifetime major depression in a multiple sclerosis population. Mult Scler. 2000;6:115–20.
3. Kessler RC, Berglund P, Demler O, et al. The Epidemiology of Major Depressive Disorder Results from the National Comorbidity Survey Replication. JAMA. 2003;289:3095–105.
4. Korostil M, Feinstein A. Anxiety disorders and their clinical correlates in multiple sclerosis patients. Mult Scler. 2007;13:67–72.
5. Galeazzi GM, Ferrari S, Giaroli G, et al. Psychiatric disorders and depression in multiple sclerosis outpatients: impact of disability and interferon beta therapy. Neurol Sci. 2005;26:255–62.
6. Kessler RC, McGonagle KA, Zhao S, et al. Lifetime and 12-month prevalence of DSM-III-R psychiatric disorders in the United States. Results from the National Comorbidity Survey. Arch Gen Psychiatry. 1994;51:8–19.
7. Kessler RC, Keller MB, Wittchen HU. The epidemiology of generalized anxiety disorder. Psychiatry Clin North Am. 2001;1:19–39.
8. Edwards LJ, Constantinescu CS. A prospective study of conditions associated with multiple sclerosis in a cohort of 658 consecutive outpatients attending a multiple sclerosis clinic. Mult Scler. 2004;10:575–81.
9. Benazzi F. Bipolar disorder—focus on bipolar II disorder and mixed Depression. Lancet. 2007;369:935–45.
10. Marrie RA, Fisk JD, Yu BN, Leung S, Elliott L, Caetano P. Mental comorbidity and multiple sclerosis : validating administrative data to support population-based surveillance. BMC Neurol. 2013;13:16.

11. Pioro EP, Brooks BR, Cummings J, Schiffer R, Thisted RA, Wynn D, et al. Dextromethorphan Plus Ultra Low-Dose Quinidine Reduces Pseudobulbar Affect. Ann Neurol. 2010;68:693–702.
12. Espinola-Nadurille M, Colin-Piana R, Ramirez-Bermudez J, Lopez-Gomez M, Flores J, Arrambide G, Corona T. Mental Disorders in Mexican Patients With Multiple Sclerosis. J Neuropsychiatry Clin Neurosci. 2010;22:63–9.
13. Johnson SK, DeLuca J, Natelson BH. Personality dimensions in the chronic fatigue syndrome: a comparison with multiple sclerosis and depression. J Psychiatr Res. 1996;30(1):9–20.
14. Quesnel S, Feinstein A. Multiple sclerosis and alcohol: a study of problem drinking. Mult Scler. 2004;10(2):197–201.
15. Schuckit MA. Alcohol-used disorders. Lancet. 2009;373:492–501.
16. Fishman I, Benedict RHB, Bakshi R, Priore R, Weinstock-Guttman B. Construct validity and frequency of euphoria sclerotic in multiple sclerosis. J Neuropsychiatry Clin Neurosci. 2004;16:350–6.
17. Feinstein A, O'Connor P, Gray T, Feinstein KJ. The prevalence and neurobehavioral correlates of pathological laughing and crying in multiple sclerosis. Arch Neurol. 1997;54:1116–21.
18. Bombardier CH, Blake KD, Ehde DH, Gibbons LE, Moore D, Kraft GH. Alcohol and drug abuse among persons with multiple sclerosis. Mult Scler. 2004;10:35.
19. Chwastiak L, Ehde DM, Gibbons LE, Sullivan M, Bowen JD, Kraft GH. Depressive symptoms and severity of illness in multiple sclerosis: epidemiologic study of a large community sample. Am J Psychiatry. 2002;159:1862–8.
20. Rosti-Otajärvi E, Hämäläinen P. Behavioural symptoms and impairments in multiple sclerosis: a systematic review and meta-analysis. Mult Scler. 2013;19:31.
21. Fromont A, Binquet C, Rollot F, Despalins R, Weill A, Clerc L, Bonithon-Kopp C, Moreau T. Comorbidities at multiple sclerosis diagnosis. J Neurol. 2013;260(10):2629–37.
22. Tarrants, M Oleen-Burkey M., Castelli-Haley J, Lage MJ. The impact of comorbid depression on adherence to therapy for multiple sclerosis. Mult Scler Int. 2011;2011:271321.
23. Amato MP, Ponziani G, Rossi F, Liedl CL, Stefanile C, Rossi L. Quality of life in multiple sclerosis: the impact of depression, fatigue and disability. Mult Scler. 2001;7(5):340–4.
24. Fredrikson S, Cheng Q, Jiang GX, Wasserman D. Elevated suicide risk among patients with multiple sclerosis in Sweden. Neuroepidemiology. 2003;22(2):146–52.
25. Bronnum-Hansen H. Suicide among Danes with multiple sclerosis. J Neurol Neurosurg Psychiatry. 2005;76(10):1457–9.
26. Sadovnick AD, Eisen K, Ebers GC, Paty DW. Cause of death in patients attending multiple sclerosis clinics. Neurology. 1991;41:1193–6.
27. Leray E, Morrissey S, Yaouanq J, Coustans M, Le Page E, Chaperon J, Edan G. Long-term survival of patients with multiple sclerosis in West France. Mult Scler. 2007;13(7):865–74.
28. Grytten Torkildsen N, Lie SA, Aarseth JH, Nyland H, Myhr KM. Survival and cause of death in Multiple sclerosis : results from a 50 year follow-up in Western Norway. Mult Scler. 2008;14:1191–8.
29. Feinstein A. An examination of suicidal intent in patients with multiple sclerosis. Neurology. 2002;59:674–8.
30. American Psychiatric Association. Diagnostic and statistical manual of mental disorders. 4th ed., text rev (DSM-IV-TR). American Psychiatric Association 2000.
31. Jones KH, Ford DV, Jones PA, John A, Middleton RM, Lockhart-Jones H, et al. A large-scale study of anxiety and depression in people with Multiple Sclerosis: a survey via the web portal of the UK MS Register. PLoS One. 2012;7(7), e41910.
32. Crawford JR, Henry JD, Crombie C, Taylor EP. Normative data for the HADS from a large non-clinical sample. Br J Clin Psychol. 2001;40:429–34.
33. Schiffer RB, Wineman NM. Antidepressant pharmacotherapy of depression associated with multiple sclerosis. Am J Psychiatry. 1990;147:1493–7.
34. Frye M. Bipolar disorder –a focus on depression. N Engl J Med. 2011;364:51–9.
35. Marrie RS, Horwitz R, Cutter G, Tyry T, Campagnolo D, Vollmer T. The burden of medical comorbidity in multiple sclerosis: frequent, underdiagnosed and undertreated. Mult Scler. 2009;15:385–92.

36. Fisk JD, Morehouse SA, Brown MG, Skedgel C, Murray TJ. Hospital-based psychiatric service utilization and morbidity in multiple sclerosis. Can J Neurol Sci. 1998;25(3):230–5.
37. Carta MG, Moro MF, Lorefice L, Trincas G, Cocco E, Del Giudice E, Fenu G, Colom F, Marrosu MG. The risk of Bipolar Disorders in Multiple Sclerosis. J Affect Disord. 2014;155:255–60.
38. Lacovides A, Androulakis E. Bipolar disorders and resembling special psychopathological manifestations in Multiple Sclerosis: a review. Curr Opin Psychiatry. 2011;24:336–340.
39. Zephir H, De Seze J, Stojkovic T, Delisse B, Ferriby D, Cabaret M, Vermersch P. Multiple sclerosis and depression: influence of interferon beta therapy. Mult Scler. 2003;9(3):284.
40. Diaz-Olavarrieta C, Cummings JL, Velazquez J. Garcia de la Cadena C. Neuropsychiatric manifestations of multiple sclerosis. J Neuropsychiatry Clin Neurosci. 1999;11:51–7.
41. Sanfilipo MP, Benedict RH, Weinstock-Guttman B, Bakshi R. Gray and white matter brain atrophy and neuropsychological impairment in multiple sclerosis. Neurology. 2006;66:685–92.
42. Parvizi J, Archiniegas DB, Bernardini GL, Hoffmann MW, Mohr JP, Rapoport MJ, et al. Diagnosis and management of pathological laughter and crying. Mayo Clin Proc. 2006;81:1482–6.
43. Ghaffar O, Chamelian L, Feinstein A. Neuroanatomy of pseudobulbar affect: a quantitative MRI study in multiple sclerosis. J Neurol. 2008;255:406–12.
44. Seliger GM, Hornstein A. Serotonin, fluoxetine, and pseudobulbar affect. Neurology. 1989;39(10):1400.
45. Udaka F, Yamao S, Nagata H, et al. Pathologic laughing and crying treated with levodopa. Arch Neurol. 1984;41:1095–6.
46. Panitch HS, Thisted RA, Smith RA, et al. Randomized controlled trial of dextromethorphan/quinidine for pseudobulbar affect in multiple sclerosis. Ann Neurol. 2006;59(5):780–7.
47. Van Os J, Kapur S. Schizophrenia. Lancet. 2009;374:635–45.
48. Blanc F, Berna F, Fleury M, Lita L, Ruppert E, Ferriby D, et al. Inaugural psychotics events in Multiple Sclerosis. Rev Neurol. 2010;166:39–48.
49. Patten SB, Svenson LW, Metz LM. Psychotic disorders in MS : population-based evidence of an association. Neurology. 2005;65:1123–5.
50. Benedikt RHB, Priore RL, Miller CMunschauer F, Jacobs L. Personality disorders in Multiple sclerosis correlates with cognitive impairment. J Neuropsychiatry Clin Neurosci. 2001;13:70–6.
51. De Assis Aquino Gondim F, Thomas F. Episodic hyperlibidinism in multiple sclerosis. Mult Scler. 2001;7:67.
52. Allanson J, Bass C, Wade DT. Characteristics of patients with persistent severe disability and medically unexplained neurological symptoms: a pilot study. J Neurol Neurosurg Psychiatry. 2002;73:307–9.
53. Caplan LR, Nadelson T. Multiple sclerosis and hysteria. Lessons learned from their association. JAMA. 1980;243(23):2418–21.
54. Whiteford HA, Degenhardt L, Rehm J, Baxter AJ, Ferrari AJ, Erskine HE, et al. Global burden of disease attributable to mental and substance use disorders: findings from the Global Burden of Disease Study 2010. Lancet. 2013;382(9904):1575–86.
55. Syed YY, McKeage K, Scott LJ. Delta-9-tetrahydrocannabinol/cannabidiol (sativex(®)): a review of its use in patients with moderate to severe spasticity due to multiple sclerosis. Drugs. 2014;74(5):563–78.
56. Clark AJ, Ware MA, Yazer E, et al. Patterns of cannabis use among patients with multiple sclerosis. Neurology. 2004;62:2098–100.
57. Chong MS, Wolff K, Wise K, Tanton C, Winstock A, Silber E. Cannabis use in patients with multiple sclerosis. Mult Scler. 2006;12(5):646–51.
58. Honarmand K, Tierney MC, O'Connor P, Feinstein A. Effects of cannabis on cognitive function in patients with multiple sclerosis. Neurology. 2011;76(13):1153–60.
59. Ghaffar O, Feinstein A. Multiple sclerosis and cannabis: a cognitive and psychiatric study. Neurology. 2008;71:164–9.

Chapter 7
Psychiatric Presentation of Brain Inflammation

Bruno Brochet

Abstract Autoimmune encephalitis associated with antibodies targeting neural cell surface antigens have emerged in the past 10 years as a major cause of encephalitis. Those associated with antibodies against the N-methyl-D-aspartate receptor (NMDAR) is the most frequent. This newly recognized disease is characterized by a stereotyped clinical phenotype. The clinical course usually begins in the majority of cases by psychiatric symptoms, sometimes preceded by prodromal symptoms. The psychiatric stage is usually followed by severe fluctuations in consciousness with neurologic involvement with cognitive impairment, speech impairment, movement disorders, seizures, and behavioral problems. Typically the disease affects young women, and an ovarian teratoma is frequently associated, but cases have been reported in children, in men, and in patients without tumors. The treatment consists of early immunotherapy and, if necessary, tumor removal. The outcome is good in many cases if the treatment is started early, but severe sequelae or death is possible. The management of psychiatric symptoms can be difficult.

Keywords Encephalitis • Autoimmune • Antibody • Neural cell surface antigen • N-Methyl-D-aspartate receptor • Psychosis • Seizures • Limbic encephalitis • Dystonia • Teratoma • Paraneoplastic

Introduction

The first central nervous system (CNS) conditions associated with the presence of autoantibodies have been recognized in the 1980s [1]. The onconeural antibodies (Ab), which target neuronal epitopes within the cytoplasm or the nucleus, were described in patients with peripheral and CNS syndromes associated with cancer (paraneoplastic syndromes) [2]. The antibodies include anti-Hu, anti-Yo, and many

B. Brochet, MD (✉)
Department of Neurology, Centre Hospitalier Universitaire Pellegrin,
INSERM U 862, Université de Bordeaux, Place Amélie Raba Léon, Bordeaux 33076, France
e-mail: bruno.brochet@chu-bordeaux.fr

B. Brochet (ed.), *Neuropsychiatric Symptoms of Inflammatory Demyelinating Diseases*, Neuropsychiatric Symptoms of Neurological Disease,
DOI 10.1007/978-3-319-18464-7_7

others, which are used as biological diagnostic markers [2, 3]. The pathogenic roles of these Ab have been questioned, as their targets are intracellular proteins and because immunotherapy is rarely useful in these cases [3, 4]. It is considered, nowadays, that T cell cytotoxicity plays a more important role than Ab in the lesions that occurs in these encephalopathies [3, 4]. In the past 15 years, several CNS disorders, with encephalopathy, have been characterized by their association to Ab that bind to membrane-associated epitopes on neuronal cells [4, 5]. These disorders are frequently not associated with cancer, and immunotherapy could be effective, suggesting that these Ab are pathogenic. Table 7.1 summarizes the main CNS disorders associated with Ab targeting neuronal cell surface antigens [5]. The most frequent clinical syndrome seen during these disorders is limbic encephalopathy (LE) which

Table 7.1 Encephalitis and other disorders associated with antibodies targeting neuronal cell surface antigens [5]

Antigen	Clinical data	Sex/age	Tumor	Outcome
NMDAR	Psychiatric symptoms, memory and language deficits, seizures, movement disorders, autonomic instability, and decreased level of consciousness	80 % female Median age 21 (12–85)	Teratoma (40–50 %)	Good outcome with timely immunotherapy (and tumor removal if required) Sequelae 25 %
LGI1 (VGKC complex[a])	LE Faciobrachial dystonic seizures	65 % male Median age 60 (30–80)	Rare	Good response to immunotherapy but absent or poor response with AED in faciobrachial dystonic seizures
CASPR2 (VGKC complex[a])	Neuromyotonia, LE, Morvan syndrome, ataxia	85 % male Median age 60 (46–77)	Thymomas, SCLC (uncommon)	Good outcome, but can be complicated by tumor
AMPAR	LE Prominent psychiatric manifestations Sometimes isolated neuropsychiatric phenotype	90 % female Median age 60 (38–78)	SCLC, thymoma, or breast cancer (70 %)	50 % of relapses, even in the absence of tumor
GABAbR	LE, seizures	50 % female Median age 62 (24–75)	SCLC (50 %)	Good outcome; relapses are rare
GlyR	Progressive encephalomyelitis, rigidity, and myoclonus Stiff person syndrome	60% male Median age 46 (1–70)	Thymoma, Hodgkin lymphoma (rare)	Good outcomes Relapses possible

[a]Rare reports of pure psychiatric phenotype associated with VGKC complex antibodies (no target identified)

is an inflammation of the limbic system, including the hippocampus, thalamus, hypothalamus, and amygdala [6]. LE is characterized by subacute development of short-term memory loss, behavioral change, and epileptic seizures. Magnetic resonance imaging (MRI) typically shows signal abnormalities on T2/FLAIR sequences in these regions. Cerebrospinal fluid is usually inflammatory [6]. Psychiatric manifestations have been described in newly identified autoimmune encephalopathies and are the main features of encephalitis associated with anti-N-methyl-D-aspartate receptor (NMDAR) Ab [4, 5, 7–9]. Autoimmune encephalitis associated with other Ab usually rarely includes psychiatric symptoms [10]. Pure psychiatric forms have been reported in encephalitis associated with anti-α-amino-3-hydroxy-5-methyl-4-isoxazolepropionic acid receptor (AMPAR) or with anti-voltage-gated potassium channels (VGKC) complex Ab [10]. However, LE is the usual presentation of encephalitis with anti-AMPAR Abs and those with anti-leucine-rich, gliomainactivated1 (LGI1) (part of VGKC complex) [5], although patients with anti-contactin-associated protein-like 2 (Caspr2) antibodies (also part of VGKC complex) have neuromyotonia or LE or both (Morvan syndrome) [11]. Confusion and paranoia have also been reported in cases of encephalitis with anti- γ-aminobutyric acid receptor (GABAbR) [10] which are usually presenting with LE. Psychiatric symptoms could also occur but rarely during viral encephalitis with enterovirus, herpes simplex virus 1, varicella-zoster virus, or West Nile virus [12]. We will focus this review on anti-NMDAR autoimmune encephalitis.

Anti-NMDAR Autoimmune Encephalitis

The first cases of encephalitis associated with Ab directed toward neuronal cell surface antigens were two cases of LE associated with anti-VGKC Ab [13], which was later attributed to anti-LGI1 [14]. Anti-NMDAR GluR ∈2 subunit autoantibodies were, first, detected in patients with Rasmussen encephalitis [15]. Two years later, the cases of four patients with severe paraneoplastic encephalitis, affecting young women with ovarian teratomas, were described [16]. The clinical syndrome associated acute psychiatric symptoms, seizures, memory impairment, hypoventilation, and decreased level of consciousness. The same group described other cases associated with Ab to unknown antigens (a subgroup of neuropil antigens) predominantly expressed in the cell membrane of hippocampal neurons [17]. The demonstration of the association of these encephalitic cases with anti-NMDAR antibodies was published in 2007 [18]. Since these seminal observations, many cases have been reported, including men, women, and children, and frequently without evidence of cancer [4, 5, 19, 20].

Glutamatergic neurotransmission is mediated via several receptors including the ionotropic NMDAR and plays a critical role in the modulation of synaptic plasticity, mood, pain transmission and regulation, cognitive processes, and motricity [21]. The hyperactivation of NMDAR could mediate acute neuronal death and is thought to play a role in chronic neurodegenerative diseases like Alzheimer's disease and amyotrophic lateral sclerosis [21]. In contrast, severe hypofunction of NMDAR is able to produce a clinical syndrome similar to schizophrenic exacerbation [22]. It has been observed that anesthetic drugs acting on NMDAR such as phencyclidine and ketamine may induce schizophrenia-like symptoms [21, 22]. Indeed, increasing evidence suggests the implication of NMDAR dysfunction in the pathogenesis of schizophrenia and, in particular, in negative and cognitive symptoms [23]. Several polymorphisms of genes controlling the NMDA receptor's pathway have been found associated with increase susceptibility for schizophrenia [23].

Encephalitis associated with anti-NMDAR Ab is a recently recognized entity and its prevalence is largely unknown. From 2005 to 2011, Dalmau and co-workers [19] identified 419 cases. Several studies attempted to measure the frequency of anti-NMDAR Ab in different populations of patients with encephalitis: from 1 % of patients with encephalitis of unknown etiology admitted to the intensive care unit of one center [24] to 4 % in a multicenter prospective study of patients with encephalitis made in England during 2 years [25] and in a prospective cohort in California [26]. In the California Encephalitis Project, who included patients aged <30 years, anti-NMDAR encephalitis was identified .4 times as frequently as herpes simplex virus-1, West Nile virus, or varicella-zoster virus [26]. It appears that NMDAR Ab-associated encephalitis is not rare and this diagnosis has to be considered in patients presenting with psychiatric symptoms [9, 10].

Clinical Presentation

Several studies have established the clinical syndrome associated with anti-NMDAR Ab [20, 27–30].

The disease was initially described exclusively in female patients, but it could also occur in less than 10 % of cases, in male patients [29, 30]. The age at onset varies considerably, and cases have been reported in children [30], adults, and geriatric patients, but the mean age is in the second decade [20, 29]. In a majority of patients (up to 86 % of cases), the disease is preceded by prodromal symptoms: headache, fever, nausea, vomiting, diarrhea, or upper respiratory tract symptoms. These symptoms last from a few days (median 5) to 2 weeks before the onset of the psychiatric symptoms [19, 20, 17, 28, 29]. The psychiatric presentation is, by far, the most frequent, occurring in nearly 80 % of cases [9, 10, 19, 29]. These symptoms include anxiety, insomnia, fear, delusions, perceptual disturbances, hyperreligiosity, disorganized thoughts and behaviors, agitation, and paranoid ideation. Also possible are social withdrawal and stereotypical behavior [9, 10]. In children, the disease presents with behavioral or personality change, and sleep dysfunction, hyperactivity, and hypersexuality are seen [9, 10, 20, 30]. The psychiatric symptoms usually last from 1 to 3 weeks.

Due to this usual psychiatric presentation, many patients are seen initially by psychiatrists.

Neurological symptoms are as frequent as psychiatric symptoms but are frequently underestimated [19]. Typically they followed psychiatric symptoms. These neurological symptoms are mainly cognitive impairment with short-term memory loss and speech problems [29]. Progressive decline in speech and language, including alogia, echolalia, perseveration, mumbling, and mutism, is characteristic [19, 29]. Seizures are present in a large majority of patients, frequently generalized tonic-clonic seizures, but sometimes partial motor or complex and other types are seen [19, 29]. Although seizures can been seen at all stages of the illness, the frequency of the seizures usually decreases as the disease evolves. In children, the first symptom to be recognized is often nonpsychiatric—e.g., seizures, status epilepticus, dystonia, verbal reduction, or mutism [30].

Dyskinesis (especially orofacial), dystonic posturing, and choreic-like movements of the limbs and spastic rigidity occur also very frequently but predominate usually in a second stage of the disease when the psychiatric symptoms decreased and are followed by decreased responsiveness, global alterations in consciousness sometimes progressing to a catatonic-like state with mutism and eyes open, or sometimes to a stage of agitation. During this second stage, autonomic symptoms (hyperthermia, urinary incontinence, cardiac arrhythmia, hypo- or hypertension, and central hypoventilation) are common and can require intubation or pacemakers [19, 29].

Spontaneous neurological improvement has been reported but appears to be slow and inconstant.

Paucisymptomatic cases have been reported, associated with only seizures [31] or dystonia [32].

Diagnostic

Magnetic resonance imaging could show signal hyperintensity on T2/FLAIR sequences in the cerebral cortex, mainly the medial temporal lobe and less frequently in cerebellar or brainstem regions or basal ganglia [29]. Cortical or meningeal contrast enhancement is possible but rare. The MRI abnormalities are not specific and could be absent in about 50 % of cases. Brain atrophy could occur in untreated cases [19].

Electroencephalograms (EEG) are abnormal in more than 90 % of cases showing nonspecific, slow, and disorganized activity, and paroxysmal activities are detected in about 20 % of cases [29]. Video-EEG could be helpful [30].

Cerebrospinal fluid (CSF) analysis is essential for diagnosis. At onset, the CSF is abnormal in 80 % of patients and becomes abnormal later in the disease in most other patients. Lymphocytic pleocytosis is common. Oligoclonal bands are positive in 60 % of cases [19, 20]. The detection of NMDAR Ab in the CSF and in sera is essential for the diagnosis. Ab are identified in CSF in 100 % of cases and in sera in 85 % [33]. Ab titers in CSF and serum were higher in patients with poor outcome or teratoma than in patients with good outcome or no tumor. Over time there was a decrease of Ab titers regardless of outcome, and after recovery, the majority of samples remains positive. However, relapses were associated with an increase in titer more often in CSF than in serum [33].

Anti-NMDAR antibodies could be detected by indirect immunofluorescence on cryopreserved sections or primary cell cultures of the rodent brain or in vitro enzyme-linked immunosorbent assay examination or using a cell-based immunoassay of culture cells (i.e., HEK cells) transfected with the complementary DNA (cDNA) representing the single or assembled NR1–NR2 subunits [20]. The cell assay is more specific.

Brain biopsies are not helpful, showing normal or nonspecific findings, including perivascular inflammation with B and T cell infiltrates and microglial activation [19].

The differential diagnosis includes viral encephalitis, acute psychosis, and mania eventually with psychotic features, drug abuse, and neuroleptic malignant syndrome.

Association with Tumors and Gender

The first reported cases were all women presenting with teratoma [16]. In later reports it appears that the disease occurs in 80 % of women and is being more frequently recognized in younger teenagers and children [30]. In a large series of 100 cases, tumors were diagnosed in 58 cases and included 53 teratomas in women, one immature teratoma of the testis and one small-cell lung cancer in two men [29]. The detection of an underlying tumor is dependent of age, sex, and ethnic background. Analysis of 400 patients confirms that tumors are less likely to be found in younger patients and that teratomas are found in more than half of women older than 18 years. Black women are more likely to have an underlying ovarian teratoma than are patients of other ethnic groups [19].

The screening procedures for ovarian teratomas include MRI, CT scan, and ultrasound. Serological tumor markers are not helpful. Exploratory laparoscopies and blind oophorectomies have been helpful in some cases [19].

Outcomes and Treatments

Immunotherapy and tumor resection are the two main aspects of the management of anti-NMDAR Ab encephalitis. Although no controlled trial has been done, there is some evidence of the efficacy of these treatments. Retrospective studies reported 4 % of mortality and a good recovery, complete or with mild sequelae in 75 % of patients, but all other patients remain severely disabled [29, 30]. Immunotherapy includes corticosteroids and intravenous immunoglobulin or plasma exchange, but they work best when an underlying tumor has been removed

[11, 19, 29]. In the absence of tumors, a second-line immunotherapy is frequently needed with mycophenolate mofetil or azathioprine [29].

Management of psychiatric symptoms is complex and not standardized [9]. It has been suggested initiating treatment with quetiapine in patients with psychotic symptoms and agitation or Thorazine in patients who refuse oral medications [9]. High-potency antipsychotics must be avoided. Valproic acid could be helpful in patients with mood symptoms, emotional lability, and/or mania [9].

References

1. Graus F, Delattre JY, Antoine JC, Dalmau J, Giometto B, Grisold W, et al. Recommended diagnostic criteria for paraneoplastic neurological syndromes. J Neurol Neurosurg Psychiatry. 2004;75:1135–40.
2. Honnorat J, Cartalat-Carel S, Ricard D, Camdessanche JP, Carpentier AF, Rogemond V, et al. Onco-neural antibodies and tumour type determine survival and neurological symptoms in paraneoplastic neurological syndromes with Hu or CV2/CRMP5 antibodies. J Neurol Neurosurg Psychiatry. 2009;80:412–6.
3. Graus F, Saiz A, Dalmau J. Antibodies and neuronal autoimmune disorders of the CNS. J Neurol. 2010;257:509–17.
4. Zuliani L, Graus F, Giometto B, Bien C, Vincent A. Central nervous system neuronal surface antibody associated syndromes: review and guidelines for recognition. J Neurol Neurosurg Psychiatry. 2012;83:638–45.
5. Coutinho E, Harrison P, Vincent A. Do neuronal autoantibodies cause psychosis? A neuroimmunological perspective. Biol Psychiatry. 2014;75:269–75.
6. Gultekin SH, Rosenfeld MR, Voltz R, Eichen J, Posner JB, Dalmau J. Paraneoplastic limbic encephalitis: neurological symptoms, immunological findings and tumour association in 50 patients. Brain. 2000;123:1481–94.
7. Dalmau J, Tüzün E, Wu HY, Masjuan J, Rossi JE, Voloschin A, et al. Paraneoplastic anti-N-methyl-D-aspartate receptor encephalitis associated with ovarian teratoma. Ann Neurol. 2007;61:25–36.
8. Dalmau J, Lancaster E, Martinez-Hernandez E, Rosenfeld MR, Balice-Gordon R. Clinical experience and laboratory investigations in patients with anti-NMDAR encephalitis. Lancet Neurol. 2011;10:63–74.
9. Kayser MS, Dalmau J. Anti-NMDA receptor encephalitis in psychiatry. Curr Psychiatry Rev. 2011;7:189–93.
10. Kayser MS, Dalmau J. The emerging link between autoimmune disorders and neuropsychiatric disease. J Neuropsychiatry Clin Neurosci. 2011;23:90–7.
11. Irani SR, Alexander S, Waters P, Kleopa KA, Pettingill P, Zuliani L, et al. Antibodies to Kv1 potassium channel-complex proteins leucine-rich, glioma inactivated 1 protein and contactin-associated protein-2 in limbic encephalitis, Morvan's syndrome and acquired neuromyotonia. Brain. 2010;133:2734–48.
12. Gable MS, Sheriff H, Dalmau J, Tilley DH, Glaser CA. The frequency of autoimmune N-methyl-D-aspartate receptor encephalitis surpasses that of individual viral etiologies in young individuals enrolled in the California Encephalitis Project. Clin Infect Dis. 2012;54:899–904.
13. Buckley C, Oger J, Clover L, Tüzün E, Carpenter K, Jackson M, et al. Potassium channel antibodies in two patients with reversible limbic encephalitis. Ann Neurol. 2001;50:73–8.
14. Lai M, Huijbers MG, Lancaster E, Graus F, Bataller L, Balice-Gordon R, et al. Investigation of LGI1 as the antigen in limbic encephalitis previously attributed to potassium channels: a case series. Lancet Neurol. 2010;9:776–85.

15. Takahashi Y, Mori H, Mishina M, Watanabe M, Fujiwara T, Shimomura J, et al. Autoantibodies to NMDA receptor in patients with chronic forms of epilepsia partialis continua. Neurology. 2003;61:891–6.
16. Vitaliani R, Mason W, Ances B, Zwerdling T, Jiang Z, Dalmau J. Paraneoplastic encephalitis, psychiatric symptoms, and hypoventilation in ovarian teratoma. Ann Neurol. 2005;58:594–604.
17. Ances BM, Vitaliani R, Taylor RA, Liebeskind DS, Voloschin A, Houghton DJ, et al. Treatment-responsive limbic encephalitis identified by neuropil antibodies: MRI and PET correlates. Brain. 2005;128:1764–77.
18. Dalmau J, Tüzün E, Wu HY, Masjuan J, Rossi JE, Voloschin A, et al. Paraneoplastic anti-N-methyl-D-aspartate receptor encephalitis associated with ovarian teratoma. Ann Neurol. 2007;61:25–36.
19. Dalmau J, Lancaster E, Martinez-Hernandez E, Rosenfeld MR, Balice-Gordon R. Clinical experience and laboratory investigations in patients with anti-NMDAR encephalitis. Lancet Neurol. 2011;10:63–74.
20. Miya K, Takahashi Y, Mori H. Anti-NMDAR autoimmune encephalitis. Brain Dev. 2014;36:645–52.
21. Rosenthal-Simons A, Durrant AR, Heresco-Levy U. Autoimmune-induced glutamatergic receptor dysfunctions: conceptual and psychiatric practice implications. Eur Neuropsychopharmacol. 2013;23:1659–71.
22. Farber NB. The NMDA receptor hypofunction model of psychosis. Ann N Y Acad Sci. 2003;1003:119–30.
23. Gruber O, Chadha Santuccione A, Aach H. Magnetic resonance imaging in studying schizophrenia, negative symptoms, and the glutamate system. Front Psychiatry. 2014;5:32.
24. Prüss H, Dalmau J, Harms L, Höltje M, Ahnert-Hilger G, Borowski K, et al. Retrospective analysis of NMDA receptor antibodies in encephalitis of unknown origin. Neurology. 2010;75(19):1735–9.
25. Granerod J, Ambrose HE, Davies NW, Clewley JP, Walsh AL, Morgan D, et al. Causes of encephalitis and differences in their clinical presentations in England: a multicentre, population-based prospective study. Lancet Infect Dis. 2010;10:835–44.
26. Gable MS, Sheriff H, Dalmau J, Tilley DH, Glaser CA. The frequency of autoimmune N-methyl-D-aspartate receptor encephalitis surpasses that of individual viral etiologies in young individuals enrolled in the California Encephalitis Project. Clin Infect Dis. 2012;54:899–904.
27. Sansing LH, Tüzün E, Ko MW, Baccon J, Lynch DR, Dalmau J. A patient with encephalitis associated with NMDA receptor antibodies. Nat Clin Pract Neurol. 2007;3:291–6.
28. Iizuka T, Sakai F, Ide T, Monzen T, Yoshii S, Iigaya M, et al. Anti-NMDA receptor encephalitis in Japan: long-term outcome without tumor removal. Neurology. 2008;70:504–11.
29. Dalmau J, Gleichman AJ, Hughes EG, Rossi JE, Peng X, Lai M, et al. Anti-NMDA Receptor encephalitis: case series and analysis of the effects of antibodies. Lancet Neurol. 2008;7:1091–8.
30. Florance NR, Davis RL, Lam C, Szperka C, Zhou L, Ahmad S, et al. Anti-N-methyl-D-aspartate receptor (NMDAR) encephalitis in children and adolescents. Ann Neurol. 2009;66:11–8.
31. Niehusmann P, Dalmau J, Rudlowski C, Vincent A, Elger CE, Rossi JE, et al. Diagnostic value of N-methyl-D-aspartate receptor antibodies in women with new-onset epilepsy. Arch Neurol. 2009;66:458–64.
32. Rubio-Agustí I, Dalmau J, Sevilla T, Burgal M, Beltrán E, Bataller L. Isolated hemidystonia associated with NMDA receptor antibodies. Mov Disord. 2011;26:351–2.
33. Gresa-Arribas N, Titulaer MJ, Torrents A, Aguilar E, McCracken L, Leypoldt F, et al. Antibody titres at diagnosis and during follow-up of anti-NMDA receptor encephalitis: a retrospective study. Lancet Neurol. 2014;13:167–77.

Chapter 8
Drug Management of Psychiatric Co-morbidity in Multiple Sclerosis

Pierre-Michel Llorca and Ludovic Samalin

Abstract Multiple sclerosis (MS) is associated with a higher risk of psychiatric comorbidities that have an impact on the evolution, prognosis, and quality of life of patients. Despite this observation, evidence-based data on the treatment of these psychiatric conditions are rather sparse. Selective serotonergic recapture inhibitors and serotonin norepinephrine reuptake inhibitors can be considered as the first-choice treatment for depression or anxiety disorder in MS patients. Second-generation antipsychotics are of interest in MS patients suffering of bipolar disorder, compared to lithium or anticonvulsants. They also have an efficacy on psychotic symptoms observed in MS patients. Adherence to treatment is also an important topic in these patients and needs to be evaluated and improved using psychoeducation programs. According to the consequences of psychiatric comorbidities, research on the efficacy of psychotropic drugs in MS patients must be developed.

Keywords Multiple sclerosis • Depression • Bipolar disorder • Anxiety • Psychotropics

Introduction

Multiple sclerosis (MS) is a relatively common chronic disabling central nervous system disease affecting 1 in 1,000 people in western countries [1]. Patients with MS appear to have higher lifetime prevalence rates of psychiatric symptoms and disorders compared with the general population (Table 8.1) [2–13].

Corticosteroids and beta interferon that are used to treat MS are also associated with an increased risk of neuropsychiatric side effects, from mood disorders to psychotic symptoms. Those side effects induce the need for patient education about psychiatric comorbidities and for regular psychiatric evaluation.

P.-M. Llorca, MD, Ph.D. (✉) • L. Samalin, MD
Department of Psychiatry B, CHU Clermont-Ferrand, University of Auvergne, EA 7280, 58 rue Montalembert, Clermont-Ferrand 63000, France
e-mail: pmllorca@chu-clermontferrand.fr; lsamalin@chu-clermontferrand.fr

B. Brochet (ed.), *Neuropsychiatric Symptoms of Inflammatory Demyelinating Diseases*, Neuropsychiatric Symptoms of Neurological Disease,
DOI 10.1007/978-3-319-18464-7_8

Table 8.1 Lifetime prevalence rates of psychiatric disorders in MS patients and the general population [2–13]

	Lifetime prevalence rates	
	Multiple sclerosis	General population
Major depressive disorder	36–54 %	16.2 %
Bipolar disorder	13 %	1–4.5 %
Anxiety disorder	35.7 %	28.8 %
Psychotic disorders	2–3 %	1.8 %

Psychiatric disorders in MS patients are associated with decreased adherence to treatment, impaired functional status, and quality of life. The reported rates of completed suicide in persons with MS are also high, and suicidality seems to be associated with psychiatric disorders [14].

This context highlights the importance of screening, diagnosing, and treating MS patients suffering from psychiatric comorbidities.

In this chapter, we will focus on the pharmacological treatments for each of the psychiatric conditions.

Drug Treatment of Depression in Multiple Sclerosis

Depression is the commonest psychiatric disorder in MS patients but remains underdiagnosed and undertreated. Major depressive disorder (MDD) in patients with MS does not relate directly to disability progression or to longer disease duration. The reported risk factors are female sex, age below 35 years, family history of major depression, and high stress levels. Multiple sclerosis patients also experience fatigue and cognitive dysfunction, both of which can worsen depression and be worsened by depression [14].

Despite the high prevalence of MDD in MS and although antidepressant (AD) use is common among patients with MS, the literature on the effectiveness of antidepressants in MS is limited.

Three double-blind controlled studies evaluated the impact of desipramine (a tricyclic antidepressant, TCA), sertraline, and paroxetine (two widely used selective serotonergic recapture inhibitors, SSRIs) in depressed MS patients [15–17]. Only the last one was placebo-controlled and the three samples were relatively small ($n=28$, $n=22$ and $n=42$, respectively). The antidepressant effect was moderate for sertraline (effect size: $d=0.46$, $p=0.047$) (findings presented in Minden et al. [2]), inconsistent for desipramine (according to the depression scales that were used), and comparable to the placebo for paroxetine. Desipramine is associated with a high level of side effects (postural hypotension, dry mouth, and constipation) that patients are not always able to tolerate.

A few open-label trials using SSRIs (sertraline, fluvoxamine, and fluoxetine), serotonin norepinephrine reuptake inhibitors (SNRIs) (duloxetine), reversible inhibitors of monoamine oxidase A (RIMAs) (moclobemide), and mirtazapine were also published [18,19]. Those studies showed the effects on depressive symptoms, evaluated with specific depression scales, and demonstrated better tolerance of SSRIs compared to TCA among MS patients.

According to those studies, antidepressants seem to reduce depressive symptoms in MS patients and should be considered for treating MDD in this population. However in an evidence-based perspective, the available literature provides insufficient evidence to support or refute the efficacy and use of TCAs and SSRIs for depressive symptoms and MDD in MS patients [2]. In this specific population, the evaluation of the balance between efficacy and risks has to be very cautious.

In their guidelines for the management of patients with mood disorders and selected comorbid medical conditions [20], the Canadian Network for Mood and Anxiety Treatments (CANMAT) considered that the use of antidepressants in this population should be strongly considered (recommendation level 2). However, due to issues with fatigue, orthostatic hypotension, balance, cognitive issues, and bladder problems, antidepressants with significant sedating or anticholinergic side effects should be avoided (recommendation level 3).

Electroconvulsive therapy (ECT) is an option in the case of treatment-resistant depression. In depressed MS patients treated with ECT, the neurological status of those patients can deteriorate, raising the question of whether ECT is a risk factor for disease exacerbation [21].

Summary

- Antidepressants are largely used in MS patients with depressive symptoms or MDD, nevertheless empirical data are lacking.
- Selective serotonergic recapture inhibitors and SNRIs can be considered as a first-line option according to the efficacy/risk balance.
- In treatment-resistant patients, ECT should be considered with caution related to the risk of neurological deterioration in the presence of active disease.
- Antidepressants combined with cognitive behavioral therapy (CBT) is considered to be of interest to meet the particular needs of each individual patient whenever possible [22].

Drug Treatment of Bipolar Disorder in Multiple Sclerosis

In a study using standardized diagnostic tools and a case–control design, compared with controls, MS patients had a significantly higher lifetime prevalence of bipolar disorder type I ($p = 0.005$) and bipolar disorder type II ($p < 0.0001$) [23]. Because of this high prevalence, the CANMAT considered that people with MS should be monitored for hypomanic and manic symptoms while they are being treated with antidepressant medications (recommendation level 4) [20].

Despite the impact of this co-morbidity, no specific clinical trials of pharmacologic interventions for these patients can be found in the literature. A few anecdotal reports underline interest in the various strategies for the treatment of bipolar disorder including benzodiazepines [24], second-generation antipsychotics [25], or lithium [26].

For most of the patients, the choice of treatment must be based on the recommendations used in general psychiatry. For the treatment of acute mania, risperidone, aripiprazole (two second-generation antipsychotics), and valproate (an anticonvulsant) are considered as a *grade 1 recommendation* (*Category A evidence and good risk–benefit ratio*) [27]. Lithium (formerly considered to be a "gold standard") is only a *grade 2 recommendation*. For the treatment of acute bipolar depression, quetiapine is the only drug considered to be a *grade 1 recommendation* [28]. Lithium, lamotrigine (an anticonvulsant), aripiprazole, and quetiapine are considered as a *grade 1 recommendation* for long-term treatment of bipolar disorder [29].

For Ameis and Feinstein [18], according to different case reports, anticonvulsants should be less effective on psychiatric symptoms in MS patients, compared with their usual efficacy. This has to be confirmed by specific studies.

In MS patients, the balance between efficacy and risk must be considered [18]:

- Lithium is not always well tolerated: the increase of diuresis induced by this compound, coupled with bladder dysfunction observed in MS, can make the patient incontinent.
- Second-generation antipsychotics are better tolerated than neuroleptics in terms of extrapyramidal side effects, but MS patients may be more sensitive to neurological side effects and must be regularly monitored.
- Anticonvulsants may induce sedation, dizziness, headaches, ataxia, tremors, nausea, constipation, and weight gain that may increase the disability associated with MS.

One of the specificities of MS patients is that some of them develop mania or hypomania secondary to treatment with steroids/ACTH prescribed for exacerbations of neurological symptoms [30]. Patients with a previous history of depression or a family history of depression must be considered to be at risk. Clinicians must consider using reduced doses of steroids/ACTH or adding lithium prophylaxis when treating high-risk patients.

Summary

- Multiple sclerosis patients are at high risk of developing bipolar disorder.
- Lithium, anticonvulsants, and second-generation antipsychotics must be considered for the treatment of acute phases (mania and depression) and for long-term maintenance treatment of bipolar disorder in MS patients.

- The specificity of the different side-effect profiles has to be taken into account.
- The iatrogenic effect of steroids/ACTH on mood has to be considered, high-risk patients have to be identified, and a prophylactic use of a mood stabilizer has to be considered.

Drug Treatment of Anxiety in Multiple Sclerosis

Generalized anxiety disorder appears to be the most common anxiety disorder in MS patients with 18 % of patients meeting the criteria for this disorder and more than half of these patients not receiving any treatment [31]. Panic disorder and obsessive–compulsive disorder may also be much more common. The advent of injectable disease-modifying treatments for MS induces an increase in "self-injection anxiety." If specific cognitive behavioral therapies have been developed for "self-injection anxiety," no studies on the pharmacological approach of anxiety have been published.

The main pharmacological agents that can be used for anxiety disorders in the general population [32] must be considered in MS patients. Selective serotonergic recapture inhibitors, SNRIs, noradrenergic and specific serotonergic antidepressants (NaSSAs), TCAs, monoamine oxidase inhibitors (MAOIs), and RIMAs have demonstrated their efficacy in the treatment of anxiety disorders. Selective serotonergic recapture inhibitors and SNRIs are usually preferred as initial treatments, since they are generally safer and better tolerated. Benzodiazepines may be useful as adjunctive therapy, particularly for acute anxiety or agitation or while waiting for the onset of adequate efficacy of SSRIs. Due to concerns about possible dependency, sedation, and cognitive impairment, those compounds should be restricted to short-term use. Several anticonvulsants and atypical antipsychotics have demonstrated efficacy in some anxiety and related disorders, but are generally recommended as second-line, third-line, or adjunctive therapies.

The choice of medication should take into consideration the evidence for efficacy and safety/tolerability specifically in MS patients. Selective serotonergic recapture inhibitors and SNRIs may induce sexual dysfunction, drowsiness, and fatigue that can be particularly disabling for these patients.

Summary

- Anxiety disorders are frequent in MS patients and have to be screened.
- Selective serotonergic recapture inhibitors and SNRIs can be considered as the cornerstone of the pharmacological treatment of anxiety disorders.
- The choice of the compound must rely on the efficacy and safety profile.

Drug Treatment of Psychosis in Multiple Sclerosis

The prevalence of psychotic disorders is higher in MS patients, with epidemiologic evidence of an association between MS and psychotic disorders in the general population [13].

There are only a few case reports describing the efficacy of neuroleptics or second-generation antipsychotics in the treatment of psychosis in MS [18]. Neuroleptics induced more neurological side effects and a higher risk for tardive dyskinesia in MS patients [33]. Consideration should be given to the fact that those patients may be more sensitive than general psychiatry patients to developing antipsychotic extrapyramidal symptoms or tardive dyskinesia.

Second-generation antipsychotics have a better risk–benefit balance with a specific interest of ziprasidone, risperidone, and aripiprazole [34]. They can be considered as a first-line treatment for MS patients with psychosis. There is no data suggesting that one second-generation antipsychotic is better than another. Clozapine is not a first-line treatment owing to the risk of agranulocytosis and the need for active blood monitoring.

Summary

- Psychotic symptoms are more prevalent in MS patients compared with the general population.
- Second-generation antipsychotics must be considered as a first-line treatment in MS patients with psychosis.
- MS patients may be more sensitive to extrapyramidal side effects that have to be monitored.

General Considerations

Adherence to Treatment

Multiple sclerosis patients with psychiatric comorbidities are more likely not to adhere to disease-modifying drug treatment and psychotropic treatment [14]. This can have a dramatic effect in terms of outcome and quality of life. There is a need to optimize adherence to psychotropic treatment:

- Clinicians must assess adherence at every consultation.
- Psychoeducation programs focusing on treatment adherence have to be developed and implemented.

Needs for Future Research

Despite the high prevalence of psychiatric disorders in MS patients and their consequences, there is a real lack of evidences to define precise pharmacological strategies. Large, methodologically rigorous, randomized, placebo-controlled studies must be conducted in this population to evaluate pharmacologic therapies with strong evidence of efficacy and widespread use for treating emotional disorders in individuals with MS [2]. This may include systematic examinations of combinations of pharmacologic and non-pharmacologic therapies.

Conclusion

Multiple sclerosis is frequently associated with psychiatric comorbidities. The drug management of these disorders is a challenge for clinicians because they can have an impact on the prognosis of MS, the adherence to treatment, and the quality of life of patients. The first step in the management of all patients with MS is the need to systematically assess and screen psychiatric comorbidities (especially MDD, bipolar disorder, general anxiety disorder, and psychotic disorder). The second step (pharmacological strategies) will select and introduce a compound as a first-line option according to the safety profile of the patient. Due to lack of evidence, most of the recommendations are based on guidelines for the treatment of psychiatric disorders in the general population. Specific studies and consequently clinical practice guidelines for patients with MS and psychiatric comorbidities are needed.

References

1. Hogancamp WE, Rodriguez M, Weinshenker BG. The epidemiology of multiple sclerosis. Mayo Clin Proc. 1997;72:871–8.
2. Minden SL, Feinstein A, Kalb RC, Miller D, Mohr DC, Patten SB, et al. Guideline Development Subcommittee of the American Academy of Neurology. Evidence-based guideline: assessment and management of psychiatric disorders in individuals with MS: report of the Guideline Development Subcommittee of the American Academy of Neurology. Neurology. 2014;82(2):174–81.
3. Minden SL, Schiffer RB. Affective disorders in multiple sclerosis: review and recommendations for clinical research. Arch Neurol. 1990;47:98–104.
4. Sadovnick AD, Remick RA, Allen J, Swartz E, Yee IM, Eisen K, et al. Depression and multiple sclerosis. Neurology. 1996;46:628–32.
5. Minden SL, Orav J, Reich P. Depression in multiple sclerosis. Gen Hosp Psychiatry. 1987;9:426–34.
6. Joffe RT, Lippert GP, Gray TA, Sawa G, Horvath Z. Mood disorder and multiple sclerosis. Arch Neurol. 1987;44:376–8.
7. Schiffer RB, Caine ED, Bamford KA, Levy S. Depressive episodes in patients with multiple sclerosis. Am J Psychiatry. 1983;140:1498–500.

8. Patten SB, Beck CA, Williams JV, Barbui C, Metz LM. Major depression in multiple sclerosis: a population-based perspective. Neurology. 2003;61:1524–7.
9. Schiffer RB, Wineman NM, Weitkamp LR. Association between bipolar affective disorder and multiple sclerosis. Am J Psychiatry. 1986;143:94–5.
10. Merikangas KR, Akiskal HS, Angst J, Greenberg PE, Hirschfeld M, Petukhova M, et al. Lifetime and 12-month prevalence of bipolar spectrum disorder in the National Comorbidity Survey Replication. Arch Gen Psychiatry. 2007;64:543–52.
11. Korostil M, Feinstein A. Anxiety disorders and their clinical correlates in multiple sclerosis patients. Mult Scler. 2007;13:67–72.
12. Kessler RC, Berglund P, Demler O, Jin R, Merikangas KR, Walters EE. Lifetime prevalence and age-of-onset distributions of DSM-IV disorders in the National Comorbidity Survey Replication. Arch Gen Psychiatry. 2005;62:593–602.
13. Patten SB, Svenson LW, Metz LM. Psychotic disorders in MS: population-based evidence of an association. Neurology. 2005;65(7):1123–5.
14. Fragoso YD, Adoni T, Anacleto A, da Gama PD, Goncalves MV, Matta AP, et al. Recommendations on diagnosis and treatment of depression in patients with multiple sclerosis. Pract Neurol. 2014;14(4):206–9.
15. Schiffer RB, Wineman NM. Antidepressant pharmacotherapy of depression associated with multiple sclerosis. Am J Psychiatry. 1990;147:1493–7.
16. Mohr DC, Boudewyn AC, Goodkin DE, Bostrom A, Epstein L. Comparative outcomes for individual cognitive-behavior therapy, supportive-expressive group psychotherapy, and sertraline for the treatment of depression in multiple sclerosis. J Consult Clin Psychol. 2001;69:942–9.
17. Ehde DM, Kraft GH, Chwastiak L, Sullivan MD, Gibbons LE, Bombardier CH, et al. Efficacy of paroxetine in treating major depressive disorder in persons with multiple sclerosis. Gen Hosp Psychiatry. 2008;30(1):40–8.
18. Ameis SH, Feinstein A. Treatment of neuropsychiatric conditions associated with multiple sclerosis. Expert Rev Neurother. 2006;6(10):1555–67.
19. Solaro C, Bergamaschi R, Rezzani C, Mueller M, Trabucco E, Bargiggia V, et al. Duloxetine is effective in treating depression in multiple sclerosis patients: an open-label multicenter study. Clin Neuropharmacol. 2013;36(4):114–6.
20. Ramasubbu R, Taylor VH, Samaan Z, Sockalingham S, Li M, Patten S, et al. The Canadian Network for Mood and Anxiety Treatments (CANMAT) task force recommendations for the management of patients with mood disorders and select comorbid medical conditions. Ann Clin Psychiatry. 2012;24(1):91–109.
21. Mattingly G, Baker K, Zorumski CF, Figiel GS. Multiple sclerosis and ECT: possible value of gadolinium-enhanced magnetic resonance scans for identifying high-risk patients. J Neuropsychiatry Clin Neurosci. 1992;4(2):145–51.
22. Goldman Consensus Group. The Goldman Consensus statement on depression in multiple sclerosis. Mult Scler. 2005;11(3):328–37.
23. Carta MG, Moro MF, Lorefice L, Trincas G, Cocco E, Giudice ED, et al. The risk of bipolar disorders in multiple sclerosis. J Affect Disord. 2014;155:255–60.
24. Blanc F, Berna F, Fleury M, Lita L, Ruppert E, Ferriby D, et al. Inaugural psychotic events in multiple sclerosis? Rev Neurol (Paris). 2010;166(1):39–48.
25. Sidhom Y, Ben Djebara M, Hizem Y, Abdelkefi I, Kacem I, Gargouri A, et al. Bipolar disorder and multiple sclerosis: a case series. Behav Neurol. 2014. doi:10.1155/2014/536503.
26. Kemp K, Lion JR, Magram G. Lithium in the treatment of a manic patient with multiple sclerosis: a case report. Dis Nerv Syst. 1977;38(3):210–1.
27. Grunze H, Vieta E, Goodwin GM, Bowden C, Licht RW, Moller HJ, et al. The World Federation of Societies of Biological Psychiatry (WFSBP) guidelines for the biological treatment of bipolar disorders: update 2009 on the treatment of acute mania. World J Biol Psychiatry. 2009;10(2):85–116. Erratum in: World J Biol Psychiatry. 2009;10(3):255.

28. Grunze H, Vieta E, Goodwin GM, Bowden C, Licht RW, Möller HJ, et al. The World Federation of Societies of Biological Psychiatry (WFSBP) for the biological treatment of bipolar disorders: update 2010 on the treatment of acute bipolar depression. World J Biol Psychiatry. 2010;11(2):81–109.
29. Grunze H, Vieta E, Goodwin GM, Bowden C, Licht RW, Möller HJ, et al. The World Federation of Societies of Biological Psychiatry (WFSBP) guidelines for the biological treatment of bipolar disorders: update 2012 on the long-term treatment of bipolar disorder. World J Biol Psychiatry. 2013;14(3):154–219.
30. Minden SL, Orav J, Schildkraut JJ. Hypomanic reactions to ACTH and prednisone treatment for multiple sclerosis. Neurology. 1988;38(10):1631–4.
31. Chwastiak LA, Ehde DM. Psychiatric issues in multiple sclerosis. Psychiatry Clin North Am. 2007;30(4):803–17.
32. Katzman MA, Bleau P, Blier P, Chokka P, Kjernisted K, Van Ameringen M, et al. Canadian clinical practice guidelines for the management of anxiety, posttraumatic stress and obsessive-compulsive disorders. BMC Psychiatry. 2014;14 Suppl 1:S1.
33. Pine DS, Douglas CJ, Charles E, Davies M, Kahn D. Patients with multiple sclerosis presenting to psychiatric hospitals. J Clin Psychiatry. 1995;56(7):297–306.
34. Davids E, Hartwig U, Gastpar M. Antipsychotic treatment of psychosis associated with multiple sclerosis. Prog Neuropsychopharmacol Biol Psychiatry. 2004;28(4):743–4.

Part II
Psychology

Chapter 9
Psychology of Multiple Sclerosis

Frederick W. Foley

Abstract Multiple sclerosis (MS) is a demyelinating disease of the central nervous system. It is unpredictable and a potentially disabling disease, with the severity and nature of symptoms varying. Clinically significant depression and anxiety are very common, although there is evidence these psychological conditions are undertreated. The lifetime prevalence rates for anxiety and depressive disorders are higher among MS patients compared to the general population and persons with other neurological disease. Treatment options include psychotherapy and psychopharmacological treatments, though there is limited clinical research. Demoralization and grief over MS-related losses are also common psychological reactions. MS can also have a psychological impact on family members, and they can experience similar emotional responses that patients do. Partners of MS patients experience emotional challenges, such as caregiver burden and changes in sexual relationships. However, many persons make positive psychological adjustments, use positive coping strategies, and find psychological benefits to the challenges presented by MS.

Keywords Multiple sclerosis • Anxiety • Depression • Pseudobulbar affect • Caregiver burden • Sexual dysfunction • Positive adjustment • Benefit finding • Cognitive behavioral therapy • Psychopharmacological treatment

Multiple sclerosis (MS) is a disease of the central nervous system (CNS) characterized by an immune-mediated attack on the myelin and oligodendrocytes, resulting in demyelination, astrocytic scarring, and axonal loss [1]. It is estimated that more than 2.3 million people worldwide suffer from multiple sclerosis [2]. The symptoms of MS vary in severity and nature and include changes in cognition, psychiatric status, gait, balance, sensation, sexual function, vision, and bladder and/or bowel function. As such, MS is an unpredictable and potentially disabling disease.

F.W. Foley, Ph.D. (✉)
Ferkauf Graduate School of Psychology, Yeshiva University,
1300 Morris Park Avenue, Bronx, NY 10461, USA

The Multiple Sclerosis Center, Holy Name Medical Center, Teaneck, NJ, USA
e-mail: ffoley1@aol.com

B. Brochet (ed.), *Neuropsychiatric Symptoms of Inflammatory Demyelinating Diseases*, Neuropsychiatric Symptoms of Neurological Disease,
DOI 10.1007/978-3-319-18464-7_9

Disorders of Mood and Affect

Mood is the subjective experience of one's inner emotional state. Affect is the observable expression of emotion. Both disorders of mood and affect occur in MS. However, there is a gradation of severity ranging from normal to extreme daily variations in mood. For example, adjustment disorders have disturbances of mood and/or behavior that cause distress and impairment but which are relatively short-lived. There is anecdotal evidence and ample clinical observation that they frequently occur following exacerbations, but the person eventually adjusts and mood returns to normal. Following the continuum from normal to severe, disorders of more severe mood and affect such as major depression, bipolar disorder, and pseudobulbar affect (involuntary emotional expressive disorder) occur in MS.

Clinical Depression

Major depression is the most common mood disorder in MS [3], which has been associated with decreased adherence to disease-modifying therapy protocols and possibly increased suicidal risk. One study in Canada that evaluated death records in MS found that suicide rates in MS patients were up to 7.5 times greater than in the general population [4]. However, other record linkage studies have reported only approximately a twofold increase in suicide in MS as compared to the general population [5, 6]. These findings are supported by other studies that show that current suicidal ideation (SI) and suicidal intent among individuals with MS were approximately 28 % and 18 %, respectively [7, 8], which is much higher than in the general population.

The 12-month prevalence of depression is approximately 25 % [9]. Point-prevalence rates range from 14 to 57 % [10] compared to general population of 1.3–3.7 % [11]. Depression in MS is found to be more common than in other chronic medical/neurological conditions, such as ALS [10, 12]. However, many features and predictors of depression in MS occur in other illnesses, such as neurovegetative symptoms, physical disability, and disease progression.

Although numerous studies estimate that between 36 and 60 % of MS patients will experience an episode of major depression at some point during the course of their life [4, 10, 13–17], there is evidence that this disorder is undertreated. In one study, 260 patients with MS who were treated by 35 neurologists in a large healthcare system in the United States were identified. Sixty-seven (25.8 %) of patients had a diagnosis of major depressive disorder, which was confirmed with a standard psychiatric interview. Of this group of patients, 65.6 % received no antidepressant treatment, 4.7 % received subthreshold doses, and only 3.1 % of patients were receiving doses exceeding the minimal threshold dose [18]. This study highlights the importance of routine systematic screening for depression in the clinic setting, as well as the importance of proper dosing with antidepressant medicines.

There are many depression screening tools available that have been validated in MS, including the Beck Depression Inventory II [19] and the seven-item Beck Depression Fast Screen [20], with one study finding that asking two questions had excellent sensitivity and specificity for detecting major depression in MS [21].

There is some evidence that depression in MS is associated with peripheral in vitro markers of inflammatory immunological activation [22, 23], although the nature of the relationship is unclear (e.g., cause and effect). However, two studies found that treatment of depression in MS were associated with decreased in vitro markers of immune activation [23, 24]. An additional study in patients that were not clinically depressed found that patients randomized to receive meditation plus cognitive behavior therapy had fewer prospective new enhancing lesions in the CNS, although this effect was not found 6 months after the cessation of treatment [25].

MS brain lesions are associated with depression and moderate the efficacy of treatments for depression [26]. Patients with higher lesion load have been found to be less responsive to either pharmacotherapy or psychotherapy treatments. In terms of adjusting to living with MS, depression and fatigue interfere with the patient's ability to adjust to MS [27]. In the latter study, depression in the patient also interfered with significant other adjustment to the MS situation.

A history of having a major depressive episode, even following recovery, may impact the perception of the quality of life in MS. One study of persons with MS who had a former depressive episode but currently did not have major depression reported lower quality of life (QOL) scores on the Multiple Sclerosis Quality of Life-54 scale (MSQOL-54) in domains of Energy, Mental Health, Cognitive Function, General QOL, Sexual Function, and Emotional Role Limitations [28]. Depression and anxiety also have a moderating (worsening) effect on the impact of severity of neurological disability as measured by the Expanded Disability Status Scale (EDSS) on quality of life measures [29].

However, treatment of depression has been found to be associated with improvement in self-reported quality of life scores. In one treatment study, 60 patients with definite MS were randomized to one of three treatments for depression. The MSQOL-54 was used both pre- and posttreatment. Controlling for initial scores and severity of neurological impairment, treatment of depression was associated with improved QOL scores, especially psychological well-being [30].

Although a lot has been learned about the nature of major depression in MS, gaps in knowledge remain. There is little information about remission rates, duration of depressive episodes, or risk of relapse or reoccurrence.

Less is known about other mood disorders in MS. Bipolar disorders are characterized by manic episodes (one episode defines bipolar I disorder) or hypomanic episodes (bipolar II disorders) or substance- or medication-induced manic episodes. In MS, much less is understood about bipolar disorders. Prevalence estimates from clinic populations suggest up to 13 % may be at risk for a manic episode compared to only 1–4 % in the general population [13, 31]. However, better studies need to be conducted, and it is not known how often manic episodes are triggered by substances, such as manic-like symptoms that are associated with administration of high-dose intravenous steroids.

Dysthymic disorder is a chronic, mild depression of at least a 2-year duration. No epidemiological or treatment data are available in MS. In the general population, there is only a modest response to antidepressants, and the primary treatment is psychotherapy.

Pseudobulbar Affect

Pseudobulbar affect, or involuntary emotional expressive disorder, involves episodes of involuntary crying and/or laughing that are discordant with the person's internal mood state. The lifetime prevalence in MS has been estimated in one epidemiology study at 10 % [32]. There is greater risk for pseudobulbar affect when cognitive impairment is present, particularly frontal-type deficits (e.g., impairments in executive control, impulsivity) [33]. However, no definitive MRI correlates have been identified to date.

Several studies have reported successful treatment of pseudobulbar affect, including one that administered low-dose amitriptyline [34], others that administered SSRIs [35, 36], or the best controlled study that administered a combination of dextromethorphan and quinidine [37], with the evidence for the latter treatment sufficient to attain FDA approval for this disorder in the United States.

Anxiety

Anxiety is also common among patients with MS and is more common than depression [33, 38, 39]. The prevalence rates of clinically significant symptoms of anxiety vary from 12 to 90 %, with most studies indicating rates of 30–50 % [40, 41]. Recent studies found 41 and 43 % of MS patients endorsed clinically significant anxiety as measured by the Hospital Anxiety and Depression Scale Anxiety subscale (HADS-A) [42, 43].

The prevalence rates of anxiety disorders in MS ranges from 14 to 41 % [40, 44]. Compared to the general population, patients with MS significantly report more anxiety. For example, the lifetime prevalence of generalized anxiety disorder (GAD) in MS is 18.6 %, while it is 5.7 % in the general population [44, 45]. Social anxiety is also common in MS, with one study finding 30.6 % of patients meeting criteria for social phobia (the fear and avoidance of social situations) using the Social Phobia Inventory [46]. Clinical correlates for anxiety disorders within the MS population include the female gender, disability, limited social support, problem-solving deficits, fatigue and sleep disturbances, pain, time since diagnosis, and comorbid diagnosis of depression [38, 44, 47, 48]. Anxiety has been found not to be strongly related to MRI-revealed pathology, so it is likely a reactive process to the unpredictable nature of MS and uncertainty that persons with MS live with [39].

Anxiety can pose a problem in the treatment of MS for some patients. For example, injection phobia is common [49]. This can be problematic for patients who receive their disease-modifying treatments by injections, which requires a schedule of either intramuscular or subcutaneous injections that occur weekly to as frequently as every other day. The recent developments in approved oral disease-modifying medicines circumvent this obstacle to treatment adherence.

Psychological Impact on Partners

The patient is not the only individual who is affected by MS. Family members, such as caregivers and partners, can also be affected emotionally. They can experience similar emotions that patients do, such as guilt, anger, anxiety, and grief [50, 51].

Caregiver Burden

Often referred to as the "invisible patients" [52], caregivers are usually family members who provide a significant amount of care for someone who needs help with activities of daily living (ADLs) and independent ADLs [53]. Individuals caring for loved ones with MS are at risk for caregiver burden, which is "a multidimensional response to physical, psychological, emotional, social, and financial stressors associated with the caregiving experience" [54]. What an individual perceives as the burden can be objective or subjective. For example, it may be the number of tasks they perform or the amount of time dedicated to caregiving (objective burden) or the emotional response to caregiving (subjective burden). Aronson [55] found that among caregivers of MS patients, subjective burden was associated with reduced quality of life and depression. In addition to depression and decreased quality of life, spousal caregivers can also have reductions in their social activities, physical health, and financial security [56].

Sexual Dysfunction

Given the frequency of depression and anxiety in MS, as well as the potential impact on partners, it is not a surprise that sexual dysfunction is highly prevalent in MS. MS is highly associated with sexual dysfunction, which can range from approximately 40–80% of patients [57]. Sexual dysfunctions can be characterized in as primary, secondary, or tertiary [58]. *Primary sexual dysfunction* is defined as MS-related neurologic change that may directly affect sexual feelings and/or sexual response, such as impaired genital sensation, erectile dysfunction, orgasm dysfunction, decreased vaginal lubrication, and loss of libido.

Secondary sexual dysfunction referred to MS-related physical changes, which affect the sexual response indirectly. Secondary sexual dysfunction is caused by MS symptoms that do not directly include nervous system pathways related to the genital system. These symptoms most commonly include fatigue, muscle tightness, weakness, spasticity, bladder and bowel dysfunction, incoordination, difficulty with mobility, adverse effects from MS medications, cognitive difficulties, and numbness, pain, burning, or discomfort in nongenital areas of the body.

Tertiary sexual dysfunction referred to the psychological, emotional, social, and cultural aspects of MS that impact upon sexuality. Tertiary sexual symptoms may include negative changes in self-image, mood, or body image and depression and anger, feeling less sexy or attractive, feeling less masculine or feminine, feeling less confident about one's sexuality, fear of being rejected sexually, worries about sexually satisfying one's partner, and difficulty communicating with one's partner. Fears of isolation and abandonment, guilt, changing gender roles, and feelings of dependency may also impact intimate relationships, affecting the way in which sexual feelings are expressed and experienced.

Because sexual dysfunction in MS is so complex and prevalent, but goes largely untreated [59], it is not surprising that both patients and intimate partners are negatively impacted. A recent study found that sexual dysfunction was far more associated with lower psychological quality of life than physical disability [59]. Dissatisfaction within marital relationships has occurred in up to a third of MS couples, with spouses showing the greatest dissatisfaction [60]. A more recent study found greater rates of dissatisfaction among carers/partners, with only 27.7 % reporting they were satisfied with their sex lives [61]. However, an intervention consisting of 12 educational/counseling sessions was developed by one group that significantly improved marital satisfaction in persons with MS and their partners. This intervention also resulted in improvements in affective and instrumental communication in both partners and persons with MS, as well as sexual satisfaction for both [62]. There were no interaction effects, which indicated the intervention demonstrated equal efficacy for both persons with MS and their partners.

Another aspect that MS creates relationship strain is the changing role demands associated with disability. Caretaking activities in partners change the dynamics of the relationship, with increasing dependency of the person with MS creating strain for both [63]. Changes in family and societal roles secondary to disability can affect both the person with MS and the partner's capacity for intimacy. The person with MS who has difficulty fulfilling his or her designated work and household roles may no longer feel like an equal partner. The partner of a severely disabled individual may feel overburdened by additional caregiving, household, and employment responsibilities. The intimate relationship can be threatened by the growing tension that results from these feelings [64].

In addition, the caregiving partner (either male or female) may have trouble switching from the nurturing role of caretaker to the more sensual role of lover. As a sexual partner of a woman (or man) with a disability, a man may begin to think of his partner as too fragile or easily injured or as a "patient" who is ill and therefore unable to be sexually expressive. If it is practical or culturally acceptable, having nonfamily members perform caretaking activities helps minimize this "role conflict." When caretaking must be performed by the sexual partner, separating caretaking activities from times that are dedicated to romantic and sexual activities can help minimize this conflict.

Accompanying these role changes may be an increasing sense of isolation in the relationship and less understanding of the partner's struggles and perspectives. The diminishing capacity to understand and work through these issues creates greater isolation and misunderstanding, leading to increasing resentments.

Cultural Expectations Regarding Sexual Behavior

The religious, cultural, and societal influences in our lives help shape our thoughts, views, and expectations about sexuality. One of the notions about sexuality that prevails in Western culture is a "goal-oriented" approach to sex. In this approach, the sexual activity is done with the goal of having penile-vaginal intercourse, ultimately leading to orgasm. Here, the sexual behaviors labeled as foreplay, such as erotic conversations, touching, kissing, and genital stimulation, are seen as steps that inevitably lead to intercourse rather than as physically and emotionally satisfying sexual activities in their own right. Hence, couples are not thought to be having "real" sex until they are engaging in coitus, and sex is typically not considered "successfully completed" until orgasm occurs [64].

This Western view of sexuality leads to spending a great deal of time and energy worrying about the MS-related barriers to intercourse and orgasm ("the goal") rather than seizing the opportunity to explore physically and emotionally satisfying alternatives to intercourse. The capacity to discover new and fulfilling ways to compensate for sexual limitations requires that couples be able to let go of preconceived notions of what sex *should* be and focus instead on openly communicating their sexual needs and pleasures without fear of ridicule or embarrassment. However, difficulties talking about sexuality are very common in both patients and partners, which serve as an impediment to effective problem-solving.

MS-Related Emotional Challenges

The MS experience is frequently associated with emotional challenges, including grief, demoralization, and clinical depression [65]. These emotional struggles may temporarily dampen interest in sex or the ability to give and receive sexual pleasure. Coping with emotional changes to enhance sexuality has several aspects: assessment, education, professional treatment, and coping interventions. Assessment of clinical depression can be done by a mental health professional who is familiar with MS. Treatment that involves antidepressant medications and psychotherapy typically offers symptom relief, including the restoration of sexual interest. It is important to select an antidepressant that will minimally impact sexual function, as sexual dysfunction is a common side effect of many antidepressants.

Positive Adjustment in MS

Persons with MS frequently are challenged in coping positively with the complexities associated with their illness. As previously discussed, patients have a high prevalence rate of depression. They may be less likely to use positive adjustment coping styles and have a depressive attributional style [66, 67]. However, an early intervention that focuses on strengthening the individual's adaptability has been found to aid in developing effective coping strategies [68].

An individual's coping strategy is an important factor associated with positive adjustment and resiliency. Three general types of coping have been identified in the literature: problem-focused, emotion-focused, and avoidance [69]. Problem-focused strategies actively involve approaching problems systematically, while emotion-focused strategies attempt to regulate distressing emotional responses. Avoidance strategies attempt to avoid thinking about MS and avoiding anything that reminds one of the MS situation. It has been found that a more active coping style is associated with psychological resilience [70, 71], as well as a higher QOL [72]. Actively "focusing on the positive" as a coping strategy has been found to be a predictor of positive physical and psychological QOL in MS, while escaping the reality of one's illness (avoidance) has been associated with a negative psychological impact [69]. Patients with MS that use a particular emotion-focused coping strategy, such as "wishful thinking," are more at risk for poor psychological adjustment and lower QOL [73]. However, individuals who were diagnosed with MS later in life were less likely to use wishful thinking as a coping strategy [73] and therefore are more likely to experience a more positive psychological state. Compared with the general population, MS patients have been found to be less likely to seek social support as a coping strategy or use problem-focused coping skills [74].

Other factors associated with psychological resilience include emotional awareness, successfully managing one's emotions, learning to face one's fears, spirituality, hardiness, social support, cognitive flexibility, realistically placing the blame rather than placing it on oneself or others, positively reframing events, acceptance, and previous exposure to mild stressful events [70, 71]. In MS, improvements on a measure of resiliency were noted among patients who participated in both social discussion and psychotherapeutic groups, and the authors noted that supportive group setting may have contributed [66]. Over time, persons with MS have been found to adapt to their condition and reach out to others for support [67].

Benefit Finding

In MS, a significant predictor of adjustment is the concept of benefit finding, which has been called "stress-related growth" and "posttraumatic growth," which refers to the individual's ability to find benefits in an adverse situation [75]. While individuals with MS can experience numerous negative effects of their disease, they can also find several benefits resulting from MS [76]. The use of benefit finding has been associated with positive adjustment outcomes and meaning-based coping strategies [75]. Individuals who have identified at least one benefit of their illness, as well as high acceptance and decreased helplessness, have better long-term outcomes [72]. In Pakenham's study [76], more than one-third of participants reported increased personal growth from their MS. In Hart, Vella, and Mohr's study [77], in which MS patients underwent telephone-based psychotherapies (CBT and supportive emotion-focused therapies), improvements in depression posttreatment predicted more benefit finding in the participants. The same study also found that benefit finding was associated with greater positive affect and optimism.

One benefit finding measure specifically for MS is the Benefit Finding in Multiple Sclerosis Scale (BFiMSS). Factor analysis revealed seven factors: mindfulness, compassion/empathy, personal growth, new opportunities, family relations growth, spiritual growth, and lifestyle gains [75]. However, a limitation of examining benefit finding in a chronic illness such as MS is a social desirability response bias, which future studies should measure and control for. Pakenham and Cox [75] found that the factors and total score of the BFiMSS were weakly correlated. It is possible that positive affect and cognitions, such as those associated with benefit finding, may be factors that promote resiliency in MS [77], although prospective studies need to examine this. In addition, future studies need to examine whether benefit finding strategies can be systematically taught to persons with MS and whether the interventions result in long-term psychological resiliency.

References

1. Slimp JC. Neurophysiology of multiple sclerosis. In: Geisser BS, editor. Primer on multiple sclerosis. New York: Oxford University; 2011. p. 31–46.
2. National Multiple Sclerosis Society: MS Prevalence. http://www.nationalmssociety.org/About-the-Society/MS-Prevalence. Accessed 15 Dec 2014.
3. Wallin MT, Wilken JA, Turner AP, Williams RM, Kane R. Depression and multiple sclerosis: review of a lethal combination. J Rehabil Res Dev. 2007;43:45–62. doi:10.1682/JRRD.2004.09.0117.
4. Sadovnick AD, Eisen K, Ebers GC, Paty D. Cause of death in patients attending multiple sclerosis clinics. Neurology. 1991;41:1193–6. doi:10.1212/WNL.41.8.1193.
5. Stenager EN, Stenager E, Koch-Henriksen N, Bronnum-Hansen H, Hyllested K, Jensen K, Bille-Brahe U. Suicide and multiple sclerosis: an epidemiological investigation. J Neurol Neurosurg Psychiatry. 1992;55:542–5. doi:10.1136/jnnp.55.7.542.

6. Fredrikson S, Cheng G, Jiang GX, Wasserman D. Elevated suicide risk among patients with multiple sclerosis in Sweden. Neuroepidemiology. 2003;22:146–52. doi:10.1159/000068746.
7. Feinstein A. An examination of suicidal intent in patients with multiple sclerosis. Neurology. 2002;59:674–8. doi:10.1212/WNL.59.5.674.
8. Foley FW, Feinstein A, Mohr D, Patten S, editors. Affective disorders in multiple sclerosis. Symposium conducted at the 21st Annual Meeting of the Consortium of Multiple Sclerosis Centers; 2007 June 1. Washington, DC: Institute of Medicine.
9. Patten SB, Beck CA, Williams JVA, Barbui C, Metz LM. Major depression in multiple sclerosis. A population-based perspective. Neurology. 2003;61:1524–7. doi:10.1212/01.WNL.0000095964.34294.B4.
10. Minden SL, Orav J, Reich P. Depression in multiple sclerosis. Gen Hosp Psychiatry. 1987;9:426–34. doi:10.1016/0163-8343(87)90052-1.
11. Anthony JC, Folstein M, Romanoski AJ, Von Korff M, Nestadt GR, Chahal R, Merchant A, Brown CH, Shapiro S, Kramer M, Gruenberg EM. Comparison of the lay Diagnostic Interview Schedule and a standardized psychiatric diagnosis. Experience in Eastern Baltimore. Arch Gen Psychiatry. 1985;42:667–75. doi:10.1001/archpsyc.1985.01790300029004.
12. Schiffer RB, Babigian HM. Behavioral disorders in multiple sclerosis, temporal lobe epilepsy, and amyotrophic lateral sclerosis. An epidemiologic study. Arch Neurol. 1984;41:1067–9. doi:10.1001/archneur.1984.04050210065016.
13. Joffe RT, Lippert GP, Gray TA, Sawa G, Horvath Z. Mood disorder and multiple sclerosis. Arch Neurol. 1987;44:376–8. doi:10.1001/archneur.1987.00520160018007.
14. Lyon-Caen O, Jouvent R, Hauser S, Chaunu MP, Benoit N, Widlöcher D, Lhermitte F. Cognitive function in recent-onset demyelinating diseases. Arch Neurol. 1986;43:1138–41. doi:10.1001/archneur.1986.00520110034010.
15. Millefiorini E, Padovani A, Pozzilli C, Loriedo C, Bastianello S, Buttinelli C, Di Piero V, Fieschi C. Depression in the early phase of MS: influence of functional disability, cognitive impairment, and brain abnormalities. Acta Neurol Scand. 1992;86:354–8. doi:10.1111/j.1600-0404.1992.tb05100.x.
16. Schiffer RB, Caine ED, Bamford KA, Levy S. Depressive episodes in patients with multiple sclerosis. Am J Psychiatry. 1983;140:1498–500. doi:10.1176/appi.ajp.159.11.1862.
17. Sullivan MJL, Weinshenker B, Mikail S, Edgley K. Depression before and after diagnosis of multiple sclerosis. Mult Scler. 1995;1:104–8. doi:10.1177/135245859500100208.
18. Mohr DC, Hart SL, Fonareva I, Tasch ES. Treatment of depression for patients with multiple sclerosis in neurology clinics. Mult Scler. 2006;12(2):204–8. doi:10.1191/135248506ms1265oa.
19. Kim S, Foley FW, Picone M, Halper J, Bongardino M, Zemon V. Depression levels and interferon treatment in patients with multiple sclerosis. Int J MS Care. 2012;14(1):12–6.
20. Benedict RHB, Fishman I, McClellan MM, Bakshi R, Weinstock-Guttman B. Validity of Beck Depression Inventory-Fast Screen in multiple sclerosis. Mult Scler. 2003;9(4):393–6. doi:10.1191/1352458503ms902oa.
21. Mohr DC, Hart SL, Julian L, Tasch ES. Screening for depression among patients with multiple sclerosis: two questions may be enough. Mult Scler. 2007;13(2):215–9. doi:10.1177/1352458506070926.
22. Foley FW, Miller AH, LaRocca NG, Traugott U, Bedell JR, Scheinberg LC. Psychoimmunological dysregulation in multiple sclerosis. Psychosomatics. 1988;29.
23. Mohr DC, Goodkin DE, Islar J, Hauser SL, Genain CP. Treatment of depression is associated with suppression of nonspecific and antigen-specific T(H)1 responses in multiple sclerosis. Arch Neurol. 2001;58:1081–6. doi:10.1001/archneur.58.7.1081.
24. Foley FW, Traugott U, LaRocca NG, Smith C, Scheinberg LC, Caruso L, Perlman K. A prospective study of depression and immune regulation in MS. Arch Neurol. 1992;49:238–44. doi:10.1001/archneur.1992.00530270052018.
25. Mohr DC, Lovera J, Brown T, Cohen B, Neylan T, Henry R, Siddigue J, Jin L, Daikh D, Pelletier D. A randomized trial of stress management for the prevention of new brain lesions in MS. Neurology. 2012;79:412–9. doi:10.1212/WNL.0b013e3182616ff9.

26. Mohr DC, Goodkin DE, Nelson S, Cox D, Weiner M. Moderating effects of coping on the relationship between stress and the development of new brain lesions in multiple sclerosis. Psychosom Med. 2002;64:803–9. doi:10.1097/01.PSY.0000024238.11538.EC.
27. King KE, Arnett PA. Predictors of dyadic adjustment in multiple sclerosis. Mult Scler. 2005;11:700–7. doi:10.1191/1352458505ms1212oa.
28. Wang JL, Reimer MA, Metz LM, Patten SB. Major depression and quality of life in individuals with multiple sclerosis. Int J Psychiatry Med. 2000;30:309–17.
29. Janssens ACJW, van Doorn PA, de Boer JB, Kalkers NF, van der Meche FGA, Passchier J, Hintzen RQ. Anxiety and depression influence the relation between disability status and quality of life in multiple sclerosis. Mult Scler. 2003;9:397–403. doi:10.1191/135245850 3ms930oa.
30. Hart S, Fonareva I, Merluzzi N, Mohr DC. Treatment for depression & its relationship to improvement in quality of life and psychological well-being in multiple sclerosis patients. Qual Life Res. 2005;14:695–703. doi:10.1007/s11136-004-1364-z.
31. Merikangas KR, Akiskal HS, Angst J, Greenberg PE, Hirschfeld RMA, Petukhova M, et al. Lifetime and 12-month prevalence of bipolar spectrum disorder in the National co-morbidity Survey Replication. Arch Gen Psychiatry. 2007;64:543–52. doi:10.1001/archpsyc.64.5.543.
32. Feinstein A, Feinstein K, Gray T, O'Connor P. Prevalence and neurobehavioral correlates of pathological laughing and crying in multiple sclerosis. Arch Neurol. 1997;54:1116–21. doi:10.1001/archneur.1997.00550210050012.
33. Feinstein A, O'Connor P, Gray T, Feinstein K. The effects of anxiety on psychiatric morbidity in patients with multiple sclerosis. Mult Scler. 1999;5:323–6. doi:10.1177/135245859900500504.
34. Schiffer RB, Herndon RM, Rudick RA. Treatment of pathological laughing and weeping with amitriptyline. N Engl J Med. 1985;312:1480–2. doi:10.1056/NEJM198506063122303.
35. Seliger GM, Hornstein A, Flax J, Herbert J, Schroeder K. Fluoxetine improves emotional incontinence. Brain Inj. 1992;6:267–70.
36. Sloan RL, Brown KW, Pentland B. Fluoxetine as a treatment for emotional lability after brain injury. Brain Inj. 1992;6:315–9.
37. Panitch HS, Thisted RA, Smith RA, Wynn DR, Wymer JP, Achiron A, Vollmer TL, Mandler RN, Dietrich DW, Fletcher M, Pope LE, Berg JE, Miller A. Pseudobulbar Affect in Multiple Sclerosis Study Group. Randomized, controlled trial of dextromethorphan/quinidine for pseudobulbar affect in multiple sclerosis. Ann Neurol. 2006;50:780–7. doi:10.1002/ana.20828.
38. Janssens AC, Buljevac D, van Doorn PA, van der Meche FG, Polman CH, Passchier J, Hintzen RQ. Prediction of anxiety and distress following diagnosis of multiple sclerosis: a two-year longitudinal study. Mult Scler. 2006;12:794–801. doi:10.1177/1352458506070935.
39. Zorzon M, deMasi R, Nasuelli D, Ukmar M, Mucelli RP, Cazzato G, Bratina A, Zivadinov R. Depression and anxiety in multiple sclerosis. A clinical and MRI study in 95 subjects. J Neurol. 2001;248:416–21.
40. Beiske AG, Svensson E, Sandanger I, Czujko B, Pedersen ED, Aarseth JH, Myhr KM. Depression and anxiety amongst multiple sclerosis patients. Eur J Neurol. 2008;15:239–45. doi:10.1111/j.1468-1331.2007.02041.x.
41. Bruce AS, Arnett PA. Longitudinal study of the symptom checklist 90-revised in multiple sclerosis patients. Clin Neuropsychol. 2008;22(1):46–59.
42. Farrell E, Beier M, D'Orio V, Picone MA, Foley F. Prevalence rates of anxiety, depression, and cognitive difficulties in patients with multiple sclerosis from an outpatient clinic. 32nd Annual Meeting and Scientific Sessions of the Society of Behavioral Medicine. Poster presented April 27–30, Washington DC; 2011.
43. Giordano A, Granella F, Lugaresi A, Martinelli V, Trojano M, Confalonieri P, Radice D, Solari A. Anxiety and depression in multiple sclerosis patients around diagnosis. J Neurol Sci. 2011;307:86–91. doi:10.1016/j.jns.2011.05.008.
44. Korostil M, Feinstein A. Anxiety disorders and their clinical correlates in multiple sclerosis patients. Mult Scler. 2007;13:67–72. doi:10.1177/1352458506071161.

45. Kessler RC, Berglund P, Demler O, Jin R, Merikangas K, Walters EE. Lifetime prevalence and age-of-onset distributions of DSM-IV disorders in the national co-morbidity survey replication. Arch Gen Psychiatry. 2005;62(6):593–602. doi:10.1001/archpsyc.62.6.593.
46. Poder K, Ghatavi K, Fisk JD, Campbell TL, Kisely S, Sarty I, Stadnyk K, Bhan V. Social anxiety in a multiple sclerosis clinic population. Mult Scler. 2009;15(3):393–8. doi:10.1177/1352458508099143.
47. Bruce JM, Arnett P. Clinical correlates of generalized worry in multiple sclerosis. J Clin Exp Neuropsychol. 2009;31:698–705. doi:10.1080/13803390802484789.
48. Janssens AC, van Doorn PA, de Boer JB, van der Meche FG, Passchier J, Hintzen RQ. Impact of recently diagnosed multiple sclerosis on quality of life, anxiety, depression and distress of patients and partners. Acta Neurol Scand. 2003;108:389–95. doi:10.1034/j.1600-0404.2003.00166.x.
49. Mohr DC, Boudewyn AC, Likosky W, Levine E, Goodkin DE. Injectable medication for the treatment of multiple sclerosis: the influence of self-efficacy expectations and injection anxiety on adherence and ability to self-inject. Ann Behav Med. 2001;23(2):125–32. doi:10.1207/S15324796ABM2302_7.
50. McDaniel SH, Hepworth J, Doherty W. Medical family therapy. New York: Basic Books; 1992.
51. Kalb R. When MS, joins the family. In: Kalb R, editor. Multiple sclerosis: a guide for families. 3rd ed. New York: Demos Medical Publishing; 2006. p. 1–10.
52. Andolsek K, Clapp-Channing N, Gehlbach S, Moore I, Proffitt VS, Sigmon A, Warshaw GA. Caregiving in elderly relatives: the prevalence of caregiving in family practice. Arch Neurol. 1988;148:2177–80. doi:10.1001/archinte.1988.00380100059013.
53. Pandya S. Caregiver in the United States. 2005. http://www.directcareclearinghouse.org/download/AARP%20Family%20caregivers%20fact%20sheet.pdf. Accessed 18 Dec 2014.
54. Buhse M. Assessment of caregiver burden in families of persons with multiple sclerosis. J Neurosci Nurs. 2008;40(1):25–31.
55. Aronson KJ. Quality of life among persons with multiple sclerosis and their caregivers. Neurology. 1997;48(1):74–80.
56. McKeown L, Porter-Armstrong A, Baxter D. The needs and experiences of caregivers of individuals with multiple sclerosis: a systematic review. Clin Rehabil. 2003;17:234–58.
57. Foley FW, Werner M. Sexuality and intimacy. In: Kalb RC, editor. Multiple sclerosis: the questions you ask, the answers you need. 2nd ed. New York: Demos Vermonde Press; 2000.
58. Foley FW, Zemon V, Campagnolo D, Tyry T, Vollmer T, Marrie R, Farrell E, Beier M, Schairer L. Reliability and validity of the Multiple Sclerosis Intimacy and Sexuality Scale: re-validation with a large United States sample. Mult Scler. 2013;19(9):1197–203.
59. Schairer L, Foley FW, Zemon V, Campagnolo D, Tyry T, Marrie R, Gromisch ES, Schairer D. The impact of sexual dysfunction on health-related quality of life in people with multiple sclerosis. Mult Scler. 2014;20:610–6. doi:10.1177/1352458513503598.
60. Dupont S. Sexual function and ways of coping in patients with multiple sclerosis and their partners. J Sex Marital Ther. 1996;11:359–72. doi:10.1080/02674659608404450.
61. O'Connor EJ, McCabe M, Firth L. The impact of neurological illness on marital relationship. J Sex Marital Ther. 2008;34:115–32. doi:10.1080/00926230701636189.
62. Foley FW, LaRocca NG, Sorgen A, Zemon V. Rehabilitation of intimacy and sexual dysfunction in couples with multiple sclerosis. Mult Scler Clin Lab Res. 2001;7(6):417–21.
63. Green G, Todd J, Pevalin D. Biographical disruption associated with multiple sclerosis: using propensity scoring to assess the impact. Soc Sci Med. 2007;65(3):524–35.
64. Foley FW. Sexuality. In: Kalb RC, editor. Multiple sclerosis and the family. 3rd ed. New York: Demos Vermonde Press; 2005.
65. Foley FW, Reznikoff M, LaRocca NG, Bedell JR, Scheinberg LC. The efficacy of stress inoculation training on coping with multiple sclerosis. J Consult Clin Psychol. 1987;55:919–22.
66. Rigby SA, Thorton EW, Young CA. A randomized group intervention trial to enhance mood and self-efficacy in people with multiple sclerosis. Br J Health Psychol. 2008;13:619–31.

67. Johnson SK, Lange G, Tiersky L, DeLuca J, Natelson BH. Health-related personality variables in chronic fatigue syndrome and multiple sclerosis. J Chronic Fatigue Syndrome. 2001;8:41–52.
68. McReynolds CJ, Koch LC, Rumrill PD. Psychosocial adjustment to multiple sclerosis: implications for rehabilitation professionals. J Vocat Rehabil. 1999;12:83–91.
69. McCabe M. A longitudinal study of coping strategies and quality of life among people in multiple sclerosis. J Clin Psychol Med Settings. 2006;13:369–79. doi:10.1007/s10880-006-9042-7.
70. Ballenger-Browning K, Johnson DC. Key facts on resilience. San Diego: Naval Center for Combat and Operational Stress Control. 2010. Accessed from www.nccosc.navy.mil
71. Armstrong AR, Galligan RF, Critchley CR. Emotional intelligence and psychological resilience to negative life events. Pers Indiv Differ. 2011;51:331–6. doi:10.1016/j.paid.2011.03.025.
72. Russell CS, White MB, White CP. Why me? Why now? Why multiple sclerosis?: making meaning and perceived quality of life in a Midwestern sample of patients with multiple sclerosis. Fam Syst Health. 2006;24:65–81.
73. McCabe MP, Stokes M, McDonald E. Changes in quality of life and coping among people with multiple sclerosis over a 2 year period. Psychol Health Med. 2009;14:86–96. doi:10.1080/13548500802017682.
74. McCabe MP, McKern S, McDonald E. Coping and psychological adjustment among people with multiple sclerosis. J Psychosom Res. 2004;56:355–61.
75. Pakenham KI, Cox S. The dimensional structure of benefit finding in multiple sclerosis and relations with positive and negative adjustment: a longitudinal study. Psychol Health. 2009;24:373–93. doi:10.1080/08870440701832592.
76. Pakenham KI. The nature of benefit finding in multiple sclerosis. Psychol Health Med. 2007;12:190–6.
77. Hart SL, Vella L, Mohr DC. Relationships among depressive symptoms, benefit-finding, optimism, and positive affect in multiple sclerosis patients after psychotherapy for depression. Health Psychol. 2008;27:230–8. doi:10.1037/0278-6133.27.2.230.

Chapter 10
Coping and Multiple Sclerosis

Valentina Bianchi and Carlo Pozzilli

Abstract Coping, considered as a process, is characterized by dynamics and changes that are a function of continuous appraisals. Two coping strategies have been identified: *problem-focused* strategies (involving dealing with internal or environmental demands that create a threat) and *emotion-focused* strategies, which are associated with emotional regulation (involving efforts to modify the distress that accompanies threat).

Coping is known to be an important mediator between multiple sclerosis (MS) and well-being. In MS people live with their disease for many years, usually increasing disabilities and reducing quality of life. In this situation, patients' abilities to cope with the disease is an important factor for reducing levels of stress and finding the correct strategy to face the daily problem of the disease.

Keywords Coping • Multiple sclerosis • Depression • Anxiety • Cognitive performances • Cognitive impairment • Coping questionnaire • Problem-focused coping strategies • Emotion-focused coping strategies • Quality of life

---

V. Bianchi, PsyD
MS Centre, S. Andrea Hospital, Neurological Sciences, Sapienza University,
Via di Grottarossa 1035, 00189 Rome, Italy

Centro Interdipartimentale Medicina Sociale (CIMS), Sapienza University,
Piazzale Aldo Moro, 00185 Rome, Italy
e-mail: valentina.bianchi@uniroma1.it

C. Pozzilli, M.D., Ph.D. (✉)
MS Centre, S. Andrea Hospital, Neurological Sciences, Sapienza University,
Via di Grottarossa 1035, 00189 Rome, Italy

Department of Neurology and Psychiatry, Sapienza University,
Viale dell'Università 30, 00185 Rome, Italy
e-mail: carlo.pozzilli@uniroma1.it

B. Brochet (ed.), *Neuropsychiatric Symptoms of Inflammatory Demyelinating Diseases*, Neuropsychiatric Symptoms of Neurological Disease,
DOI 10.1007/978-3-319-18464-7_10

Coping and Multiple Sclerosis

Definition of Coping

It's possible to define coping as the cognitive and behavioral efforts to manage specific external and/or internal demands appraised as taxing or exceeding the resources of the individual. With this definition, we can suppose that coping is process-oriented; it makes no a priori judgment about the quality of coping processed, and it makes a distinction between coping and automatic adaptive behaviors.

A process-oriented approach to coping is directed toward what an individual think. Coping, when considered as a process, is characterized by dynamics and changes that are a function of continuous appraisals, and it is defined as realistic and flexible thought and acts that solve problems. Judging the quality of coping, it is possible to evaluate the effectiveness of a given coping strategy contextually; in this way, there is the possibility that one coping strategy is adaptive in one context and maladaptive in another. Coping versus automatic behavior implies that coping refers only to those adaptive activities that involve effort [1].

Therefore, this concept includes behavioral as well as cognitive strategies, used to cope with or to counteract a difficult situation.

Two coping strategies have been identified: problem-focused strategies, which imply the efforts made by the subject to overcome his difficulty, and emotion-focused strategies, which are associated with emotional regulation [2].

Problem-solving functions involve dealing with internal or environmental demands that create a threat, such as studying for an exam. Emotion regulation functions involve efforts to modify the distress that accompanies threat, for example, by denying that the threat exists or by drinking in excess [3].

Problem-focused coping strategies are efforts to do something active to alleviate stressful situation; in contrast, emotion-focused coping involves efforts to regulate the emotional consequences of stressful events. In the first way, the subject tries to find the strategy that concretely solves the specific problem, seeking the correct information or searching for an adequate explanation. In the second way, the subject searches for the solution in spiritual ideas, for example, turning to religion or resignation.

Coping strategies are strongly related to chronic disease. In MS, for example, people live with their disease for many years, usually increasing disabilities and reducing quality of life. In this situation, patients' abilities to cope with the disease is an important factor for reducing levels of stress and finding the correct strategy to face the daily problem of the disease. In chronic disease, we can use the term coping as being adaptive or nonadaptive and successful or unsuccessful strategy and we can describe coping as a mediating factor between chronic disease and adjustment to the disease. In chronic disease, the two different strategies of coping (problem-focused and emotion-focused) are both important and can have important beneficial consequences for physical and mental health [4]. Studies on coping and MS have shown that in the first stage of the disease, patients tend to use emotion-focused coping and that this strategy of coping is strongly related to attacks of the disease. On the other hand, when patients have lived with the disease for several years and have experienced of disabling symptoms, they tend to use problem-focused coping [5].

Coping Questionnaires

There are many tools used to assess and explore coping strategies in MS (Table 10.1).

Ways of Coping Questionnaire (WCQ) [6] comprises 66 items representing different behaviors a person may use to cope with a stressful situation; subjects give a score on each item depending on the frequency with which they used the described behavior. Subjects refer to the most stressful MS-related situation they experienced in the last 2 weeks in the context of work, family, or social life. The different coping strategies explored by WCQ are:

- "Confrontive" describes aggressive efforts to alter the situation and suggests some degree of hostility and risk-taking.
- "Distancing" describes cognitive efforts to detach oneself and minimize the significance of the situation.
- "Self-controlling" describes efforts to regulate one's feelings and actions.
- "Seeking social support" describes efforts to seek informational support, tangible support, and emotional support.
- "Accepting responsibility" describes the acknowledgement of one's own role in the problem with a concomitant theme of trying to put things right.
- "Escape-avoidance" describes wishful thinking and behavioral efforts to escape or avoid the problem.
- "Planful problem-solving" describes deliberate problem-focused efforts to alter the situation, coupled with an analytic approach to solving the problem.
- "Positive reappraisal" describes efforts to create positive meaning by focusing on personal growth. It also has a religious dimension.

Coping style can also be measured using a shortened 30-item version of Folkman and Lazarus's [1] *Way of Coping Questionnaire* (WOQ) [6]. This questionnaire comprises a list of different cognitive and behavioral strategies a person may use to cope with a stressful situation. The five subscales calculated from this questionnaire were:

- Problem-focused strategies – 10 items (e.g., "I try to analyze the problem in order to understand it better")
- Detachment – 6 items (e.g., "I try to forget the whole thing")
- Wishful thinking – 5 items (e.g., "Hope a miracle will happen")
- Seek social support – 5 items (e.g., "Talk to someone about how I am feeling")
- Focusing on the positive – 4 items (e.g., "I'm changing or growing as a person in a good way")

Patients are asked to rate the frequency with which they would use each strategy on a four-point Likert-type scale, ranging from 0 = not used to 3 = used a great deal.

Table 10.1 Questionnaires used to explore coping strategies in MS [6–10]

Questionnaire	Coping strategies explored
Ways of Coping Questionnaire (WCQ) [6]	Confrontive
	Distancing
	Self-controlling
	Seeking social support
	Accepting responsibility
	Escape-avoidance
	Planful problem-solving
	Positive reappraisal
Way of Coping Questionnaire (WOQ) [6]	Problem-focused
	Detachment
	Wishful thinking
	Seek social support
	Focusing on the positive
Ways of Coping Checklist (WCC) [7]	Problem-focused
	Emotional-focused
The Coping with Health, Injuries, and Problems Scale (CHIP) [8]	Diversion
	Palliative
	Instrumental
	Emotional coping
The Ways of Coping (WOC) [9]	Emotion-centered coping:
	Emotion-focused
	Emotional respite
	Passive-avoidant
	Escape-avoidance
	Problem-centered coping:
	Problem-focused
	Cognitive reframing
	Active-constructive coping
	Planful problem-solving
COPE Dispositional Coping Styles Scale [10]	Problem-focused coping:
	Planning
	Restraint coping
	Seeking social support for instrumental reasons
	Emotion-focused coping:
	Seeking social support for emotional reasons
	Positive reinterpretation and growth
	Denial
	Behavioral disengagement
	Mental disengagement

Ways of Coping Checklist (WCC) [7] assesses two coping strategies: problem-focused coping and emotional-focused coping. It has been widely used in literature. Subject thinks about a stressful situation met in the previous months and selects the strategies he/she used to cope with it.

The Coping with Health, Injuries, and Problems Scale (CHIP) [8] is a specific tool for subjects suffering from somatic disease. The subject indicates which coping strategy he uses to cope with the disease. This scale identifies four strategies:

- "Diversion" (e.g., to dream of agreeable things)
- "Palliative" (e.g., to spare his energy)
- "Instrumental" (e.g., to look for efficient treatments)
- "Emotional coping" (e.g., to feel frustrated)

The Ways of Coping (WOC) [9] consists of 65 items. Patients should indicate the coping strategies they use to deal with MS. The items of this scale describe numerous coping strategies which people use to deal with stressful situations. Patients rate the frequency with which they would use each strategy using a scale from 1 = not used to 4 = used a great deal. This questionnaire indicates four measures of emotion-centered coping:

- Emotion-focused coping
- Emotional respite
- Passive-avoidant coping
- Escape-avoidance

and four measures of problem-centered coping:

- Problem-focused coping
- Cognitive reframing
- Active-constructive coping
- Planful problem-solving

The COPE is a short form of *COPE Dispositional Coping Styles Scale* [10]. This questionnaire is a fine-grained dispositional measure of individual differences in coping. This scale has been used in studies concerning coping with chronic disease [11]. In this questionnaire, patients should indicate the ways to deal with MS problems. COPE consists of 15 subscales, with four items each, including problem-focused strategies and emotional-focused strategies.

The subscales measuring problem-focused coping are:

- Planning
- Restraint coping
- Seeking social support for instrumental reason

The subscales measuring emotion-focused coping are:

- Seeking social support for emotional reason
- Positive reinterpretation and growth
- Focus on and venting of emotion

- Denial
- Behavioral disengagement
- Mental disengagement

Each of these subscales has four items. For example, "planning" in problem-focused strategies consists of the four items:

- "I try to come up with a strategy about what to do"
- "I make a plan of action"
- "I think hard about what steps to take"
- "I think about how I might best handle the problem"

In the case of "denial" in emotion-focused problems, the four items are:

- "I refuse to believe that it has happened"
- "I pretend that it hasn't really happened"
- "I act as though it hasn't even happened"
- "I say to myself this isn't real"

Patients rate using a scale from 0 = not at all to 5 = a lot. A sum score is calculated for each subscale.

Coping and MS

Given the debilitating nature of MS, patients face uncertainty about their future health. The symptoms associated with this neurological disorder may lead to negative emotional responses. The extent to which this occurs may be associated with the way in which people with MS cope with the changing symptoms associated with the illness. It is relevant to understand the level of adjustment and the coping styles of people with MS compared to the general population.

Because people with MS live with uncertainty as to the course of the disease, they need to cope with unpredictable deteriorating health, changing social and intimate relationships, and increasing support needs. Daily needs can be outside an individual's control and can lead to negative reactions. A perception of a lack of control over the illness may make people feel depressed. It is important to understand the reason why different people with MS respond in different ways to the same situation and why they adopt different coping strategies. It has been observed that MS patients with more impaired response are also more debilitating or have had the illness for a longer period [12]. Devins et al. also found that decreased psychological well-being and increased levels of distress (poorer psychological functioning) were associated with increasing levels of disability among people with MS [13]. On the other hand, other authors found no association between mood and specific characteristics of the disease such as its duration or the disability level, suggesting denial as other coping strategies may mediate the level of negativity experienced by respondents [14, 15]. Problem-focused-based strategy of coping are related to better adjustment, while emotion-focused strategy are related to poorer adjustment including depression and distress [16]. As confirmed, Pakenham in a longitudinal study found that social adjustment and less distress at 12 months were predicted by greater use of problem-focused coping and less use of emotional-focused coping [5]. Jean et al. suggest that emotion-focused strategies were used during the most difficult moment of the illness and this interpretation can explain the link between this strategy and poorer adjustment [17]. Hickey and Green have investigated the relationship between gender and coping style [18]. They found that women in terms of depression or hopelessness were no different from a psychiatrically depressed group of women, while men presented less depression and hopelessness than the group of psychiatrically depressed men. Mohr et al. [19] found that adaptive coping (problem-focused coping) was related to higher levels of "benefit finding," has deepening of relationships, and increase in spiritual interest. Teaching coping skills to people with MS was associated with a greater number of aspects of well-being than a program that involved telephone support from peers [20]. Lode et al. explored the correlation between the quality of perceived disease information and later coping styles applied by MS patients in stress situations related to their disease [4]. All patients were informed about the disease by a neurologist. The information consisted mainly of facts about the disease, and how it may affect daily living. The authors developed an instrument for the study to examine the quality of perceived information. The instrument was based on the literature and the content of the information at the time of diagnosis. The authors asked patients to indicate how they experienced the information they were given at the time of their diagnosis. The questions were: "Was the information about the disease MS satisfactory?" "Did they inform you about the treatment for the disease?" "Did you get information about how to handle your job, family, and friends?" "Did you get information about what it implies to live with MS?" In addition, the patients indicated the statement about their experience of information, for example, if the neurologist encouraged them to ask questions at the time of diagnosis. Approximately 43 % of patients were dissatisfied or very dissatisfied with the information at the time of diagnosis. The authors found that the most frequently employed strategies of MS-related coping were "positive reinterpretation and growth," "planning," and "restraint coping," while "denial" was the strategy most infrequently used by patients. The authors demonstrated that the correct style of information at the time of diagnosis may indicate positive coping styles producing a better adaptation to living with MS.

Many factors are associated with coping strategies in MS. In a recent study Mosson et al. have explored the impact of physical activity on adopting strategies [21]. The authors found that patients with a high or moderate level of physical activity used more active coping strategies than those who had lower level of physical activity. In addition, these patients were also able to analyze their emotions much better.

Madan and Pakenham [22] examined the effect of global hope on changes in adjustment to MS (the outcomes were anxiety, depression, positive effect, positive states of mind, and life satisfaction) caregiving over 12 months. Predictors were stress, hope, agency, and pathways. Results showed that greater hope was associated with better adjustment. The authors found that hope is an important protective resource for coping with MS caregiving.

Senders et al. [23] studied the role of mindfulness in health and well-being in MS. they examined the relationship between trait mindfulness and perceived stress, coping, and resilience in MS patients. Mindful consciousness can alter the impact of stressful events and has the potential to improve health outcomes in MS. Greater trait mindfulness was associated with decreased psychological stress and better coping skills. Mindfulness training enhances psychological resilience and improves well-being for MS patients.

Personality Traits and Coping Styles

Stability of coping styles is based on multiple psychological and social factors. Bolger, not in recent years, thought that personality is one of the major factors affecting psychological functioning, potentially exerting influence on both coping choice (problem-focused or emotion-focused) and coping effectiveness [24]. Speculating that certain personality dimensions and traits could predispose people to cope with stress in different ways; extroversion, optimism, hardiness, self-esteem, and locus of control have been related to functional coping strategies such as problem-focused coping. On the other hand, less advantageous qualities like neuroticism have been correlated with emotional-focused coping [25]. Those who think that personality traits are correlated with coping strategies emphasize the importance of a *five-factor personality model*. The five-factor model of personality is a hierarchical organization of personality traits in terms of five basic dimensions:

Neuroticism (a tendency to easily experience unpleasant emotions such as anxiety, anger, or depression), *Extraversion* (energy and the tendency to seek stimulation and the company of others), *Openness to experience* (appreciation for art, emotion, adventure, and unusual ideas, imaginative, and curious), *Agreeableness* (a tendency to be compassionate and cooperative rather than suspicious and antagonistic toward others), and *Conscientiousness* (a tendency to show self-discipline, act dutifully, and aim for achievement). The importance of these five factors remained hidden from most personality psychologists throughout the 1960s and 1970s. In the 1980s, however, researchers from many different traditions were led to conclude that these factors were fundamental dimensions of personality. The relationship between the five-factor personality model and coping has been explored by several authors [26].

Neuroticism describes the differences in the levels of disability, somatization, and pain in a person with similar impairments [27]. Extraversion has been found to explain the concentration on social support-seeking and problem-focused coping. Openness to experience has been found to be related to problem-solving strategies, probably due to flexibility, intellectuality, and imagination [25]. Agreeableness is the only personality trait weakly linked to coping strategies, and conscientiousness has been linked to problem-focused strategy, and it is considered as predictor of this type of coping [28]. The importance of the five-personality domain in explaining adaptive and maladaptive coping with chronic disease and with disease-related distress is unclear. The symptoms in MS are different, and any of the symptoms produce social and psychical sub-stressors, and the distinction between disease-related stressors and general stressors in a specific situation is complicated.

Coping and Mood Alteration

Coping styles are linked to mood alteration [29]. Anxiety is reported to affect 23.5–41 % [30] of MS patients, while the depression range in people with MS is 10–41.8 % [31].

Depression may occur at any time over the course of MS, even in the early phase of the disease. Depression was more frequent in the secondary progressive course of MS than in remitting relapsing and primary progressive courses [32]. Different results on the incidence of mood alteration in MS make difficult the comparison among the studies. However, a better understanding of the link existing between depression, anxiety, and coping strategies is very relevant for managing MS.

Aikens et al. [33] in a prospective study found that the only coping strategy able to predict future mood alteration in MS was the escape-avoidance strategy. Intervention programs among people with MS that focused on improving skills were more effective at reducing depression than interventions that focused on improving insight [34]. McCabe et al. [12] examined psychological adjustment among people with MS. In their study, they evaluate the impact of coping on levels of depression, anxiety, anger, fatigue, and confusion. They hypothesized that patients who used emotion-focused coping strategies such as “wishful thinking” and fewer problem-focused strategies such as “seeking social support,” who had experience of severe disability and lower social support, would experience lower levels of adjustment than other people with MS. Equally, patients who had the experience of severe illness for a short time presented a lower level of adjustment. The authors also speculated that people with MS usually used emotion-focused coping strategies and that they used less problem-focused coping strategies than the general population. The coping strategy most consistently associated with poor psychological adjustment was “wishful thinking” suggesting that people with MS who used passive approaches to their illness and who expected that the problem can go away without a concrete and active behavior are more likely to experience higher levels of depression and confusion. Women and men used different types of coping strategies [12]. Women (with MS or not) were more inclined to use coping strategies that involved seeking social support and focusing on the positive than men. Men with MS may evidence better adjustment if they attempt to solve the problem and use an oriented approach, whereas women with MS may evidence better adjustment if they interact socially, seeking social support from family and friends. Emotion-focused coping strategies like wishful thinking are used both by men and women when there is poor adjustment and avoidance. The extent to which men and women continued to be engaged in their day-to-day activities was associated with levels of depression; therefore, it is important for people with MS to continue to be involved in their daily activities, as much as possible.

Lynch et al. [35] explored the link between disability and depression in MS and the coping strategies adopted by patients with MS. The authors found that depression was significantly related to disability, uncertainty concerning illness, level of hope, and the use of emotion-focused strategies. In particular, these variables were significant independent predictors of depression, together accounting for approximately 40 % of the variance in patients' self-reported depression. Depression was not significantly related to problem-focused coping strategies.

In a recent research, Ozura and Sega [36] proposed a new approach to explore depression and capacity for coping with stress in MS. They focused on the profile of depression, capacity for coping with stress, and experienced distress in patients with MS measured by a performance-based method for personality assessment – the Rorschach Inkblot Method (RIM) – scored by the Exner Comprehensive system. In comparison with healthy controls, patients had lower ability for coping with stress, complexity of information processing, body image, willingness to process emotional stimulation, and interpersonal interest. The author's interpretation is that depression in MS patients can be described in terms of negative symptoms such as emotional withdrawal and apathy.

Nielsen et al. [37] explored volitional coping competencies in relation to mood disturbance in a cross-sectional study of 121 participants with clinically defined MS. They found that daily stress situation, stimulation of self-access, stimulation of volitional inhibition, self-motivation, and emotional perseverance/state orientation after failure appear to be valuable predictors of depression. Their interpretation is that personality-accentuated volitional coping competencies elicited by daily stressful situations could be a relevant factor for depressive mood states in individuals with MS.

Few research studies have explored coping strategies related to mood alteration in patients with early MS. In a recent longitudinal study, Bianchi et al. [38] investigated mood alterations and coping strategies in patients with clinically isolated syndrome (CIS) and early relapsing-remitting MS (RRMS). The hypothesis of the authors was that a decrease of psychological distress over time was associated with active coping strategies. A reduction in anxiety and depression and in "seeking social support" and an increase in "planful problem-solving" were observed over the 24-month follow-up. Patients with a higher acceptance of illness ("accepting responsibilities") complained mood alterations, and both CIS and RRMS patients showed less reliance on "confrontive coping" strategy compared to healthy controls.

The study suggests that a depressive reaction involves the use of coping strategies based on accountability rather than escape. The finding of a realistic approach to the disease that initially is related to depression and anxiety can be considered a relevant factor in determining the disappearance of depressive symptoms during the follow-up in most of the patients and can lead to the use of active coping strategies.

Arnett et al. [39] explored the relationship between depression, cognitive dysfunction, and coping strategies adopted by patients with MS. To test this concept, 55 patients with definite MS were administered a neuropsychological battery and measures of depression and coping. Consistent with the author's hypotheses, regression analyses showed that coping significantly moderated the relationship between cognitive dysfunction and depression. In particular, patients with cognitive damage and depression presented a high use of avoidance coping style and a low use of problem-focused coping. Conversely, patients with cognitive difficulties did not show depression and presented low levels of avoidance coping style and a high level of problem-focused coping strategies. It is possible that cognitive damage can lead to depression just when patients use avoidance coping strategies. In contrast, also when patients reported cognitive difficulties, the use of problem-focused strategies may make patients more resistant to depression. In this hypothesis, coping strategies are protective factors for experiencing depression. These results suggest that intervention for depression in MS involves helping patients develop more active coping strategies, reducing the use of emotion-focused coping such as avoidance. An alternative formulation is that depression can lead to cognitive dysfunction in MS just when patients use avoidance or emotion-focused strategies. The greater use of emotion-focused coping strategies can be an example of poor adjustment in MS patients and can be linked to the presence of high levels of depression. Apathy, fatigue, and low motivation characterizing depression make it difficult for depressed patients to use active coping strategies. One more hypotheses is the importance of lesion damage in the brain. Disruption in frontal and subcortical system which commonly affects MS patients has been shown to be associated with depression. Additionally, this damage in the brain that is typical in MS is also related to difficulty in working memory and in the ability to plan new active strategies for solving problems.

Coping and Cognitive Performance

MS symptoms vary among people with MS and include disturbance in strength, balance and vision, sexual dysfunction, fatigue, and cognitive impairment [40].

Cognitive impairment (CI) is a relevant symptom in MS, but only during the last two decades have clinicians become increasingly aware of its prevalence in MS [41]. CI can have a dramatic impact on quality of life, affecting the ability to maintain employment and social life [42].

In neuropsychological studies, 30 % of early MS patients show CI [43, 44], and the deficits domain-specific [45] involves memory, information processing speed, executive functions, and attention [44].

To determine the role of psychological and neuropsychological variables in the prediction of coping with disease-related stressors and satisfaction with coping effort, Jean et al. [17] explored 56 MS patients in terms of their coping strategies and cognitive performances. No significant relationships between neuropsychological variables and coping strategies were observed. The authors concluded that the absence of links between cognitive performance and coping strategies suggests that CI will not pose a major obstacle to successful therapy. However, more recent studies suggest that cognitive deficit may impair abilities to use adaptive coping strategies, leaving the subjects more likely to use maladaptive strategies. Deficit of sustained attention and executive functions may negatively affect the patient's ability to develop active coping strategies [46–49]. Memory is one of the most consistently impaired cognitive functions in MS. It is possible that the deficits in learning new information are associated with executive dysfunction and slow processing speed [50] and this can be linked to the difficulty in using problem-focused coping strategies. In fact, in problem-focused coping strategies, it is necessary to analyze the problem and maintain and manipulate information in the brain for a short period. Processing speed is also important for solving a problem, and processing speed deficits are observed on even the most basic task in MS patients [51]. Montel et al. [52] explored coping strategies in MS patients with frontal cognitive disorder. The authors divided 135 MS patients into two groups: "mild cognitive impairments" and "without frontal cognitive impairments." Patients with mild cognitive impairments tended to use more emotion-focused coping strategies, in particular "self-blame," than patients without cognitive impairments. No other differences were found between the two groups as to coping strategies and mood.

Disease-related variables that can influence the style of coping in MS are poorly understood, but the hypothesis that sustained attention and executive functions have an important role in a patient's choice to assume active or non-active coping strategies is always a valid assumption. This hypothesis emphasizes the importance of a comprehensive assessment of MS patients, including mood disorder, coping strategies, personality traits, and cognitive performance.

Quality of Life and Coping Strategies

Until now, depression has been considered the single most important predictor of quality of life (QOL) [53]; however, coping has proved to be important for adjusting to the adaptive demands of chronic disease.

McCabe and MCKern [54] showed that coping strategies are predictors of QOL; in particular, the "wishful thinking" strategy is considered as the best predictor of poor QOL.

Aikens et al. [33] found a positive link between problem-focused coping strategies and QOL. Emotion-focused coping strategies such as avoidance, wishful thinking, and self-accusation were linked to poor QOL [12].

Montel and Bungener [2] found a relationship between emotion-focused coping strategies and poorer QOL. In this study, the authors compared coping strategies and QOL according to the disease course. The study showed that the only difference between various MS courses was the higher score of emotion-focused coping strategies in the SPMS group of patients who presented the worst QOL.

QOL in MS patients is related to disability levels. Mohr et al. [55] studied the relationship between disability, depression, and coping in MS patients. They found in 101 patients with clinically definite MS that the depression was related to disability; patients who presented high disability showed also high levels of depression. Emotional-focused coping strategies such as escape-avoidance were positively related to levels of depression. Problem-focused coping strategies (planful problem-solving and cognitive reframing) were negatively related to depression. Recently Mikula et al. [56] explored the importance of QOL in MS patients and their coping strategies. The authors aimed to find out whether there is a link between problem-focused coping, coping focused on getting support, and coping focused on stopping unpleasant emotions and different levels of physical and mental QOL. The sample consisted of 113 consecutive MS patients. Coping is significantly associated with mental QOL, but not with physical QOL. All types of coping are associated positively with the mental component of QOL. Stopping unpleasant emotions seems to be the most important type of coping in MS patients, and it seems to be very adaptive for patients with MS.

Goretti et al. [57] explored the impact of psychological features in the choice of coping strategies in MS patients and their link to QOL. In the study's sample, MS patients were less likely to use problem-focused strategies compared to the general population, and they adopted avoidance coping strategies more frequently. This strategy is usually adopted by patients with high levels of depression and in particular by patients with RRMS course of disease, although these patients were younger and less disabled than progressive subjects and their daily living activity had a lower impact. Positive attitude and planning activity were linked to lower disease duration and socialization. In this sample, the link between shorter disease duration and positive attitude can be due to higher levels of optimism when facing the disease in its early stages. In this study, the use of social support and positive attitude strategies influenced QOL positively, and MS patients prevalently adopted avoiding strategies rather than positive attitude strategies that are related to better QOL when compared to the general population.

Conclusion

Coping can be defined as cognitive and behavioral efforts to manage specific demands.

Two coping strategies have been identified: problem-focused and emotion-focused strategies. The first implies the efforts made by the subject to overcome his difficulty, and the second strategy is associated with emotion regulation and involves efforts to modify the distress that accompanies threat.

It is possible to describe coping as being a mediating factor between chronic disease and adjustment to the disease. Coping strategies are strongly related to chronic disease and to MS in particular. People live many years with the disease increasing disabilities and reducing quality of life. In MS both coping strategies are important and can have crucial beneficial consequences for physical and mental health.

Studies on coping and MS have shown that in the first stage of the disease, patients tend to use emotion-focused coping and that this strategy is more related to attacks of the disease. On the other hand, other authors found no association between mood and specific characteristics of the disease such as its duration or disability levels.

Some studies found that personality is one of the major factors affecting psychological functioning, potentially exerting influence on both coping choices.

Coping styles are also linked to mood alteration. A depressive reaction in the first phase of disease can be considered a relevant factor in determining the disappearance of depressive symptoms during the follow-up in most of the patients and can lead to use of active coping strategies.

Neuropsychological dysfunctions that affected people with MS, such as deficit of complex attention, may negatively affect a patient's ability to develop active coping strategies, but the link between cognitive impairment and coping is not clear.

Coping is relevant in terms of QOL. Emotion-focused coping strategies such as avoidance, wishful thinking, and self-accusation were linked to poor QOL in MS.

Coping strategies are crucial in chronic disease and in MS in particular, because they can be active efforts to do something to alleviate stressful situations, increasing adjustment to the disease.

References

1. Folkman S, Lazarus RS. Manual of ways of coping questionnaire. Palo Alto: Consulting Psychology Press; 1988.
2. Montel SR, Bungener C. Coping and quality of life in one hundred and thirty five subjects with multiple sclerosis. Mult Scler. 2007;13:393–401.
3. Kasl SV, Cooper CL. Stress and health: issue in research methodology. Chichester: Wiley; 1987.
4. Lode K, Larsen JP, Bru E, Klevan G, Myhr KM, Nyland H. Patient information and coping styles in multiple sclerosis. Mult Scler. 2007;13:792–9.
5. Pakenham KI. Adjustment to multiple sclerosis: application of a stress and coping model. Health Psychol. 1999;18:383–92.
6. Scherer RF, Luther DC, Wiebe FA, Adams JS. Dimensionality of coping: factor stability using the Ways of Coping Questionnaire. Psychol Rep. 1998;62:763–70.
7. Vitaliano PP, Russo J, Carr JE, Maiuro RD, Becker J. The ways of coping check-list: revision and psychometric properties. Multivariate Behav Res. 1985;20:3–26.
8. Endler NS, Parker JDA, Summerfelt LJ. Coping with health problems: developing a reliable and valid multidimensional measure. Psychol Assess. 1998;10:195–205.
9. Folkman S, Lazarus RS. If it changes it must be a process: study of emotion and coping during three stages of a college examination. J Pers Soc Psychol. 1985;48:150–70.
10. Carver CS, Scheier MF, Weintraub JK. Assessing coping strategies: a theoretically based approach. J Pers Soc Psychol. 1989;56:267–83.
11. Karlsen B. Coping with diabetes. A study of factors influencing psychological well-being and coping in adults with diabetes. Bergen: University of Bergen; 2004.

12. McCabe MP, McKern S, McDonald E. Coping and psychological adjustment among people with multiple sclerosis. J Psychosom Res. 2004;56:355–61.
13. Devins GM, Styra R, O'Connor P, Gray T, Seland TP, Klein GM, et al. Psychological impact of illness intrusiveness moderated by age in multiple sclerosis. Psychol Health Med. 1996;1:179–91.
14. Ford H, Trigwell P, Johnson M. The nature of fatigue in multiple sclerosis. J Psychosom Res. 1998;45:33–8.
15. Noy S, Achiron A, Gabbay U, Barak Y, Rostein Z, Laor N, et al. A new approach to affective symptoms in relapsing-remitting multiple sclerosis. Compr Psychiatry. 1995;36:390–5.
16. Pakenham KI, Stewart CA, Rogers A. The role of coping in adjustment to multiple-sclerosis-related adaptive demands. Psychol Health Med. 1997;2:197–211.
17. Jean V, Paul RH, Beatty WW. Psychological and neurological predictors of coping pattern by patients with multiple sclerosis. J Clin Psychol. 1999;55:21–6.
18. Hickey A, Greene SM. Coping with multiple sclerosis. Ir J Psychol Med. 1989;6:118–24.
19. Mohr DC, Dick LP, Russo D, Pinn J, Boudewyn AC, Likosky W, et al. The psychosocial impact of multiple sclerosis: exploring the patient's perspective. Health Psychol. 1999;18:376–82.
20. Schwartz CE. Teaching coping skills enhances quality of life more than peer support: result of a randomized trial with multiple sclerosis patients. Health Psychol. 1999;18:211–20.
21. Mosson M, Peter L, Montel S. Impact of physical activity level on alexithymia and coping strategies in an over-40 multiple sclerosis population: a pilot study. Rev Neurol. 2014;170(1):19–25.
22. Madan S, Pakenham KL. The stress-buffering effects of hope on change in adjustment to caregiving in multiple sclerosis. J Health Psychol. 2013; Epub ahead of print.
23. Senders A, Bourdette D, Hanes D, Yadav V, Shinto L. Perceived stress in multiple sclerosis: the potential role of mindfulness in health and well-being. J Evid Based Complementary Altern Med. 2014;19(2):104–11.
24. Bolger N, Zuckerman A. A framework for studying personality in the stress process. J Pers Soc Psychol. 1995;69:890–902.
25. Watson D, Hubbard B. Adaptational styles and dispositional structure: coping in the context of the five-factor model. J Pers. 1996;64:737–74.
26. McCormick RA, Dowd ET, Quirk S, Zegarra JH. The relationship of NEO-PI performance to coping styles, patterns of use, and triggers for use among substance abusers. Addict Behav. 1998;23:497–507.
27. Russo J, Katon W, Lin E, Von-Korff M. Neuroticism and extraversion as predictors of health outcome in depressed primary care patients. Psychosomatics. 1997;38:339–48.
28. Hooker K, Frazier LD, Monahan DJ. Personality and coping among caregivers of spouses with dementia. Gerontologist. 1994;34:386–92.
29. Arnett PA, Randolph JJ. Longitudinal course of depression symptoms in multiple sclerosis. J Neurol Neurosurg Psychiatry. 2006;77(5):606–10.
30. Suhn Y, Motl RW, Mohr DC. Physical activity, disability, and mood in the early stage of multiple sclerosis. Disabil Health J. 2010;3:93–8.
31. Wood B, Van Der Mei IA, Ponsonby AL, Pittas F, Quinn S, Dwyer T, et al. Prevalence and concurrence of anxiety, depression and fatigue over time in multiple sclerosis. Mult Scler. 2012;19(2):217–24.
32. Bakshi R, Czarnecki D, Shaikh ZA, Priore RL, Janardhan V, Kaliszky Z, et al. Brain MRI lesions and atrophy are related to depression in multiple sclerosis. Neuroreport. 2000;11:1153–8.
33. Aikens JE, Fisher JS, Namey M, Rudick RA. A replicated prospective investigation of life stress, coping, and depressive symptoms in multiple sclerosis. J Behav Med. 1997;20:433–45.
34. Mohr DC, Goodkin DE. Treatment of depression in multiple sclerosis: review and meta-analysis. Clin Psychol Sci Pract. 1999;6:1–9.

35. Lynch SG, Kroencke DC, Denney DR. The relationship between disability and depression in multiple sclerosis: the role of uncertainty, coping, and hope. Mult Scler. 2001;7:411–6.
36. Ozura A, Sega S. Profile of depression, experienced distress and capacity for coping with stress in multiple sclerosis patients – a different perspective. Clin Neurol Neurosurg. 2013;115 Suppl 1:S12–6.
37. Nielsen PJ, Saliger J, Guldenberg V, Breier G, Karbe H. Stress-stimulated volitional coping competencies and depression in multiple sclerosis. J Psychosom Res. 2013;74(3):221–6.
38. Bianchi V, De Giglio L, Prosperini L, Mancinelli C, De Angelis F, Barletta V, et al. Mood and coping in clinically isolated syndrome and multiple sclerosis. Acta Neurol Scand. 2014;129(6):374–81.
39. Arnett PA, Higginson CI, Voss VD, Randolph JJ, Grandey AA. Relationship between coping, cognitive dysfunction and depression in multiple sclerosis. Clin Neuropsychol. 2002;16:341–55.
40. Dennison L, Moss-Morris R, Siliber E, Galea I, Chalder T. Cognitive and behavioral correlates of different domains of psychological adjustment in early-stage multiple sclerosis. J Psychosom Res. 2010;69:353–61.
41. Amato MP, Zipoli V, Portaccio E. Multiple sclerosis-related cognitive changes: a review of cross-sectional and longitudinal studies. J Neurol Sci. 2006;245:41–6.
42. Amato MP, Zipoli V. Clinical management of cognitive impairment on multiple sclerosis: a review of current evidence. Int MS J. 2003;10(3):72–83.
43. Amato MP, Portaccio E, Goretti B, Zipoli V, Iudice A, Della Pina D, et al. Relevance of cognitive deterioration in early relapsing-remitting MS: a 3-year follow-up study. Mult Scler. 2010;16(12):1474–82.
44. Reuter F, Zaaraoui W, Crespy L, Faivre A, Rico A, Malikova I, et al. Frequency of cognitive impairment dramatically increases during the first 5 years of multiple sclerosis. J Neurol Neurosurg Psychiatry. 2011;82:1157–9.
45. Glanz B, Holland C, Gauthier S, Amunwa EL, Liptak Z, Houtchens MK, et al. Cognitive dysfunction in patients with clinically isolated syndromes or newly diagnosed multiple sclerosis. Mult Scler. 2007;13:1004–10.
46. Ehrensperger MM, Grether A, Romer G, Berres M, Monsch AU, Kappos L, et al. Neuropsychological dysfunction, depression, physical disability, and coping processes in families with a parent affected by multiple sclerosis. Mult Scler. 2008;14:1106–12.
47. Rabinowitz AR, Arnett PA. A longitudinal analysis of cognitive dysfunction, coping, and depression in multiple sclerosis. Neuropsychology. 2009;23:581–91.
48. Goretti B, Portaccio E, Zipoli V, Razzolini L, Amato MP. Coping strategies, cognitive impairment, psychological variables and their relationship with quality of life in multiple sclerosis. Neurol Sci. 2010;31 Suppl 2:S227–30.
49. Goretti B, Portaccio E, Zipoli V, Hakiki B, Siracusa G, Sorbi S, et al. Impact of cognitive impairment on coping strategies in multiple sclerosis. Clin Neurol Neurosurg. 2010;112:127–30.
50. Chiaravalloti ND, DeLuca J. Cognitive impairment in multiple sclerosis. Lancet Neurol. 2008;7:1139–51.
51. Guimaraes J, Sa MJ. Cognitive dysfunction in multiple sclerosis. Front Neurol. 2012;3(74):1–8.
52. Montel S, Spitz E, Bungener C. Coping strategies in multiple sclerosis patients with frontal cognitive disorders. Eur Neurol. 2012;68(2):84–8.
53. Siegert RJ, Abernethy DA. Depression in multiple sclerosis: a review. J Neurol Neurosurg Psychiatry. 2005;76:469–75.
54. McCabe MP, MCKern S. Quality of life and multiple sclerosis: comparison between people with multiple sclerosis and people from the general population. J Clin Psychol Med Settings. 2002;9:287–95.
55. Mohr DC, Goodkin DE, Gatto N, Van der Wende J. Depression, coping and level of neurological impairment in multiple sclerosis. Mult Scler. 1997;3(4):254–8.

56. Mikula P, Nagyova I, Krokavcova M, Vitkova M, Rosenberger J, Szilasiova J, Gdovinova Z, Groothoff JW, van Dijk JP. Coping and its importance for quality of life in patients with multiple sclerosis. Disabil Rehabil. 2014;36(9):732–6.
57. Goretti B, Portaccio E, Zipoli V, Hakiki B, Siracusa G, Sorbi S, Amato MP. Coping strategies, psychological variables and their relationship with quality of life in multiple sclerosis. Neurol Sci. 2009;30:15–20.

Chapter 11
Fatigue

Vikram Bhise and Lauren B. Krupp

Abstract Fatigue is the most frequently reported symptom among individuals with MS. It can be assessed using self-report instruments and/or performance-based measures. Fatigue may be primary or secondary to other disorders in MS, such as depression, sleep disturbance, or pain. Brain imaging and neurophysiology studies have helped demonstrate multiple areas and aspects of brain dysfunction, implicating a disruption of physiological "networks." Disorders of energy metabolism, immune regulation, the hypothalamic-pituitary axis, and the autonomic nervous system are also theorized to participate in the pathogenesis. Other markers of fatigue include findings on polysomnography and tests of vigilance and attention. Nonpharmacologic treatments include energy conservation, cooling, exercise, rehabilitation therapy, cognitive behavioral therapy, electromagnetic maneuvers, and mindfulness techniques. Medication trials have met with mixed results. Ideally future studies will better help link elements of fatigue to study findings and enable improved therapeutic modalities.

Keywords Fatigue • Fatigability • Scales • Cognition • Neuroimaging • Neuropsychology

Introduction

Fatigue is one of the most challenging symptoms of multiple sclerosis (MS). In one US population sample, 83 % of patients indicated fatigue as their number one symptom associated with MS [1]. Other studies have similarly reported that over 75 % of

V. Bhise, M.D. (✉)
Department of Pediatrics, Neurology, Rutgers – Robert Wood Johnson Medical School, Child Health Institute of New Jersey, 89 French Street, Suite 2200, New Brunswick, NJ 08901, USA
e-mail: bhisevi@rwjms.rutgers.edu

L.B. Krupp, M.D.
Department of Neurology, Stony Brook Medicine, HSC T12 020, Stony Brook, NY 11794-8121, USA
e-mail: lauren.krupp@stonybrook.edu

B. Brochet (ed.), *Neuropsychiatric Symptoms of Inflammatory Demyelinating Diseases*, Neuropsychiatric Symptoms of Neurological Disease,
DOI 10.1007/978-3-319-18464-7_11

patients experience it as among their top three most troubling problems [2]. Fatigue alone can also be the presenting symptom of MS [3]. The socioeconomic consequences of fatigue are severe. It is closely linked to unemployment and decreased quality of life (QOL) [2, 4–6]. Though the pathophysiology of this MS complication is not fully understood, current research is shedding light on its possible etiologies. Fatigue is considered a primary symptom in MS, stemming from dysfunction in the central nervous system; however, additional problems present in MS such as sleep disorders and depression can also cause or exacerbate fatigue [7, 8]. Both pharmacological and nonpharmacological methods can address this troublesome symptom [9–12]. Overall our understanding of fatigue in MS remains incomplete; nonetheless, numerous research endeavors continue to further investigate its mechanism and best management.

Definition

The definition of fatigue may be approached either subjectively or objectively. Fatigue can be the subjective feeling described by the patient or objectively measured as performance decrement in a given activity over time [13]. The 1998 Multiple Sclerosis Counsel for Clinical Practice Guideline consensus statement defined fatigue as a "subjective lack of physical and/or mental energy that is perceived by the individual or caregiver to interfere with usual and desired activities [14]." Fatigue may be either chronic or acute. Acute fatigue is new onset fatigue occurring within a 6-week time frame, limiting functional activity or impairing quality of life. Chronic fatigue, on the other hand, is defined as any amount of fatigue occurring over more than 50 % of days for more than a 6-week time frame that also limits functional activity or quality of life. Fatigue may be primary – due to the disease MS itself – or secondary to other illnesses such as depression. MS fatigue may be further sub-defined as either asthenia, fatigability, or worsening of other symptoms [15]. Asthenia represents increased fatigue at times of rest. Fatigability refers to the inability to sustain a specific task or activity over time. Notably, this fatigability need not only be physical, but can be cognitive in nature. Individuals with cognitive fatigue suffer from impaired ability to sustain mental tasks. Worsening of other symptoms such as spasticity may contribute to increasing baseline fatigue. Other secondary causes of fatigue include the increased tiredness associated with the afternoon, possibly due to circadian rhythm fluctuations experienced by everyday individuals, as well as deconditioning due to lack of exercise and/or prolonged immobility, neuromuscular fatigue, and fatigue due to other medical issues such as depression, anemia, or hypothyroidism. The variety of causes thus creates difficulty for patients and practitioners to best define fatigue and complicates research and treatment (Table 11.1).

Table 11.1 Categorization of fatigue

Onset
Acute – less than 6 weeks
Chronic – greater than 6 weeks
Activity
Asthenia
Fatigability
Worsening of symptoms
Type
Normal fatigue
Cognitive fatigue
Physical fatigue
Neuromuscular fatigue and deconditioning
Fatigue due to other medical problems
Pathophysiological changes and fatigue
Physical exertion → physical fatigue
Mental exertion → cognitive fatigue
Neurochemical → depression
Neuromuscular → physical fatigue

Consequences

Fatigue in MS has far-reaching consequences affecting multiple aspects of a person's well-being. These problems include early retirement or the need to reduce work hours [6, 16]. Patients with moderate or severe fatigue compared to their non-MS non-fatigued colleagues are at increased risk of becoming unemployed [5, 16]. Fatigued patients require increased healthcare resources and engender increased socioeconomic burden. MS patients with fatigue also seek increased outpatient visits including rehabilitation [17]. Multiple studies demonstrate a lower quality of life overall for patients with fatigue [4, 18–20].

Demographics

Studies on the demographics of fatigue have been inconsistent. In the NARCOMS study, more severe fatigue was associated with male gender and older patients [2]; however, this finding was not seen in other studies [21–24]. Other better developed associations include more fatigue in patients who have less education [2, 21, 24].

Clinical Features

Primary MS fatigue is distinct from other forms of fatigue. Unlike everyday fatigue experienced by healthy individuals without chronic illness, MS fatigue interferes with the quality of life and activities of daily living [3]. The difference between MS fatigue and other disorders is a notable increase with and sensitivity to heat, known clinically as Uhthoff's phenomenon [25]. Patients often also commonly describe diurnal symptoms with fatigue worsening as the day proceeds and maximal in the afternoon, unlike depression where fatigue can be worse in the morning.

Fatigue is seen in all subtypes of MS, including progressive and relapsing. It is known to be highest among patients with secondary progressive MS [26]. In addition, patients who are gait impaired tend to have greater fatigue, as do patients who have relapsing disease with progressive impairment compared to those with relapsing yet stable disease [2]. Fatigue appears to be correlated with disability, but this correlation is greatly attenuated after controlling for depression [27]. Surprisingly, fatigue severity is not associated with disease duration [28].

One common concern for patients is that disease-modifying therapy may cause or exacerbate fatigue. In a study of 320 patients and the NARCOMS survey of 9,205 respondents, no difference in fatigue was found significantly associated with disease-modifying therapy [2, 29]. The Cognitive Impairment in MS (COGIMUS) study examined interferon-beta 1a in 331 patients and found no association [30], though a small study of interferon-beta 1a did find some benefits on fatigue and cognitive measures [31]. Benefits are more convincingly seen with glatiramer acetate and are thought to be due to improvements in the quality of life [32]. These improvements seen with the use of glatiramer acetate remained stable over a 2-year time frame; and improvements seen at the 6-month time point predicted outcome at 2 years [33]. A concern with interferon-beta lies in the side effects of depression and flu-like symptoms which may exacerbate underlying fatigue. Indeed one group reported that patients noted lower levels of fatigue when switching from interferon-beta to glatiramer [2]. Reduced fatigue is also seen in patients who switched to natalizumab. Cross-sectional studies also suggest lower fatigue levels among those on natalizumab relative to other disease-modifying therapies [34–36].

Unfortunately, fatigue can be a problem which persists in most patients and correlates with pain, mood, and neurological impairment [26]. Table 11.2 notes problems commonly associated with fatigue in MS.

Biomarkers

Evoked potentials may provide a surrogate electrophysiological marker [37]. Fatigue in MS in one study correlated with increased P100 latency, decreased P100 amplitude, and increased interocular P100 latency on visual evoked potential testing, as well as increased V component latency, increased I-III-V interlatency, and

Table 11.2 Problems commonly associated with fatigue in MS

Impaired quality of life
Increased need for healthcare resources
Reduced work hours
Higher risk of becoming unemployed
Depression
Anxiety
Sleep disorders
Pain
Cognitive impairment
Decreased motor sustainability

decreased V component amplitude on brainstem auditory evoked potential testing. These findings are not pronounced, however, and tend to be most obvious in the most fatigued patients.

Furthermore, higher fatigue on the Fatigue Severity Scale (FSS) predicted conversion from clinically isolated syndrome to clinically definite MS, with a hazard ratio of 2.6 when fatigue was represented as a dichotomous variable for FSS scores greater than or equal to 5 [38].

Depression

Fatigue and depression are intricately intertwined. As fatigue can lead to situational depression, depression can be the cause for fatigue [39–41]. Consequently, many studies have found a correlation between the two [7, 42–45]. These studies have also found that fatigue is associated with depression, even when correcting for fatigue in the measures used to assess depression [21, 46]. Notably symptoms of depression may mimic fatigue namely apathy, sleep disturbances, decreased energy, and the inability to complete tasks often seen in patients with depression. Fatigue may further lead to depression as symptoms of loss of control occurring with disease progression can trigger depressive episodes [7].

Low sense of control over one's self or environment may predict later MS fatigue [7, 47–50]. Mood problems such as anxiety or depression are interlinked with fatigue as well. For example, depression may predict later fatigue and anxiety, while anxiety and fatigue may predict later depression [26, 51, 52]. On the other hand, it should be emphasized that an individual with MS may have severe fatigue but entirely lack symptoms of depression, corroborating that primary MS fatigue is intrinsic to the disease itself [21].

While fatigue and depression may have separate elements altogether, treatment of depression may help severity of fatigue by improving symptoms of mood as opposed to vegetative symptoms such as changes in energy, appetite, or sleep [53].

Anxiety

Anxiety is also a fairly common MS symptom [54]. The correlation between anxiety and fatigue is not nearly as much studied as that between depression and fatigue; however, available research supports a consistent connection between anxiety and fatigue [21, 43, 49, 55, 56]. It may be that anxiety better correlates with self-reported cognitive fatigue than physical measures [43, 49].

Sleep Disorders

Sleep disorders represent a major problematic symptomatology in patients with MS. The sleep disorders occurring more commonly among those with MS relative to the general population include sleep apnea, insomnia, circadian rhythm disorder, restless legs syndrome, periodic limb movements, rapid eye movement (REM), sleep behavior disorder, and narcolepsy [57, 58]. Some sleep disorder symptoms correlate with objective findings, such as cerebellar lesions and periodic limb movements [59]. Objective measures on polysomnography include decreased sleep efficiency, greater wake time after sleep onset, decreased sleep latency, and increased total arousal index in MS patients with fatigue compared to those without, ultimately depicting a decreased sleep efficiency [60, 61]. One study discovered that 20 % of their patient population had a formal diagnosis of obstructive sleep apnea (OSA) and that over half of their population were at risk for a diagnosis of OSA [62]. The questions used for screening sleep apnea also significantly correlated with fatigue.

Treatment medications may also be implicated in symptoms of hypersomnia or insomnia [63]. Additionally, other disease-related symptoms may play a role in sleep disorders. Impaired sleep is a common complaint in patients who are depressed or anxious. Other symptoms such as nocturia may necessitate frequent nighttime awakening leading to fragmented sleep [57, 64]. Over half of patients with MS with insomnia can have middle insomnia, significantly correlating with fatigue [65]. Nocturnal pain, paresthesias, and muscle spasms are other problems which may also disrupt sleep [66].

Pain

Pain may be a primary symptom in patients with MS or may be due to other sensory-motor disturbances such as neuralgias, paresthesias, and painful muscle spasms. The symptoms in turn can lead to disrupted sleep, exacerbate depression, or progress to immobility causing physical deconditioning. All together, these factors conspire to increase daytime somnolence and overall fatigue, creating a progressive downward spiraling of the disease [49, 65].

Cognition

Over the past 20–25 years, research has aimed to dissect the interaction between fatigue and cognitive impairment. Early studies demonstrated that standard neuropsychological testing does not show a significant association between self-reported fatigue and cognition [7]. This lack of association between routine cognitive tests and fatigue has been confirmed in larger subsequent samples [67–69]. Specifically, the majority of reports demonstrate little association between fatigue and tests of verbal and visual memory, working memory, or cognitive processing speed. In contrast, performance on tests of vigilance and alertness is significantly correlated with self-reported fatigue [70–72]. Most tests of vigilance assess performance decrement over an extended task and hence can be considered measures of cognitive fatigability. Fatigability as a change in performance over time, distinct from self-reported fatigue, is further discussed in the fatigue measurement section.

Motor Function and Fatigue

Fatigued patients with MS often describe physical motor fatigue as part of their symptomatology. These deficits have been identified physiologically as an inability to efficiently recruit motor pathways, sustain muscle contraction, and maintain appropriate levels of metabolites needed for muscle activity during periods of exertion [73–75]. However, these assessments of motor function do not well predict self-reported measures of fatigue.

EEG studies demonstrate hyperactivity in sensorimotor areas during movement execution in MS patients with fatigue, with failure of inhibition after movement's end [76]. Transcranial magnetic stimulation (TMS) studies in patients complaining of self-reported fatigue show decreased motor cortex inhibition compared to MS patients without fatigue [77]. The time to normalization of the motor threshold correlates with fatigue severity. A complementary study of evoked potentials found failure in normal reduction of amplitudes in fatiguing repetitive muscle stimulation [78]. These findings would suggest either aberrant hyperexcitability of these pathways and circuits or perhaps a compensatory increase in pathways to maintain the same level of effort. Comparing healthy controls to MS patients, one study found no changes in motor evoked potentials or central conduction time to suggest motor pathway conduction blockade during a voluntary fatiguing motor exercise, thus postulating that MS fatigue is instead due entirely to an impaired cortical drive [79].

Fatigue Measurement

Fatigue is measured in two ways, either by questionnaires that are validated or by measures of performance of motor or cognitive tasks. Patients' experience of fatigue tends to be a more important factor for patients' quality of life [14].

Self-Reported Measures

Most studies employ questionnaires based on self-report methods. Questionnaires consist of a variety of forms ranging from visual analogue scales, unidimensional scales (e.g., the Fatigue Severity Scale (FSS)) to longer multidimensional scales, such as the Modified Fatigue Impact Scale (MFIS). These measures are frequently used in MS clinical trials (particularly the MFIS) and longitudinal research settings, showing sensitivity to changes in fatigue progression [25, 80, 81]. They are fast, easy to administer, and well validated. Although the 9-item FSS is designed as a unidimensional scale, Rasch analysis demonstrated that removal of four of the nine items increases the unidimensionality of the scale [82]. Conversely, the MFIS uses a multidimensional approach separating the different components into cognitive, physical, and psychosocial dimensions. The physical dimension appears to be the most associated with neurologic impairment, disability, and the EDSS. Unfortunately, neuropsychological measures do not correlate well with the cognitive dimension. Another often used multidimensional scale is the Fatigue Descriptive Scale (FDS) which evaluates three modalities: asthenia, fatigability, and worsening symptoms [15]. There is also the Fatigue Scale [83, 84] which uses two dimensions – a physical and a mental component. Fatigue may also be measured as a component of a larger inventory, such as the inverse of the vitality subscale within the SF-36 [85, 86], the Fatigue-Inertia Scale of the Profile of Mood States (POMS) [87], or portions of the Sickness Impact Profile (SIP) [88].

These questionnaires are available to the practitioner and can be quickly administered in the office setting. All self-reported measures are subject to limitations of recall bias and other time-based factors [89, 90].

In the past several years, other fatigue measures which incorporate current scaling techniques recommended by the FDA have been developed. Two examples are the Neurological Fatigue Index (NFI-MS) [91, 92] and the NIH-supported PROMIS fatigue measure [93, 94]. Proposed values for clinically meaningful change on the NFI-MS are available [92]. Furthermore, to develop assessments tailored for specific research purposes, the PROMIS website provides access to a databank of 95 items (available at http://www.nihpromis.org/default.aspx) that can be selected in different combinations as needed. Nonetheless, neither the NFI-MS nor the PROMIS measure has been widely used in MS. Further study of these fatigue-related outcome measures is needed for clinical trials focused on fatigue. Table 11.3 lists some commonly used fatigue measurement scales.

Performance-Based Measures

Performance measures have been developed to provide investigators with more objective measurements of fatigue [95]. Therefore, fatigue can be defined as a quantitative change in performance over a specified period of time. This can be applied to both motor function and cognitive function.

Table 11.3 Commonly used fatigue measurement scales

Commonly used fatigue measurement scales
Modified Fatigue Impact Scale (MFIS)
Fatigue Severity Scale (FSS)
Fatigue Impact Scale (FIS)
Fatigue Assessment Instrument (FAI)
Fatigue Rating Scale (FRS)
Fatigue Descriptive Scale (FDS)
Fatigue Scale
Fatigue Scale for Motor and Cognitive Functions (FSMC)
Rochester Fatigue Diary

Motor Fatigability To measure motor fatigability, an isometric strain gauge can be used to determine motor fatigue by determining the degree of force produced through muscle contraction to create a fatigue index, i.e., the ratio between muscle strength decay over time and maximal voluntary contraction. This index is consistently elevated in MS patients particularly those with motor tract dysfunction [96, 97]. Fatigue may also otherwise be described in terms of change in motor unit firing rate or muscle metabolism [73–75, 79, 95, 96, 98]. Patients with MS demonstrate more fatiguing of sustained contraction than healthy controls [95].

Mental Fatigability The objective fatigue measures for cognitive fatigue include neuropsychological testing techniques that contain sustained effortful cognitive tasks. Performance on neuropsychological testing significantly declined in MS patients compared to healthy controls demonstrating cognitive fatigability [99]. These objective changes did not correlate with self-reported fatigue scales such as the FSS or MFIS [99].

Another study demonstrated that participants with MS versus controls had increases in the variability of cognitive processing speed across multiple repeated administrations of the Symbol Digit Modality Test (SDMT). The authors interpreted this increased variability as an indicator of cognitive fatigability [100]. Other neurocognitive studies of fatigability-inducing tasks in MS have shown decreases in working memory performance over time [15, 101]. In one report self-reported fatigue was linked to a challenging cognitive task, though findings were highly variable depending on specific tasks and scoring techniques [102].

Neuroimaging

Fatigue does not have a discreet anatomic location within the central nervous system. Studies using conventional MRI did not find any single region in the brain that was most associated with fatigue. Initial studies also did not find a correlation with demyelinating lesion burden [103, 104]; however, newer studies with larger sample sizes and better mathematical modeling do indeed show that fatigue correlates with increased gray and white matter atrophy as well as higher lesion load [105, 106]. A longitudinal study demonstrated that increases in fatigue in the first 2 years can be predictive of greater whole-brain atrophy 6 years later [105, 106]. Neuroimaging studies have also provided insight into potential pathogenic mechanisms linking fatigue to cerebral dysfunction and structural damage pertaining to energy metabolism, cognitive fatigability, and motivation. Additional imaging methods applied to studies of fatigue include positron emission tomography (PET), magnetic resonance spectroscopy (MRS), and functional MRI (fMRI) [107–109].

Fatigue and Energy Metabolism

Focal reduction in glucose metabolism has been demonstrated in fatigued MS using fluorodeoxyglucose PET, linked to multiple brain regions: prefrontal cortex, basal ganglia, internal capsule, posterior parietal cortex, and temporo-occipital gyri [107]. Reduced energy metabolism in the frontal cortex and basal ganglia may indicate that fatigue is due to their failure to interconnect, possibly due to disruption of a "dorsolateral prefrontal circuit [107, 110]."

N-Acetylaspartate (NAA) is a marker of neuronal activity, and increased levels are believed to be indicative of axonal damage. Proton MRS measuring these levels found an association between the Fatigue Severity Scale and a decreased NAA-to-creatinine ratio, implicating widespread axonal dysfunction as part of the pathophysiology of fatigue [108]. Reductions in NAA could also be due to mitochondrial dysfunction within intact neurons which would suggest that, at the cellular level, energy depletion contributes to MS fatigue.

Fatigue and Complex Attention

It is possible that pathways responsible for complex attention overlap with those underlying fatigue. Associations between self-reported fatigue in MS patients have been correlated with lesion burden in the right parietal temporal region white matter and left frontal region white matter [110]. The right parietotemporal region is important for attention and plays a role in the alerting and orienting networks. Fatigue was also associated with gray matter atrophy in the left superior frontal gyrus and bilateral middle frontal gyri. An emerging model for cognitive fatigue has implicated the striato-thalamic-frontal network, which connects the anteromedial thalamus, basal ganglia, and frontal lobes via the anterior limb of the internal capsule [111]. Damage to these thalamocortical fibers could lead to cognitive fatigability, along with additional specific involvement of the caudate in fatigued MS patients.

Fatigue and Motivation

Multimodal imaging comparing whole-brain volume and regional brain volumes demonstrates more atrophy in fatigued MS patients compared to non-fatigued patients in multiple areas that include the nucleus accumbens, a dopaminergic structure implicated in motivation, reward, and the regulation of effort. Other regions with increased atrophy among fatigued vs. non-fatigued individuals with MS include the right inferior temporal gyrus, left superior frontal gyrus, and the forceps major [112]. Lower fractional anisotropy values have also been seen in the forceps major, left inferior fronto-occipital fasciculus, and right anterior thalamic radiation.

Furthermore, the basal ganglia, or rather their cortical connections, have been implicated as critical pathways underlying both fatigue and lack of motivation. Individuals with MS fatigue have abnormalities in the basal ganglia that include atrophy, decreased blood flow, and metabolic changes [107, 113, 114]. Functional MRI studies have demonstrated increased activation in the basal ganglia during cognitively demanding tasks in MS patients with fatigue. Taken together, these studies suggest abnormalities in MS fatigue can be linked to cortical-basal ganglionic circuits showing decreased functional connectivity with the medial prefrontal cortex and decreased connectivity between the anterior cingulate cortex and the caudate nucleus [115]. The medial prefrontal cortex is involved in motivation; while lesions of the anterior cingulate are known to lead to decreased effort and lethargy. These studies implicate dysregulation of dopamine as well.

DTI studies confirm the importance of cerebral pathways involved in motivation. Disruption in the forceps major occurs in MS patients with both fatigue and depression, whereas fatigued-only patients also have damage to the right anterior thalamic radiation [112]. The anterior thalamic radiation provides excitatory fibers to the dorsolateral prefrontal cortex, while the forceps major connects the dorsolateral prefrontal cortex, including the middle and superior frontal gyri. Additionally, the forceps major carries fibers from the corpus callosum, which is also believed to be affected in fatigued MS patients [116].

Functional neuroimaging studies have described cortical functional reorganization in brains of MS patients [117]. Compared to healthy controls, MS patients show increased cortical activation in both ipsilateral and contralateral brain regions during repetitive motor tasks [109] and a fatiguing task [118]. These findings may also represent later disease stages that show rather increased functional connectivity with the motor cortex, as a compensatory mechanism for the increased effort required to maintain the same level of functionality.

Pathogenesis

The true pathogenesis of fatigue in MS remains elusive. Multiple mechanisms have been proposed including structural, endocrine, inflammatory, metabolic, and neurochemical processes.

Neuroimmune Mechanisms

Fatigue is seen not only in MS but in other autoimmune conditions as well, such as systemic lupus erythematosus and rheumatoid arthritis. Therefore, it has been suggested that the autoimmune process itself may participate in the pathogenesis of fatigue. That disease-modifying therapies can induce fatigue, most notably interferon-beta, supports the contention that cytokines and immune dysregulation are contributory to fatigue. Elevations in proinflammatory cytokines are also found in other disorders afflicted with fatigue, such as chronic fatigue syndrome, cancer, and viral infections [119–122]. MS patients with fatigue show higher expression of TNF-α mRNA in peripheral mononuclear blood cells, suggesting that TNF-α may be the neurochemical mediator of fatigue [122]. Other studies however have been inconsistent in linking circulating cytokines to fatigue [122–124].

Neuroendocrine Mechanisms

Some studies suggest that hyperactivity of the hypothalamic-pituitary-adrenal axis may be responsible for fatigue [125]. Significantly elevated levels of ACTH are associated with MS patients with fatigue [125]. The role of neuroendocrine factors and fatigue could be due to secondary effects of proinflammatory cytokines affecting corticoid receptor signaling [125]. Use of antidepressants may affect corticoid receptor function and may be part of the mechanism in treatment of fatigue. Other studies however have failed to corroborate this finding [126]. One study did note, though, that low levels of dehydroepiandrosterone (DHEA) and its sulfated conjugate (DHEAS) are seen more often in fatigued MS patients [127]. The low levels of DHEA may feed forward and hyperstimulate the HPA axis.

Autonomic Nervous System Dysregulation

Abnormalities in the autonomic cardiovascular system have been proposed to underlie symptoms such as generalized weakness, dizziness, and other neurocognitive complaints [128]. Up to 20 % of fatigued MS patients also had signs of autonomic failure. Autonomic testing in MS patients with fatigue found impaired adrenergic orthostatic responses, thought to be due to impaired sympathetic vasomotor activity [129]. This correlation has not been identified in other studies however [130].

Dysregulation in Vigilance and Alertness

The author of an interesting review of cognitive and fatigue studies in MS has made a compelling argument implicating deficits in the attentional/vigilance network as one explanation for MS fatigue [131]. Three elements in the vigilance network can be delineated. These include alerting functions, orienting/selecting behavior, and the executive network. Poor performance on tests of vigilance can develop due to increased cognitive load, depletion of attentional resources (loss of the ability to ignore distracting internal or external stimuli), and impaired executive function. This network is associated with activity in the midbrain, thalamus, frontoparietal areas, and the anterior cingulate. Individuals with MS show deficits on measures of vigilance and alertness which are likely due in part to an impaired ability to maintain attention on the task at hand. These performance deficits have been linked in MS to focal atrophy and reduced cortical thickness in frontal parietal regions and the anterior cingulate. Some functional MRI studies also demonstrate dysregulation in these areas.

Physical Deconditioning

Lack of exercise induced by gait impairment or ongoing pain can lead to a state of relative inactivity in MS patients and subsequent physical deconditioning [132]. This state is characterized by decreased muscle bulk and increased generalized weakness, which often prompts patients to further avoidance of exercise. Patients may become severely disabled eventually with serious risk for respiratory compromise. It is critical to reengage patients and restore physical activity to offset the associated co-morbidities.

Temperature Sensitivity

The neurophysiological correlate of the Uhthoff's phenomenon is believed to be a heat-induced central conduction motor fiber block [133]. These findings may have been missed by earlier studies using motor evoked potential testing due to insufficient sensitivity, but correlate well with patients' subjective impression of their vulnerability to increases in temperature.

Table 11.4 Medical evaluation for new onset fatigue

Complete blood count
Complete metabolic panel
Vitamin B12 level
Thyroid function tests
Urinalysis (subclinical infection)

Evaluation

The presence of fatigue is part of the comprehensive history and physical examination in patients suspected of or diagnosed with multiple sclerosis. It is helpful to ask about possible triggers such as heat, stress, and illnesses, as well as delve into the timing of onset of fatigue, current medications, and environments which appear to exacerbate or alleviate fatigue. The history should also include other symptoms that are known to contribute to fatigue, such as depression, gait impairment, physical deconditioning, nocturia, ongoing pain, heat intolerance, and more. Common medications that can induce fatigue in MS patients include anti-spasticity medications such as baclofen, tricyclic antidepressants, antihistamines, anticonvulsants, benzodiazepines, antihypertensives, and sedatives. Self-reported measures may be helpful in assessing the severity of fatigue symptoms as well as distinguishing mimics such as depression and sleep disorders [20]. Specific attention needs to be given to the diagnosis of depression due to the multiple overlapping features. Mood disorders in particular tend to be underdiagnosed in patients with multiple sclerosis.

At the time of the initial complaint, routine blood work should be performed [20] to also evaluate for other causes of fatigue, such as electrolyte imbalance, anemia, vitamin deficiency (B12), and thyroid disease. In addition, other autoimmune disorders tend to be more common in patients with multiple sclerosis than in the general population; for example, it is not uncommon to also find evidence of Hashimoto's hypothyroidism (Table 11.4).

Treatment

Treatment strategies for fatigue in MS have been varied, including both pharmacologic and nonpharmacologic interventions. These strategies have met with varied success, though none has been shown to be a primary intervention.

Complementary Therapies

Studies on exercise have demonstrated mild to moderate benefits [134]. A meta-analysis of its effects on the quality of life showed only mild improvements, the best seen with programs involving over 90 min per week of aerobic rather than resistance or isometric training [10]. Sustained benefits are typically seen in the quality of life and fatigue through exercise, though continued participation tends to be difficult for patients severely affected [11]. A second meta-analysis of exercise and MS fatigue limited to only randomized controlled trials also found similar modest benefits, but identified programs with alternative or resistance training as more effective than aerobic exercise alone [135].

Beyond exercise, other strategies such as energy conservation and heat avoidance have been partly useful in patients with MS. Improvements were seen in fatigue as measured by MFIS in patients taught energy conversation strategies compared to a wait-listed control group [12]. For patients with heat intolerance, cool water immersion and cap-and-vest cooling provide short-term improvements in fatigue, though ongoing daily therapy can provide longer-term benefits [136–138].

Other strategies have ranged from complementary therapies such as meditation and Tai Chi, to more mainstream strategies such as rehabilitation and biofeedback. A study of mindfulness training conducted over 8 weeks in a single-blind randomized trial showed statistically significant improvements in depression, anxiety, and fatigue lasting at least 8 months [139]. However, a study of Tai Chi performed with weekly 90-min sessions compared to control MS patients demonstrated only that fatigue did not progress over a 6-month time frame [140]. Vestibular rehabilitation improved fatigue after 6 weeks compared to exercise-stretching or wait-listed control groups [141]. Aquatic therapy combines the benefits of both exercise and cooling which can be particularly useful for MS patients with heat intolerance [142]. A study of 24 patients using biofeedback found significant decrements in fatigue and depression scores over 8 weeks, which remained stable over a 2-month period [143]. It is not clear that these two symptoms were independent however. A small perspective case series identified benefits in MS patients failing amantadine alone after 2 months of acupuncture sessions [144].

Targeting patient outlook can help individuals remain engaged in long-term exercise and retain enduring benefits in fatigue. Optimistic personalities appear more likely to experience beneficial outcomes [145]. Threats to self-identity such as impairment in activities of daily living, loss of employment and/or income, and loss of control may be the biggest features contributing to nonparticipation in exercise. Men and women often seek different strategies to maintain goals, men seeking self-paced activities and women tending more toward group activities. Success was seen in individuals able to readjust their tasks and goals on a continual basis.

Cognitive behavioral therapy (CBT) [9] improves fatigue among those with chronic fatigue syndrome and depression [146–149]. Both relaxation training and CBT helped fatigued MS patients more than healthy controls, bringing fatigue levels closer to those seen in healthy controls [9]. The beneficial effects were sustained up to a 6-month time frame post intervention with additional benefit seen for depression, anxiety, and stress.

Electromagnetic techniques have shown small but statistically significant improvements in fatigue among MS patients exposed to daily weak low-frequency magnetic pulses for 2 months via a portable watch-sized magnetic pulsing device. Unfortunately, these benefits were not seen with the use of shorter daily exposure

Table 11.5 Nonpharmacologic therapies evaluated for MS fatigue

Exercise – aerobic, resistance, isometric
Energy conservation
Heat avoidance
Cool water immersion
Aquatic therapy
Vestibular rehabilitation
Mindfulness training
Tai chi chuan
Yoga
Biofeedback
Acupuncture
Adaptive devices
Cognitive behavioral therapy
Transcranial magnetic stimulation
Direct transcranial electric stimulation

times [150–152]. Transcranial direct cortical stimulation (tDCS), another technique, typically consists of a constant 1–2 mA current applied to the scalp and delivered by sponge electrode with anodal application increasing cortical stimulation and cathodal application decreasing cortical stimulation. Direct low-intensity transcranial electrical stimulation in 15-min intervals daily for five consecutive days showed benefits for fatigue persisting up to 3 weeks posttreatment in a small placebo-controlled study [153]. A small study of 13 RRMS patients with chronic fatigue found improvements in fatigue lasting 3 weeks past the final 5-week session with anodal stimulation to the primary motor cortex [154]. On the hypothesis that the motor cortex is already hyperexcitable in MS patients, another study applied tDCS to the entire bilateral somatosensory strip alone and found a 28 % reduction in fatigue compared to an 8 % reduction in sham stimulation [155]. Patients were given 1.5 mA of stimulation for 15 min a day for 5 days straight during treatment sessions and assessed with MFIS outcomes. Employing a related strategy, repetitive transcranial magnetic stimulation (TMS) uses high-frequency stimulation on the primary motor cortex to provide analgesia, continuous trains of theta burst stimulation to reduce corticospinal excitability, and intermittent theta burst stimulation to increase corticospinal excitability. TMS has been applied to stroke and epilepsy as well as multiple sclerosis. Benefits in fatigue and depression are seen for TMS with motor cortex but not prefrontal cortex stimulation [156] (see Table 11.5).

Medication Therapies

Often prescription medications are tried for fatigue. While amantadine, modafinil, armodafinil, or aspirin are often prescribed, these therapies are not well supported in clinical trials. Amantadine was initially used for the treatment of influenza, but its dopaminergic and glutaminergic effects were also noted to have activating benefits [157]. While five separate randomized studies with different outcomes for fatigue have shown positive results [158–163], all studies had design limitations or analytic problems, which hampered interpretation according to a recent Cochrane review [158].

Modafinil remains one of the most commonly prescribed treatments for fatigue in patients with MS [2]. This medication has combined noradrenergic and dopaminergic properties but is not a classic sympathomimetic agent. Improvements in fatigue, dexterity, and focused attention were seen in a small randomized controlled trial of modafinil [164]. However, other trial results have been more mixed. One study with a crossover design showed improvement in the FSS, though this benefit could have been due to placebo [165]. Another study using a parallel group design found no difference between placebo- and the actively treated group except for more insomnia and gastrointestinal side effects for modafinil [166]. The study however used a dose titration schedule which led fewer patients to reach the optimal dose of 200 mg daily. Another retrospective case series collected over 5 years found better responses in fatigue scores in patients reporting excessive daytime sleepiness over 1 month, possibly denoting a subgroup with greater potential for benefit [167]. Contradictory results were seen in another recent assessment of 121 MS patients [168]. Overall the benefits for modafinil remain somewhat unclear when compared to placebo; however, it appears individual treatment responses occur and empiric trials are justified.

A small placebo-controlled study found no differences in FSS for fatigued MS patients tried on lisdexamfetamine, though improvements were seen for cognitive processing speed [169]. Studies for other stimulants such as methylphenidate or amphetamine are lacking. Nevertheless, this class of stimulants has shown benefit for fatigue due to other chronic illnesses [170]. One study noted a 9 % increase in processing speed in MS patients following a single 3 mg dose of rivastigmine, which may help individuals complaining of cognitive fatigue [171].

Aspirin has also been investigated for possible use in fatigue. A randomized controlled trial showed benefit, though some argue that these benefits were secondary to pain reduction [172]. One potential mechanism for aspirin may be altered hypothalamic output due to changes in autonomic and neuroendocrine responses. If MS fatigue is indeed cytokine induced, aspirin and other nonsteroidal anti-inflammatory medications may reduce proinflammatory cytokine activity such as those modulated by interferon.

Particular interest has been shown for the aminopyridines, which inhibit potassium channels of exposed axons and thereby improve neuronal conduction. An extended-release formulation of this compound, dalfampridine, is currently approved for use in gait impairment in patients with MS. Earlier examinations of 3,4-diaminopyridine and 4-aminopyridine suggested benefits for ambulation, fatigue, and vision, perhaps more so for the latter agent [173–176]. Studies done with 4-aminopyridine and dalfampridine show improvements in both short-term and long-term cognitive and physical fatigue, as well as for walking speed [177, 178]. These effects may be more evident in patients with higher serum levels of the medication.

Patients commonly seek nonprescription therapies as well. Two grams daily of acetyl L-carnitine given for 3 months compared to 200 mg daily amantadine demonstrated higher efficacy and tolerability for fatigue, though a subsequent study over 1 month only noted a trend toward improvement [179, 180]. A current Cochrane review reported the evidence as inconclusive and that new ongoing studies are eagerly anticipated [181]. A preliminary study of high-dose thiamine (vitamin B1) found improvements in fatigue despite normal thiamine serum levels [182]. This finding was reported for other disorders as well and likely represents a nonspecific benefit for fatigue. A small placebo-controlled study of ginseng also showed improvement in fatigue, though American ginseng extract has not been found to be beneficial [183]. Ginkgo biloba was also tested in a limited number of studies with suggestions for improvement in cognitive function and fatigue [184, 185].

One randomized double-blind placebo-controlled study of alfacalcidol, a synthetic vitamin D analogue, enrolling 158 patients found a 30 % decrease in FIS scores, with the biggest change seen in cognition [186]. As there were lower numbers of relapses in the treatment group, these benefits could be due to reduction in inflammation.

As previously mentioned, disease-modifying therapies themselves may help improve fatigue such as with glatiramer acetate. Another disease-modifying therapy, natalizumab, also appears to demonstrate improvements in fatigue. The ENER-G study showed benefits in patients with fatigue over the course of 1 year which remained stable for up to 48 weeks [187]. In the TYNERGY study, patients also had significant improvements in fatigue measured over the course of 12 months, as well as in quality of life, sleepiness, depression, and cognition [188]. Both studies were uncontrolled and these improvements in fatigue may only represent improvement in overall disease control. Interestingly fatigue was noted rather as a side effect in the original AFFIRM study with 27 % occurrence vs. 21 % placebo [189] (see Table 11.6).

Conclusion

Fatigue remains an important yet elusive symptom in the diagnosis and management of multiple sclerosis. Fatigue is not only intrinsic to the nature of MS but also affected by other MS problems, such as depression, sleep disturbance, or pain. It can respond to pharmacologic and nonpharmacologic intervention. Research on specific areas of the brain "network" points to several pathogenic mechanisms. Although some treatments have certainly been helpful for MS patients, more effective interventions are needed. Ongoing studies will be critical to better define the link between fatigue and MS to negate its deleterious effects.

Table 11.6 Pharmacologic therapies investigated for MS fatigue

Prescription	Doses examined
Amantadine	100 mg twice daily
Modafinil	200 mg twice daily
Armodafinil	250 mg daily
Rivastigmine	3 mg
3,4-Diaminopyridine	25–60 mg daily
4-Aminopyridine	0.5–1 mg/kg daily, 8 mg four times daily
Dalfampridine	10 mg twice daily
Disease-modifying therapies	
Glatiramer acetate	20 mg SQ daily
Natalizumab	300 mg IV monthly
Over the counter	
Aspirin	650 mg twice daily
L-Acetylcarnitine	2,000 mg daily
Thiamine (B1)	600–1,500 mg daily
Ginseng	250 mg twice daily
Ginkgo biloba	240 mg daily
Alfacalcidol	1 mcg daily

References

1. Minden SL, Frankel D, Hadden L, et al. The Sonya Slifka Longitudinal Multiple Sclerosis Study: methods and sample characteristics. Mult Scler. 2006;12(1):24–38.
2. Hadjimichael O, Vollmer T, Oleen-Burkey M. Fatigue characteristics in multiple sclerosis: the North American Research Committee on Multiple Sclerosis (NARCOMS) survey. Health Qual Life Outcomes. 2008;6:100.
3. Krupp LB, Alvarez LA, LaRocca NG, et al. Fatigue in multiple sclerosis. Arch Neurol. 1988;45(4):435–7.
4. Amato MP, Ponziani G, Rossi F, et al. Quality of life in multiple sclerosis: the impact of depression, fatigue and disability. Mult Scler. 2001;7(5):340–4.
5. Julian LJ, Vella L, Vollmer T, et al. Employment in multiple sclerosis. Exiting and re-entering the work force. J Neurol. 2008;255(9):1354–60.
6. Smith MM, Arnett PA. Factors related to employment status changes in individuals with multiple sclerosis. Mult Scler. 2005;11(5):602–9.
7. Schwartz CE, Coulthard-Morris L, Zeng Q. Psychosocial correlates of fatigue in multiple sclerosis. Arch Phys Med Rehabil. 1996;77(2):165–70.
8. Strober LB, Arnett PA. An examination of four models predicting fatigue in multiple sclerosis. Arch Clin Neuropsychol. 2005;20(5):631–46.
9. van Kessel K, Moss-Morris R, Willoughby E, et al. A randomized controlled trial of cognitive behavior therapy for multiple sclerosis fatigue. Psychosom Med. 2008;70(2):205–13.
10. Motl RW, McAuley E, Snook EM. Physical activity and multiple sclerosis: a meta-analysis. Mult Scler. 2005;11(4):459–63.
11. McCullagh R, Fitzgerald AP, Murphy RP, et al. Long-term benefits of exercising on quality of life and fatigue in multiple sclerosis patients with mild disability: a pilot study. Clin Rehabil. 2008;22(3):206–14.

12. Sauter C, Zebenholzer K, Hisakawa J, et al. A longitudinal study on effects of a six-week course for energy conservation for multiple sclerosis patients. Mult Scler. 2008;14(4):500–5.
13. Wessely S, Hotopf M, Sharpe D. Chronic fatigue and its syndromes. New York: Oxford University Press; 1998.
14. Guidelines MSCfCP, editor. Fatigue and multiple sclerosis: evidence-based management strategies for fatigue in multiple sclerosis. Washington, DC: Paralyzed Veterans Association of America; 1998.
15. Iriarte J, Katsamakis G, de Castro P. The Fatigue Descriptive Scale (FDS): a useful tool to evaluate fatigue in multiple sclerosis. Mult Scler. 1999;5(1):10–6.
16. Edgley K, Sullivan M, Dehoux E. A survey of multiple sclerosis: II. Determinants of employment status. Can J Rehabil. 1991;4:127–32.
17. Johansson S, Ytterberg C, Gottberg K, et al. Use of health services in people with multiple sclerosis with and without fatigue. Mult Scler. 2009;15(1):88–95.
18. Aronson KJ. Quality of life among persons with multiple sclerosis and their caregivers. Neurology. 1997;48(1):74–80.
19. Janardhan V, Bakshi R. Quality of life in patients with multiple sclerosis: the impact of fatigue and depression. J Neurol Sci. 2002;205(1):51–8.
20. Krupp LB. Fatigue in multiple sclerosis: a guide to diagnosis and management. New York: Demos Medical Publishing; 2004.
21. Chwastiak LA, Gibbons LE, Ehde DM, et al. Fatigue and psychiatric illness in a large community sample of persons with multiple sclerosis. J Psychosom Res. 2005;59(5):291–8.
22. Colosimo C, Millefiorini E, Grasso MG, et al. Fatigue in MS is associated with specific clinical features. Acta Neurol Scand. 1995;92(5):353–5.
23. Flachenecker P, Kumpfel T, Kallmann B, et al. Fatigue in multiple sclerosis: a comparison of different rating scales and correlation to clinical parameters. Mult Scler. 2002;8(6):523–6.
24. Lerdal A, Celius EG, Moum T. Fatigue and its association with sociodemographic variables among multiple sclerosis patients. Mult Scler. 2003;9(5):509–14.
25. Krupp LB, LaRocca NG, Muir-Nash J, et al. The fatigue severity scale. Application to patients with multiple sclerosis and systemic lupus erythematosus. Arch Neurol. 1989;46(10):1121–3.
26. Patrick E, Christodoulou C, Krupp LB. Longitudinal correlates of fatigue in multiple sclerosis. Mult Scler. 2009;15(2):258–61.
27. Bakshi R, Shaikh ZA, Miletich RS, et al. Fatigue in multiple sclerosis and its relationship to depression and neurologic disability. Mult Scler. 2000;6(3):181–5.
28. Freal JE, Kraft GH, Coryell JK. Symptomatic fatigue in multiple sclerosis. Arch Phys Med Rehabil. 1984;65(3):135–8.
29. Putzki N, Katsarava Z, Vago S, et al. Prevalence and severity of multiple-sclerosis-associated fatigue in treated and untreated patients. Eur Neurol. 2008;59(3–4):136–42.
30. Patti F, Amato MP, Trojano M, et al. Quality of life, depression and fatigue in mildly disabled patients with relapsing-remitting multiple sclerosis receiving subcutaneous interferon beta-1a: 3-year results from the COGIMUS (COGnitive Impairment in MUltiple Sclerosis) study. Mult Scler. 2011;17(8):991–1001.
31. Melanson M, Grossberndt A, Klowak M, et al. Fatigue and cognition in patients with relapsing multiple sclerosis treated with interferon beta. Int J Neurosci. 2010;120(10):631–40.
32. Jongen PJ, Lehnick D, Sanders E, et al. Health-related quality of life in relapsing remitting multiple sclerosis patients during treatment with glatiramer acetate: a prospective, observational, international, multi-centre study. Health Qual Life Outcomes. 2010;8:133.
33. Jongen PJ, Lehnick D, Koeman J, et al. Fatigue and health-related quality of life in relapsing-remitting multiple sclerosis after 2 years glatiramer acetate treatment are predicted by changes at 6 months: an observational multi-center study. J Neurol. 2014;261(8):1469–76.
34. Iaffaldano P, Viterbo RG, Paolicelli D, et al. Impact of natalizumab on cognitive performances and fatigue in relapsing multiple sclerosis: a prospective, open-label, two years observational study. PLoS One. 2012;7(4), e35843.

35. Putzki N, Yaldizli O, Tettenborn B, et al. Multiple sclerosis associated fatigue during natalizumab treatment. J Neurol Sci. 2009;285(1–2):109–13.
36. Yildiz M, Tettenborn B, Putzki N. Multiple sclerosis-associated fatigue during disease-modifying treatment with natalizumab, interferon-beta and glatiramer acetate. Eur Neurol. 2011;65(4):231–2.
37. Pokryszko-Dragan A, Bilinska M, Gruszka E, et al. Assessment of visual and auditory evoked potentials in multiple sclerosis patients with and without fatigue. Neurol Sci. 2015;36(2):235–42.
38. Runia TF, Jafari N, Siepman DA, et al. Fatigue at time of CIS is an independent predictor of a subsequent diagnosis of multiple sclerosis. J Neurol Neurosurg Psychiatry. 2015;86(5):543–6.
39. Feinstein A. The neuropsychiatry of multiple sclerosis. Can J Psychiatry. 2004;49(3):157–63.
40. Patten SB, Beck CA, Williams JV, et al. Major depression in multiple sclerosis: a population-based perspective. Neurology. 2003;61(11):1524–7.
41. Sadovnick AD, Remick RA, Allen J, et al. Depression and multiple sclerosis. Neurology. 1996;46(3):628–32.
42. Fisk JD, Pontefract A, Ritvo PG, et al. The impact of fatigue on patients with multiple sclerosis. Can J Neurol Sci. 1994;21(1):9–14.
43. Ford H, Trigwell P, Johnson M. The nature of fatigue in multiple sclerosis. J Psychosom Res. 1998;45(1 Spec No):33–8.
44. Kroencke DC, Lynch SG, Denney DR. Fatigue in multiple sclerosis: relationship to depression, disability, and disease pattern. Mult Scler. 2000;6(2):131–6.
45. Moller A, Wiedemann G, Rohde U, et al. Correlates of cognitive impairment and depressive mood disorder in multiple sclerosis. Acta Psychiatr Scand. 1994;89(2):117–21.
46. Vercoulen JH, Hommes OR, Swanink CM, et al. The measurement of fatigue in patients with multiple sclerosis. A multidimensional comparison with patients with chronic fatigue syndrome and healthy subjects. Arch Neurol. 1996;53(7):642–9.
47. Vercoulen JH, Swanink CM, Galama JM, et al. The persistence of fatigue in chronic fatigue syndrome and multiple sclerosis: development of a model. J Psychosom Res. 1998;45(6):507–17.
48. van der Werf SP, Evers A, Jongen PJ, et al. The role of helplessness as mediator between neurological disability, emotional instability, experienced fatigue and depression in patients with multiple sclerosis. Mult Scler. 2003;9(1):89–94.
49. Trojan DA, Arnold D, Collet JP, et al. Fatigue in multiple sclerosis: association with disease-related, behavioural and psychosocial factors. Mult Scler. 2007;13(8):985–95.
50. Jopson NM, Moss-Morris R. The role of illness severity and illness representations in adjusting to multiple sclerosis. J Psychosom Res. 2003;54(6):503–11. discussion 13–4.
51. Brown RF, Valpiani EM, Tennant CC, et al. Longitudinal assessment of anxiety, depression, and fatigue in people with multiple sclerosis. Psychol Psychother. 2009;82(Pt 1):41–56.
52. Johansson S, Ytterberg C, Hillert J, et al. A longitudinal study of variations in and predictors of fatigue in multiple sclerosis. J Neurol Neurosurg Psychiatry. 2008;79(4):454–7.
53. Mohr DC, Hart SL, Goldberg A. Effects of treatment for depression on fatigue in multiple sclerosis. Psychosom Med. 2003;65(4):542–7.
54. Feinstein A, O'Connor P, Gray T, et al. The effects of anxiety on psychiatric morbidity in patients with multiple sclerosis. Mult Scler. 1999;5(5):323–6.
55. Skerrett TN, Moss-Morris R. Fatigue and social impairment in multiple sclerosis: the role of patients' cognitive and behavioral responses to their symptoms. J Psychosom Res. 2006;61(5):587–93.
56. Iriarte J, Subira ML, Castro P. Modalities of fatigue in multiple sclerosis: correlation with clinical and biological factors. Mult Scler. 2000;6(2):124–30.
57. Fleming WE, Pollak CP. Sleep disorders in multiple sclerosis. Semin Neurol. 2005;25(1):64–8.

58. Tachibana N, Howard RS, Hirsch NP, et al. Sleep problems in multiple sclerosis. Eur Neurol. 1994;34(6):320–3.
59. Ferini-Strambi L, Filippi M, Martinelli V, et al. Nocturnal sleep study in multiple sclerosis: correlations with clinical and brain magnetic resonance imaging findings. J Neurol Sci. 1994;125(2):194–7.
60. Kaynak H, Altintas A, Kaynak D, et al. Fatigue and sleep disturbance in multiple sclerosis. Eur J Neurol. 2006;13(12):1333–9.
61. Attarian HP, Brown KM, Duntley SP, et al. The relationship of sleep disturbances and fatigue in multiple sclerosis. Arch Neurol. 2004;61(4):525–8.
62. Braley TJ, Segal BM, Chervin RD. Obstructive sleep apnea and fatigue in patients with multiple sclerosis. J Clin Sleep Med. 2014;10(2):155–62.
63. Brass SD, Duquette P, Proulx-Therrien J, et al. Sleep disorders in patients with multiple sclerosis. Sleep Med Rev. 2010;14(2):121–9.
64. Amarenco G, Kerdraon J, Denys P. Bladder and sphincter disorders in multiple sclerosis. Clinical, urodynamic and neurophysiological study of 225 cases. Rev Neurol. 1995;151(12):722–30.
65. Stanton BR, Barnes F, Silber E. Sleep and fatigue in multiple sclerosis. Mult Scler. 2006;12(4):481–6.
66. Caminero A, Bartolome M. Sleep disturbances in multiple sclerosis. J Neurol Sci. 2011;309(1–2):86–91.
67. Morrow SA, Weinstock-Guttman B, Munschauer FE, et al. Subjective fatigue is not associated with cognitive impairment in multiple sclerosis: cross-sectional and longitudinal analysis. Mult Scler. 2009;15(8):998–1005.
68. Bol Y, Duits AA, Hupperts RM, et al. The impact of fatigue on cognitive functioning in patients with multiple sclerosis. Clin Rehabil. 2010;24(9):854–62.
69. Karadayi H, Arisoy O, Altunrende B, et al. The relationship of cognitive impairment with neurological and psychiatric variables in multiple sclerosis patients. Int J Psychiatry Clin Pract. 2014;18(1):45–51.
70. Rotstein D, O'Connor P, Lee L, et al. Multiple sclerosis fatigue is associated with reduced psychomotor vigilance. Can J Neurol Sci. 2012;39(2):180–4.
71. Weinges-Evers N, Brandt AU, Bock M, et al. Correlation of self-assessed fatigue and alertness in multiple sclerosis. Mult Scler. 2010;16(9):1134–40.
72. Greim B, Benecke R, Zettl UK. Qualitative and quantitative assessment of fatigue in multiple sclerosis (MS). J Neurol. 2007;254 Suppl 2:Ii58–64.
73. Sharma KR, Kent-Braun J, Mynhier MA, et al. Evidence of an abnormal intramuscular component of fatigue in multiple sclerosis. Muscle Nerve. 1995;18(12):1403–11.
74. Kent-Braun JA, Sharma KR, Miller RG, et al. Postexercise phosphocreatine resynthesis is slowed in multiple sclerosis. Muscle Nerve. 1994;17(8):835–41.
75. Latash M, Kalugina E, Nicholas J, et al. Myogenic and central neurogenic factors in fatigue in multiple sclerosis. Mult Scler. 1996;1(4):236–41.
76. Leocani L, Colombo B, Magnani G, et al. Fatigue in multiple sclerosis is associated with abnormal cortical activation to voluntary movement – EEG evidence. Neuroimage. 2001;13(6 Pt 1):1186–92.
77. Liepert J, Mingers D, Heesen C, et al. Motor cortex excitability and fatigue in multiple sclerosis: a transcranial magnetic stimulation study. Mult Scler. 2005;11(3):316–21.
78. Perretti A, Balbi P, Orefice G, et al. Post-exercise facilitation and depression of motor evoked potentials to transcranial magnetic stimulation: a study in multiple sclerosis. Clin Neurophysiol. 2004;115(9):2128–33.
79. Sheean GL, Murray NM, Rothwell JC, et al. An electrophysiological study of the mechanism of fatigue in multiple sclerosis. Brain. 1997;120(Pt 2):299–315.
80. Schwartz JE, Jandorf L, Krupp LB. The measurement of fatigue: a new instrument. J Psychosom Res. 1993;37(7):753–62.
81. Whitehead L. The measurement of fatigue in chronic illness: a systematic review of unidimensional and multidimensional fatigue measures. J Pain Symptom Manage. 2009;37(1):107–28.

82. Mills R, Young C, Nicholas R, et al. Rasch analysis of the Fatigue Severity Scale in multiple sclerosis. Mult Scler. 2009;15(1):81–7.
83. Paul RH, Beatty WW, Schneider R, et al. Cognitive and physical fatigue in multiple sclerosis: relations between self-report and objective performance. Appl Neuropsychol. 1998;5(3):143–8.
84. Chalder T, Berelowitz G, Pawlikowska T, et al. Development of a fatigue scale. J Psychosom Res. 1993;37(2):147–53.
85. Ware Jr JE, Sherbourne CD. The MOS 36-item short-form health survey (SF-36). I. Conceptual framework and item selection. Med Care. 1992;30(6):473–83.
86. Ware Jr JE. SF-36 health survey update. Spine (Phila Pa 1976). 2000;25(24):3130–9.
87. McNair DM, Lorr M, Droppleman LF. Profile of mood states manual. San Diego: Educational and Industrial Testing Service; 1971.
88. Gilson BS, Gilson JS, Bergner M, et al. The sickness impact profile. Development of an outcome measure of health care. Am J Public Health. 1975;65(12):1304–10.
89. Krupp LB, Christodoulou C. Fatigue in multiple sclerosis. Curr Neurol Neurosci Rep. 2001;1(3):294–8.
90. Stone AA, Shiffman S. Capturing momentary, self-report data: a proposal for reporting guidelines. Ann Behav Med. 2002;24(3):236–43.
91. Mills RJ, Young CA, Pallant JF, et al. Development of a patient reported outcome scale for fatigue in multiple sclerosis: the Neurological Fatigue Index (NFI-MS). Health Qual Life Outcomes. 2010;8:22.
92. Mills RJ, Calabresi M, Tennant A, et al. Perceived changes and minimum clinically important difference of the Neurological Fatigue Index for multiple sclerosis (NFI-MS). Mult Scler. 2013;19(4):502–5.
93. Reeve BB, Hays RD, Bjorner JB, et al. Psychometric evaluation and calibration of health-related quality of life item banks: plans for the Patient-Reported Outcomes Measurement Information System (PROMIS). Med Care. 2007;45(5 Suppl 1):S22–31.
94. Christodoulou C, Junghaenel DU, DeWalt DA, et al. Cognitive interviewing in the evaluation of fatigue items: results from the patient-reported outcomes measurement information system (PROMIS). Qual Life Res. 2008;17(10):1239–46.
95. Schwid SR, Thornton CA, Pandya S, et al. Quantitative assessment of motor fatigue and strength in MS. Neurology. 1999;53(4):743–50.
96. Djaldetti R, Ziv I, Achiron A, et al. Fatigue in multiple sclerosis compared with chronic fatigue syndrome: a quantitative assessment. Neurology. 1996;46(3):632–5.
97. Krupp LB, Pollina DA. Mechanisms and management of fatigue in progressive neurological disorders. Curr Opin Neurol. 1996;9(6):456–60.
98. Schubert M, Wohlfarth K, Rollnik JD, et al. Walking and fatigue in multiple sclerosis: the role of the corticospinal system. Muscle Nerve. 1998;21(8):1068–70.
99. Krupp LB, Elkins LE. Fatigue and declines in cognitive functioning in multiple sclerosis. Neurology. 2000;55(7):934–9.
100. Holtzer R, Foley F, D'Orio V, et al. Learning and cognitive fatigue trajectories in multiple sclerosis defined using a burst measurement design. Mult Scler. 2013;19(11):1518–25.
101. Kujala P, Portin R, Revonsuo A, et al. Attention related performance in two cognitively different subgroups of patients with multiple sclerosis. J Neurol Neurosurg Psychiatry. 1995;59(1):77–82.
102. Walker LA, Berard JA, Berrigan LI, et al. Detecting cognitive fatigue in multiple sclerosis: method matters. J Neurol Sci. 2012;316(1–2):86–92.
103. Bakshi R, Miletich RS, Henschel K, et al. Fatigue in multiple sclerosis: cross-sectional correlation with brain MRI findings in 71 patients. Neurology. 1999;53(5):1151–3.
104. van der Werf SP, Jongen PJ, Lycklama a Nijeholt GJ, et al. Fatigue in multiple sclerosis: interrelations between fatigue complaints, cerebral MRI abnormalities and neurological disability. J Neurol Sci. 1998;160(2):164–70.
105. Tedeschi G, Dinacci D, Lavorgna L, et al. Correlation between fatigue and brain atrophy and lesion load in multiple sclerosis patients independent of disability. J Neurol Sci. 2007;263(1–2):15–9.

106. Marrie RA, Fisher E, Miller DM, et al. Association of fatigue and brain atrophy in multiple sclerosis. J Neurol Sci. 2005;228(2):161–6.
107. Roelcke U, Kappos L, Lechner-Scott J, et al. Reduced glucose metabolism in the frontal cortex and basal ganglia of multiple sclerosis patients with fatigue: a 18 F-fluorodeoxyglucose positron emission tomography study. Neurology. 1997;48(6):1566–71.
108. Tartaglia MC, Narayanan S, Francis SJ, et al. The relationship between diffuse axonal damage and fatigue in multiple sclerosis. Arch Neurol. 2004;61(2):201–7.
109. Filippi M, Rocca MA, Colombo B, et al. Functional magnetic resonance imaging correlates of fatigue in multiple sclerosis. Neuroimage. 2002;15(3):559–67.
110. Sepulcre J, Masdeu JC, Goni J, et al. Fatigue in multiple sclerosis is associated with the disruption of frontal and parietal pathways. Mult Scler. 2009;15(3):337–44.
111. Genova HM, Rajagopalan V, Deluca J, et al. Examination of cognitive fatigue in multiple sclerosis using functional magnetic resonance imaging and diffusion tensor imaging. PLoS One. 2013;8(11), e78811.
112. Rocca MA, Parisi L, Pagani E, et al. Regional but not global brain damage contributes to fatigue in multiple sclerosis. Radiology. 2014;273(2):511–20.
113. Tellez N, Alonso J, Rio J, et al. The basal ganglia: a substrate for fatigue in multiple sclerosis. Neuroradiology. 2008;50(1):17–23.
114. Calabrese M, Rinaldi F, Grossi P, et al. Basal ganglia and frontal/parietal cortical atrophy is associated with fatigue in relapsing-remitting multiple sclerosis. Mult Scler. 2010;16(10):1220–8.
115. Finke C, Schlichting J, Papazoglou S, et al. Altered basal ganglia functional connectivity in multiple sclerosis patients with fatigue. Mult Scler. 2014. doi:10.1177/1352458514555784.
116. Yaldizli O, Penner IK, Frontzek K, et al. The relationship between total and regional corpus callosum atrophy, cognitive impairment and fatigue in multiple sclerosis patients. Mult Scler. 2014;20(3):356–64.
117. Rocca MA, Agosta F, Colombo B, et al. fMRI changes in relapsing-remitting multiple sclerosis patients complaining of fatigue after IFNbeta-1a injection. Hum Brain Mapp. 2007;28(5):373–82.
118. White AT, Lee JN, Light AR, et al. Brain activation in multiple sclerosis: a BOLD fMRI study of the effects of fatiguing hand exercise. Mult Scler. 2009;15(5):580–6.
119. Bower JE, Ganz PA, Aziz N, et al. Fatigue and proinflammatory cytokine activity in breast cancer survivors. Psychosom Med. 2002;64(4):604–11.
120. Kerr JR, Barah F, Mattey DL, et al. Circulating tumour necrosis factor-alpha and interferon-gamma are detectable during acute and convalescent parvovirus B19 infection and are associated with prolonged and chronic fatigue. J Gen Virol. 2001;82(Pt 12):3011–9.
121. Kurzrock R. The role of cytokines in cancer-related fatigue. Cancer. 2001;92(6 Suppl):1684–8.
122. Flachenecker P, Bihler I, Weber F, et al. Cytokine mRNA expression in patients with multiple sclerosis and fatigue. Mult Scler. 2004;10(2):165–9.
123. Giovannoni G, Thompson AJ, Miller DH, et al. Fatigue is not associated with raised inflammatory markers in multiple sclerosis. Neurology. 2001;57(4):676–81.
124. Heesen C, Nawrath L, Reich C, et al. Fatigue in multiple sclerosis: an example of cytokine mediated sickness behaviour? J Neurol Neurosurg Psychiatry. 2006;77(1):34–9.
125. Gottschalk M, Kumpfel T, Flachenecker P, et al. Fatigue and regulation of the hypothalamo-pituitary-adrenal axis in multiple sclerosis. Arch Neurol. 2005;62(2):277–80.
126. Heesen C, Gold SM, Raji A, et al. Cognitive impairment correlates with hypothalamo-pituitary-adrenal axis dysregulation in multiple sclerosis. Psychoneuroendocrinology. 2002;27(4):505–17.
127. Tellez N, Comabella M, Julia E, et al. Fatigue in progressive multiple sclerosis is associated with low levels of dehydroepiandrosterone. Mult Scler. 2006;12(4):487–94.
128. Merkelbach S, Dillmann U, Kolmel C, et al. Cardiovascular autonomic dysregulation and fatigue in multiple sclerosis. Mult Scler. 2001;7(5):320–6.

129. Flachenecker P, Rufer A, Bihler I, et al. Fatigue in MS is related to sympathetic vasomotor dysfunction. Neurology. 2003;61(6):851–3.
130. Egg R, Hogl B, Glatzl S, et al. Autonomic instability, as measured by pupillary unrest, is not associated with multiple sclerosis fatigue severity. Mult Scler. 2002;8(3):256–60.
131. Hanken K, Eling P, Hildebrandt H. Is there a cognitive signature for MS-related fatigue? Mult Scler. 2015;21(4):376–81.
132. Foglio K, Clini E, Facchetti D, et al. Respiratory muscle function and exercise capacity in multiple sclerosis. Eur Respir J. 1994;7(1):23–8.
133. Humm AM, Beer S, Kool J, et al. Quantification of Uhthoff's phenomenon in multiple sclerosis: a magnetic stimulation study. Clin Neurophysiol. 2004;115(11):2493–501.
134. Rietberg MB, Brooks D, Uitdehaag BM, et al. Exercise therapy for multiple sclerosis. Cochrane Database Syst Rev. 2005;(1):CD003980.
135. Pilutti LA, Greenlee TA, Motl RW, et al. Effects of exercise training on fatigue in multiple sclerosis: a meta-analysis. Psychosom Med. 2013;75(6):575–80.
136. Beenakker EA, Oparina TI, Hartgring A, et al. Cooling garment treatment in MS: clinical improvement and decrease in leukocyte NO production. Neurology. 2001;57(5):892–4.
137. Schwid SR, Petrie MD, Murray R, et al. A randomized controlled study of the acute and chronic effects of cooling therapy for MS. Neurology. 2003;60(12):1955–60.
138. White AT, Wilson TE, Davis SL, et al. Effect of precooling on physical performance in multiple sclerosis. Mult Scler. 2000;6(3):176–80.
139. Grossman P, Kappos L, Gensicke H, et al. MS quality of life, depression, and fatigue improve after mindfulness training: a randomized trial. Neurology. 2010;75(13):1141–9.
140. Burschka JM, Keune PM, Oy UH, et al. Mindfulness-based interventions in multiple sclerosis: beneficial effects of Tai Chi on balance, coordination, fatigue and depression. BMC Neurol. 2014;14:165.
141. Hebert JR, Corboy JR, Manago MM, et al. Effects of vestibular rehabilitation on multiple sclerosis-related fatigue and upright postural control: a randomized controlled trial. Phys Ther. 2011;91(8):1166–83.
142. Kooshiar H, Moshtagh M, Sardar MA, et al. Aquatic exercise effect on fatigue and quality of life of women with multiple sclerosis: a randomized controlled clinical trial. J Sports Med Phys Fitness. 2014 [Epub ahead of print].
143. Choobforoushzadeh A, Neshat-Doost HT, Molavi H, et al. Effect of neurofeedback training on depression and fatigue in patients with multiple sclerosis. Appl Psychophysiol Biofeedback. 2015;40(1):1–8.
144. Foroughipour M, Bahrami Taghanaki HR, Saeidi M, et al. Amantadine and the place of acupuncture in the treatment of fatigue in patients with multiple sclerosis: an observational study. Acupunct Med. 2013;31(1):27–30.
145. Smith CM, Fitzgerald HJ, Whitehead L. How fatigue influences exercise participation in men with multiple sclerosis. Qual Health Res. 2015;25(2):179–88.
146. Deale A, Chalder T, Marks I, et al. Cognitive behavior therapy for chronic fatigue syndrome: a randomized controlled trial. Am J Psychiatry. 1997;154(3):408–14.
147. Prins JB, Bleijenberg G, Bazelmans E, et al. Cognitive behaviour therapy for chronic fatigue syndrome: a multicentre randomised controlled trial. Lancet. 2001;357(9259):841–7.
148. Sharpe M, Hawton K, Simkin S, et al. Cognitive behaviour therapy for the chronic fatigue syndrome: a randomized controlled trial. BMJ. 1996;312(7022):22–6.
149. Shaw BF. Cognitive-behavior therapies for major depression: current status with an emphasis on prophylaxis. Psychiatr J Univ Ott. 1989;14(2):403–8. discussion 409–12.
150. Richards TL, Lappin MS, Acosta-Urquidi J, et al. Double-blind study of pulsing magnetic field effects on multiple sclerosis. J Altern Complement Med. 1997;3(1):21–9.
151. Lappin MS, Lawrie FW, Richards TL, et al. Effects of a pulsed electromagnetic therapy on multiple sclerosis fatigue and quality of life: a double-blind, placebo controlled trial. Altern Ther Health Med. 2003;9(4):38–48.

152. Mostert S, Kesselring J. Effect of pulsed magnetic field therapy on the level of fatigue in patients with multiple sclerosis – a randomized controlled trial. Mult Scler. 2005;11(3):302–5.
153. Ferrucci R, Vergari M, Cogiamanian F, et al. Transcranial direct current stimulation (tDCS) for fatigue in multiple sclerosis. NeuroRehabilitation. 2014;34(1):121–7.
154. Saiote C, Goldschmidt T, Timaus C, et al. Impact of transcranial direct current stimulation on fatigue in multiple sclerosis. Restor Neurol Neurosci. 2014;32(3):423–36.
155. Tecchio F, Cancelli A, Cottone C, et al. Multiple sclerosis fatigue relief by bilateral somatosensory cortex neuromodulation. J Neurol. 2014;261(8):1552–8.
156. Schippling S, Paul F. TMS may reduce depression, fatigue in MS patients. Abstract #165. 29th congress of the European Committee for Treatment and Research in Multiple Sclerosis (ECTRIMS), 4 Oct 2013, Copenhagen.
157. Hayden FG. Combination antiviral therapy for respiratory virus infections. Antiviral Res. 1996;29(1):45–8.
158. Pucci E, Branas P, D'Amico R, et al. Amantadine for fatigue in multiple sclerosis. Cochrane Database Syst Rev. 2007;(1):CD002818.
159. A randomized controlled trial of amantadine in fatigue associated with multiple sclerosis. The Canadian MS Research Group. Can J Neurol Sci. 1987;14(3):273–8.
160. Cohen RA, Fisher M. Amantadine treatment of fatigue associated with multiple sclerosis. Arch Neurol. 1989;46(6):676–80.
161. Geisler MW, Sliwinski M, Coyle PK, et al. The effects of amantadine and pemoline on cognitive functioning in multiple sclerosis. Arch Neurol. 1996;53(2):185–8.
162. Krupp LB, Coyle PK, Doscher C, et al. Fatigue therapy in multiple sclerosis: results of a double-blind, randomized, parallel trial of amantadine, pemoline, and placebo. Neurology. 1995;45(11):1956–61.
163. Rosenberg GA, Appenzeller O. Amantadine, fatigue, and multiple sclerosis. Arch Neurol. 1988;45(10):1104–6.
164. Lange R, Volkmer M, Heesen C, et al. Modafinil effects in multiple sclerosis patients with fatigue. J Neurol. 2009;256(4):645–50.
165. Rammohan KW, Rosenberg JH, Lynn DJ, et al. Efficacy and safety of modafinil (Provigil) for the treatment of fatigue in multiple sclerosis: a two centre phase 2 study. J Neurol Neurosurg Psychiatry. 2002;72(2):179–83.
166. Stankoff B, Waubant E, Confavreux C, et al. Modafinil for fatigue in MS: a randomized placebo-controlled double-blind study. Neurology. 2005;64(7):1139–43.
167. Littleton ET, Hobart JC, Palace J. Modafinil for multiple sclerosis fatigue: does it work? Clin Neurol Neurosurg. 2010;112(1):29–31.
168. Moller F, Poettgen J, Broemel F, et al. HAGIL (Hamburg Vigil Study): a randomized placebo-controlled double-blind study with modafinil for treatment of fatigue in patients with multiple sclerosis. Mult Scler. 2011;17(8):1002–9.
169. Morrow SA, Smerbeck A, Patrick K, et al. Lisdexamfetamine dimesylate improves processing speed and memory in cognitively impaired MS patients: a phase II study. J Neurol. 2013;260(2):489–97.
170. Peuckmann V, Elsner F, Krumm N, et al. Pharmacological treatments for fatigue associated with palliative care. Cochrane Database Syst Rev. 2010;(11):CD006788.
171. Huolman S, Hamalainen P, Vorobyev V, et al. The effects of rivastigmine on processing speed and brain activation in patients with multiple sclerosis and subjective cognitive fatigue. Mult Scler. 2011;17(11):1351–61.
172. Wingerchuk DM, Benarroch EE, O'Brien PC, et al. A randomized controlled crossover trial of aspirin for fatigue in multiple sclerosis. Neurology. 2005;64(7):1267–9.
173. Polman CH, Bertelsmann FW, de Waal R, et al. 4-Aminopyridine is superior to 3,4-diaminopyridine in the treatment of patients with multiple sclerosis. Arch Neurol. 1994;51(11):1136–9.

174. Polman CH, Bertelsmann FW, van Loenen AC, et al. 4-aminopyridine in the treatment of patients with multiple sclerosis. Long-term efficacy and safety. Arch Neurol. 1994;51(3):292–6.
175. Romani A, Bergamaschi R, Candeloro E, et al. Fatigue in multiple sclerosis: multidimensional assessment and response to symptomatic treatment. Mult Scler. 2004;10(4):462–8.
176. Sheean GL, Murray NM, Rothwell JC, et al. An open-labelled clinical and electrophysiological study of 3,4 diaminopyridine in the treatment of fatigue in multiple sclerosis. Brain. 1998;121(Pt 5):967–75.
177. Rossini PM, Pasqualetti P, Pozzilli C, et al. Fatigue in progressive multiple sclerosis: results of a randomized, double-blind, placebo-controlled, crossover trial of oral 4-aminopyridine. Mult Scler. 2001;7(6):354–8.
178. Ruck T, Bittner S, Simon OJ, et al. Long-term effects of dalfampridine in patients with multiple sclerosis. J Neurol Sci. 2013;337(1–2):18–24.
179. Ledinek AH, Sajko MC, Rot U. Evaluating the effects of amantadin, modafinil and acetyl-L-carnitine on fatigue in multiple sclerosis – result of a pilot randomized, blind study. Clin Neurol Neurosurg. 2013;115 Suppl 1:S86–9.
180. Tomassini V, Pozzilli C, Onesti E, et al. Comparison of the effects of acetyl L-carnitine and amantadine for the treatment of fatigue in multiple sclerosis: results of a pilot, randomised, double-blind, crossover trial. J Neurol Sci. 2004;218(1–2):103–8.
181. Tejani AM, Wasdell M, Spiwak R, et al. Carnitine for fatigue in multiple sclerosis. Cochrane Database Syst Rev.. 2012;(5):CD007280.
182. Costantini A, Nappo A, Pala MI, et al. High dose thiamine improves fatigue in multiple sclerosis. BMJ Case Rep. 2013;2013.
183. Etemadifar M, Sayahi F, Abtahi SH, et al. Ginseng in the treatment of fatigue in multiple sclerosis: a randomized, placebo-controlled, double-blind pilot study. Int J Neurosci. 2013;123(7):480–6.
184. Diamond BJ, Johnson SK, Kaufman M, et al. A randomized controlled pilot trial: the effects of EGb 761 on information processing and executive function in multiple sclerosis. Explore. 2013;9(2):106–7.
185. Johnson SK, Diamond BJ, Rausch S, et al. The effect of Ginkgo biloba on functional measures in multiple sclerosis: a pilot randomized controlled trial. Explore. 2006;2(1):19–24.
186. Achiron A, Givon U, Magalashvili D, et al. Effect of Alfacalcidol on multiple sclerosis-related fatigue: a randomized, double-blind placebo-controlled study. Mult Scler. 2015;21(6):767–75.
187. Wilken J, Kane RL, Sullivan CL, et al. Changes in fatigue and cognition in patients with relapsing forms of multiple sclerosis treated with natalizumab: the ENER-G study. Int J MS Care. 2013;15(3):120–8.
188. Svenningsson A, Falk E, Celius EG, et al. Natalizumab treatment reduces fatigue in multiple sclerosis. Results from the TYNERGY trial; a study in the real life setting. PLoS One. 2013;8(3), e58643.
189. Polman CH, O'Connor PW, Havrdova E, et al. A randomized, placebo-controlled trial of natalizumab for relapsing multiple sclerosis. N Engl J Med. 2006;354(9):899–910.

Chapter 12
Psychological and Behavioral Therapies in Multiple Sclerosis

Peter A. Arnett, Dede Ukueberuwa, and Margaret Cadden

Abstract Depression in multiple sclerosis (MS) is very common, with a lifetime prevalence of around 50 %, which is much higher than the 8 % reported for the general population. In this chapter, we explore some of the psychological and behavioral depression treatments that have been studied in MS. Our review shows clearly that depression in MS is treatable, via both psychotherapeutic and behavioral interventions. Randomized clinical trials (RCTs) on psychotherapy have generally revealed very large effect sizes, with the one benchmarking study (one that examines treatment as it actually occurs in clinical settings) in the literature showing a still large but somewhat smaller effect size than RCTs. Exercise appears to be a promising potential treatment of depression in MS and has been shown to be correlated with lower depression. Additionally, RCTs of exercise have revealed small to moderate effect sizes on reducing depression. One problem with psychological and behavioral treatments of depression in MS is the salience of disability and travel issues that make involvement in such intensive treatment more challenging. Thus, even though treatments are often effective, patients may avoid them in favor of other interventions. Telephone-based interventions have been shown to be very effective in treating depression in MS and could represent a promising approach that circumvents such obstacles. Even though current treatments are effective, future research should explore why nearly half of patients do not respond to available treatments. A focus on the possibility that co-morbid conditions (e.g., anxiety or personality disorders) could interfere with standard depression treatments, as well as possible mediators of treatment (e.g., fatigue, disability levels, sleep disturbance), may be promising avenues for future research.

P.A. Arnett, Ph.D. (✉)
Department of Psychology, College of the Liberal Arts, Penn State University,
352 Bruce V. Moore Bldg., University Park, PA 16802-3105, USA
e-mail: paa6@psu.edu

D. Ukueberuwa, M.S. • M. Cadden, B.S.
Department of Psychology, Penn State University,
372 Moore Building, University Park, PA 16802, USA
e-mail: dedemu@gmail.com; margaret.cadden@gmail.com

B. Brochet (ed.), *Neuropsychiatric Symptoms of Inflammatory Demyelinating Diseases*, Neuropsychiatric Symptoms of Neurological Disease,
DOI 10.1007/978-3-319-18464-7_12

Keywords Multiple sclerosis (MS) • Depression • Anxiety • Personality • Cognitive function • Mental health • Neuropsychological • Cognitive behavioral therapy (CBT) • Physical activity • Randomized clinical trials (RCTs)

Psychological and Behavioral Therapies in MS

Depression is extremely common in MS, with a lifetime prevalence rate of about 50 % [1–3], compared with only about 8 % in the general population. MS depression is treatable, but available interventions are effective in only about 50 % of patients [4–6]. Better treatments are needed for those patients who do not respond to treatment; still, treatments should be offered to all MS patients who present with clinically significant depression. In this chapter, we will first review some of the complexities in measuring depression in MS. Then we will review some of the extant psychological and behavioral therapies for depression in MS, including interventions that involve treating depression through increasing exercise.

Measuring Depression in MS

Before depression in MS can be treated, it must be accurately assessed. An important factor complicating the measurement of depression in MS is the overlap between neurovegetative symptoms of depression in MS and disease symptoms. For example, symptoms such as fatigue, sleep disturbance, and sexual dysfunction are all considered neurovegetative symptoms of depression. However, they are also prominent MS disease symptoms. When patients report such symptoms, how is it possible to determine whether they reflect depression or disease symptoms? One suggested method is through careful interviewing to try and identify the source of such symptoms. Given the time-consuming nature of such an approach, however, some investigators have suggested simply omitting neurovegetative symptoms from consideration to get an accurate measurement of depression in MS [3]. Another suggested approach is to consider neurovegetative depression symptoms as reflecting depression in MS only when they exceed what is typically reported in nondepressed MS patients [7].

At present, the most sensitive measures for screening depression in MS do not include neurovegetative symptoms and take only a few minutes to administer. The Beck Depression Inventory-Fast Screen (BDI-Fast Screen) [8] consists of only seven items and has been validated in at least two studies of MS patients [9, 10], with an optimal cutoff score (≥4) that mirrors what is recommended in the BDI-Fast Screen manual for medical patients more generally. The Hospital Anxiety and Depression Scale (HADS) [11] has been validated in at least one MS study, and the depression scale from this measure consists of only eight items; the optimal cutoff score for screening depression in MS is ≥8. The HADS is further appealing in that the other half of the scale measures anxiety, something that is highly co-morbid with depression in MS, and can complicate treatment if depression is identified and treatment attempted without proper identification of co-morbid anxiety.

Treating Depression in MS

Because of its high prevalence, impact on quality of life [12], effect on medication compliance [13], possible impact on cognition [3], and the fact that it is unlikely to remit without intervention [14], treatment of depression is essential in MS. What follows is a review of some of the psychological and behavioral therapies that have been employed in MS.

Psychological and Behavioral Approaches

Mohr and Goodkin [14] conducted a meta-analysis of the depression treatment literature in MS about 15 years ago and concluded that psychological and behavioral therapies were at least as effective as medication. In particular, they also found that approaches that focused on skill building (e.g., improving coping strategies, general cognitive behavioral therapy (CBT) approaches) were more effective than approaches primarily focusing on insight. Since their publication, a number of additional studies have been published supporting and refining their initial review. For example, one study showed that CBT was more effective than an approach involving supportive group therapy [5]. Further supporting the CBT approach, Mohr and colleagues found that even when cognitive behavioral therapy was administered via telephone, it was effective [15, 16]. These demonstrations were extremely important, in that they illustrated that interventions could be effective even when MS patients, because of their disability or other limitations, were unable to travel to a site where in-person therapy could occur.

Further supporting the CBT approach to treating depression in MS, Cooper and colleagues [17] found in a randomized clinical trial (RCT) that a computerized CBT approach was effective in reducing BDI-II scores in 24 MS patients. One limitation of this approach, however, was that patients displayed poor adherence to the treatment protocol. However, clearly this is a promising approach with great potential in need of further research to better understand obstacles to adherence.

Although RCTs are rigorous tests of the efficacy of treatment interventions for depression, they are limited by the fact that the approaches used within the trial and the types of patients selected for them may not be representative of what truly happens in clinical practice. Because of this, benchmarking studies that examine treatment as it actually occurs in clinical settings are conducted. With this in mind, Askey-Jones and colleagues [18] conducted what appears to be the first benchmarking study on depression treatment in MS. Their study examined anxiety as well. MS patients in this study were identified for treatment if their HADS score was greater than 8, generally consistent with Honarmind and Feinstein's [11] recommendation, as noted above. In contrast to typical RCT approaches, patients were not excluded if they had co-morbid conditions, such as anxiety, if they were using medication, or if they were in the midst of an MS relapse. However, the authors did exclude those with significant cognitive problems or other severe disability that would impact their ability to attend treatment sessions. A CBT approach was used on the 29 patients enrolled in the study, and treatment was provided by MS nurse specialists. In addition to being used for participant selection, the HADS was used as the primary outcome variable. Because this study used a different outcome variable (HADS) than most published RCTs (which most often used the BDI-II), the investigators compared effect sizes between their study and the existing RCT data. The study had a 32 % dropout rate, with most non-completers reporting the primary impediment to continuing the treatment as the distance to the clinic for treatment.

Results of the study revealed significant decreases in both HADS Depression and Anxiety scale scores. The effect sizes for both were large by conventional standards (1.02 for depression and 1.18 for anxiety), though generally lower than the larger effect sizes (1.54–3.42) reported in the five RCTs to which they compared their data. The authors also highlighted their higher dropout rate compared with the computerized CBT RCT, as well as telephone-based CBT studies. Thus, although Askey-Jones et al.'s [18] study shows that depression in MS can be treated effectively in routine clinical practice, it does raise important issues involving transportation to therapy sites that may provide an impediment to treatment for some MS patients. However, because most of the non-completers in Askey-Jones et al.'s study cited travel concerns to the treatment site as their primary reason for discontinuing treatment, one possible solution to this problem might be to conduct a benchmarking study that employs telephone-based CBT.

Sleep problems are much more common in MS than the general population and are often co-morbid with depression and anxiety [19]. Over 50 % of MS patients complain of sleep onset problems or early morning awakening, compared with only about 10–15 % of those in the general population. With this in mind, Baron and colleagues examined data from a telephone-based CBT study to evaluate whether changes in depression and anxiety with treatment were associated with improvement in sleep difficulties. Their large group of MS participants ($n = 127$) received either CBT or supportive emotion-focused therapy. About half of the patients met criteria for major depression at treatment onset, and this was reduced by almost half at the treatments' conclusion. In terms of the relationship of treatment with sleep problems, patients who continued to report insomnia posttreatment were nearly six times more likely to meet criteria for major depressive disorder and over three times more likely to have elevated anxiety scores. Thus, these investigators found that persisting depression and anxiety were highly associated with continued problems with insomnia. One caveat is that these authors found that about one-third of their sample continued to have significant problems with insomnia after treatment even when they did not meet major depressive disorder criteria or have elevated anxiety.

They reasoned that these individuals' problems with sleep may have been more related to core MS symptoms (e.g., restless legs, difficulty swallowing, upper airway weakness) than affective problems.

Behavioral Approaches Involving Physical Activity and Exercise

Exercise has emerged as an effective treatment for depression in the general population, with hundreds of published studies exploring this topic. The most recently published large-scale Cochrane review on this topic concluded that exercise has a moderate ameliorative effect on depressive symptoms at time of treatment. Albeit these gains appeared to lessen over time, they continued to exist even at treatment follow-up [20]. Although this relationship has been less explored in the MS population, the existing literature is promising. This section will review the evidence of exercise as an effective behavioral treatment for depression in MS as revealed through both cross-sectional work and randomized clinical trials. Additionally, potential mediators of this relationship will be discussed. The section will conclude with a brief discussion on the limitations of exercise as a treatment for depression specific to the MS population.

Cross-Sectional Studies

Several cross-sectional studies examining the relationship between exercise and depressive symptoms in MS exist. Although these studies differ in methods of measuring exercise and depressive symptoms, they generally all come to the conclusion that higher levels of exercise are associated with lower depressive symptoms in MS.

One of the earliest studies on this topic was conducted by Stroud and Minehan [21]. Using the International Physical Activity Questionnaire (IPAQ), the authors dichotomized MS participants in their study into exercisers and non-exercisers based on the criteria of them participating in two 30-min bouts of exercise per week. They found that regular exercisers reported having lower fatigue, lower depressive symptoms on the Beck Depression Inventory (BDI), and higher quality of life compared to non-exercisers. Ensari and colleagues [22] found that level of exercise, as measured by the Godin Leisure-Time Exercise Questionnaire (GLTEQ), was one of the several factors that contributed to depression course in MS. The authors identified two course types for depression in MS; the first course was marked by low initial levels of depression that maintained or lowered over time, and the second course was marked by high initial levels of depression that increased over time. Participating in physical activity was predictive of the first course type, or, the course type in which depressive symptoms were generally less severe. Jensen and colleagues [23] further illuminated exercise as a potential treatment for depression in MS by examining how age and level of exercise (i.e., moderate versus vigorous) moderated this relationship. Using items from the 2003 Center for Disease Control and Prevention Behavioral Risk Factor Surveillance system (BRFSS) Survey Questionnaire, the authors measured how many minutes per week individuals in their study engaged in moderate and vigorous activity. After controlling for age and disability (EDSS), a moderate level of activity significantly predicted depressive symptoms, with higher activity indicative of fewer symptoms. The vigorous exercise findings were more complicated, with time spent in vigorous exercise being associated with fewer depressive symptoms in middle-aged individuals (45–64 years old) but not in younger or older individuals.

Thus far, the studies discussed used surveys as their measure of exercise. However, similar evidence for exercise as a treatment for depression exists when using more objective exercise measurements. Suh et al. [24] measured physical activity using accelerometers in a group of individuals early on in their MS course (5-year disease duration or less); depression was measured with the Hospital Anxiety and Depression Scale (HADS). These investigators found that higher levels of physical activity were associated with significantly lower depression levels. Path analysis revealed that this relationship could be completely explained by disability level; specifically, the analysis suggested that physical activity led to reduced disability, which in turn resulted in lower depression. Although literature exists which supports the idea that physical activity can reduce disability [25], it is also possible that individuals with low levels of disability are more likely to be physically active as well as less depressed. The results of this study are provocative, but it is important to keep in mind that this study, as well as others discussed, has a correlational design; therefore, causal relationships cannot be clearly inferred.

Randomized Clinical Trials

Findings such as those reviewed above that show greater levels of physical activity are associated with fewer depression symptoms have resulted in efforts to research exercise as a treatment for depression. The goal of a randomized controlled trial (RCT), in which participants are randomly assigned to a treatment or control group, is an examination of an intervention for which causal inferences about outcomes can more clearly be made. Thus far, intervention studies of the therapeutic benefits of exercise for depression show mixed results [26]. In general they indicate a small positive effect of exercise interventions in reducing depression symptoms in MS [22].

Different types of exercise programs have been examined in RCTs. Exercise may consist of strength and resistance training, such as repetitive training of the arms and legs with elastic resistance bands. Aerobic exercise may include walking, running, biking, swimming, or even climbing indoors or outdoors. Practices such as yoga and tai chi, which are thought to improve muscle relaxation, posture, and balance, have also been studied for potential therapeutic effects on mood in MS [27]. In a typical research design, assessments of patients' physical and emotional functioning are collected at baseline before administration of the intervention. At the end of the exercise intervention period, typically 8–12 weeks, outcome measures are then administered to patients to assess for changes in depression symptoms. One RCT evaluated an exercise training program as a treatment for depression symptoms in addition to physical and cognitive functioning and fatigue in patients with progressive MS [28]. The study involved 8–10 weeks of arm strengthening, rowing, or cycling, with 2–3 sessions per week, and the specific training program was tailored to the fitness level of each participant at baseline. Nearly half of participants had moderate to severe clinical depression scores at baseline, and symptoms decreased significantly in the groups with arm or bicycle training compared to a waitlist control group.

Exercise programs can be set up remotely by telephone or through a website, sometimes known as *telerehabilitation*. Patients may then engage in the activities at home or with a local physical therapist, thus expanding the reach of the trial and reducing barriers to accessibility of the program due to other responsibilities or limitations of transportation that people with MS may experience [29]. Telephone-based counseling may consist of motivational interviewing, an evidence-based technique for counseling patients to change behaviors such as engaging in an exercise program [30]. Web-based programs use a website to describe exercise activities through video or text and to offer advice to patients [31]. While a physical therapist may offer additional in-person guidance for carrying out exercises, a home-based intervention program may aid in long-term adherence [32]. In one RCT, patients in a home-based exercise program engaged in aerobic endurance training and were also provided with elastic bands for resistance training [33].

In a meta-analysis that included all of the 13 randomized controlled trials of exercise as a treatment for depression in MS, Ensari and colleagues [22] found an overall reduction in depression symptoms as a result of an exercise routine. This reduction was seen across studies, regardless of the specific method of exercise that was prescribed or the demographic or clinical characteristics of the MS samples. However, the effect of the treatment was small – these studies indicate that the treatment would have a 59 % chance of successfully reducing depression symptoms. These studies also typically used patients on a waitlist as a control group, and thus we are not able to make direct comparisons between exercise and other types of interventions for depression. Overall, the results are promising but indicate that we need more information to conclude that exercise is a consistently effective treatment for depression in MS, and additional trials of the therapeutic benefits of exercise in MS should include depression as a measure of outcome.

Mediators and Confounds

While the primary goal of these research studies reviewed in this section has been to examine the relationship between exercise and depression, changes in depression symptoms could be an indirect outcome of other effects of exercise. We are beginning to understand some factors that influence how exercise interventions could lead to changes in depression. Among adults with major depressive disorder, greater frequency and longer duration of physical activity led to increased positive affect

[34, 35]. Kratz, Ehde, and Bombardier [36] examined affective components of depression in people with MS within a randomized controlled trial and found that increases in positive affect and not decreases in negative affect mediated the effects of a physical activity intervention on reduced symptoms of depression. In this study, one-third of patients showed at least 50 % reduction in symptoms. Statistical mediation models indicated that increased physical activity leads to increased positive affect, but physical activity was not significantly related to changes in negative affect. Assignment to an intervention group – telephone-based motivational interviewing versus a waitlist control – had a direct effect on changes in both positive and negative affects, regardless of changes in physical activity. These results indicate that a counseling intervention had a general benefit for patients' affect and that physical activity could specifically improve depression symptoms in patients with MS by increasing positive affect. The authors reasoned that physical activity becomes a rewarding behavior and is associated with increased positive affect. Additional research suggests that changes in physical disability and perceived stress levels during an exercise intervention could be mediators of changes in depression symptoms [37].

In addition to these possible therapeutic mechanisms, other factors may be further examined as potential confounds in understanding the relationship between exercise interventions and changes in depression. The symptoms of fatigue and depression overlap, and fatigue is a common problem for people with MS [38]. One study, which was not an RCT, found that change in fatigue was a mediator between exercise training and change in depressive symptoms [39]. Specifically, exercise resulted in reduced fatigue, which in turn predicted lower depression. Future trials should include measures of fatigue and examine pathways leading from physical activity to depression. An additional confounding factor in these studies could be the benefits of social interaction between researchers or physical therapists and patients in the treatment group. In the cross-sectional assessment by Suh et al. [37], physical activity and social support each had independent relationships to depression. This suggests that patients in the treatment group are likely to experience a reduction of depression compared to a waitlist control group, regardless of the specific features of the treatment, such as an exercise program. These results also suggest that increasing social support could be an additional target of future depression treatment studies in people with MS, in addition to separately examining the benefits of engaging in exercise.

Limitations

Although exercise has been found to be an effective treatment for depression in the general population, it is important to consider the limitations that having MS may pose to successfully exercising. First and foremost, individuals with MS may face barriers to exercising such as physical impairments that limit their ability to participate in certain activities (such as walking if ambulation is poor). Additionally, individuals with MS may face physical barriers such as a lack of transportation to appropriate exercise facilities or a lack of accessibility to these facilities (i.e., not available in their area or not handicap accessible). According to one recent study, individuals with MS reported fatigue, impairment, and a lack of time as the three biggest barriers to exercising [40]. This same study found that the more perceived barriers to exercise individuals reported, the higher their self-reported depressive symptoms and the lower their self-reported perceived health. Therefore, special attention should be given when designing exercise regimens for individuals with MS that account for their disability, as well as physical and emotional barriers they may face. Asano et al. [40] suggested creating exercise regimens that are short in duration in order to address perceived barriers such as fatigue and lack of time. Morrison and Stuifbergen [41] found results which indicated that increasing physical and social expectations of exercise outcomes may increase the likelihood of exercise engagement in MS as well.

Summary and Conclusions

Depression is very common in MS, with lifetime prevalence rates around 50 % compared with 8 % in the general population. Because of these high prevalence rates, there has been an intense focus on identifying effective treatments for depression in MS. The results of this review show that depression in MS is treatable, via both psychotherapeutic and behavioral interventions. The effect sizes for most psychotherapy RCTs are very large; the one benchmarking study in the literature also reported large effect sizes, albeit somewhat smaller than the RCTs. Studies involving exercise show that exercise is a promising potential treatment of depression in MS. Exercise has been shown to be correlated with lower depression in MS; furthermore, RCTs involving exercise have revealed small to moderate effects of exercise on depression, with at least one study showing that the effect may be through increasing positive affect rather than reducing negative affect per se.

One problem with treating depression in MS is that patients have far more impediments to treatment, especially surrounding getting to treatment sites, than typical non-MS individuals seeking treatment. Thus, although existing psychological and behavioral treatments are often effective, patients may not seek out treatment because they are unable to get to treatment. One way of circumventing this that has been explored in the literature is by using treatment delivery systems that occur via telephone or through web-based strategies. Telephone-based CBT appears to be very effective in treating depression in MS, so it could be a viable way of providing patients with treatment when travel/disability issues are paramount.

A concern raised by psychological and behavioral treatment studies conducted thus far is that, although effect sizes on reducing depression are often large, around half of patients treated do not respond to treatment. Given the costs of depression to the well-being of patients and their families, more research is necessary that attempts to understand why some patients do not respond to treatment. It may be that comorbid conditions (e.g., anxiety or personality disorders), when not accurately identified and then treated, interfere with standard depression treatments. It may also be that more attention needs to be paid to possible mediators of treatment (e.g., fatigue, disability levels, sleep disturbance) so that they can be addressed, as well as more direct depression symptoms.

References

1. Sadovnick AD, Remick RA, Allen J, Swartz E, Yee IML, Eisen K, Paty DW. Depression and multiple sclerosis. Neurology. 1996;46:628–32.
2. Chwastiak L, Ehde DM, Gibbons LE, Sullivan M, Bowen JD, Kraft GH. Depressive symptoms and severity of illness in multiple sclerosis: epidemiologic study of a large community sample. Am J Psychiatry. 2002;159:1862–8.
3. Arnett PA, Barwick FH, Beeney JE. Depression in multiple sclerosis: review and theoretical proposal. J Int Neuropsychol Soc. 2008;14:691–724.
4. Ehde DM, Kraft GH, Chwastiak L, Sullivan MD, Gibbons LE, Bombardier CH, Wadhwani R. Efficacy of paroxetine in treating major depressive disorder in persons with multiple sclerosis. Gen Hosp Psychiatry. 2008;30:40–8.
5. Mohr DC, Boudewyn AC, Goodkin DE, Bostrom A, Epstein L. Comparative outcomes for individual cognitive-behavior therapy, supportive-expressive group psychotherapy, and sertraline for the treatment of depression in multiple sclerosis. J Consult Clin Psychol. 2001;69(6):1–8.
6. Baron KG, Corden M, Jin L, Mohr DC. Impact of psychotherapy on insomnia symptoms in patients with depression and multiple sclerosis. J Behav Med. 2011;34:92–101.
7. Strober LB, Arnett PA. Assessment of depression in multiple sclerosis: development of a "trunk and branch" model. Clin Neuropsychol. 2010;24:1146–66.
8. Beck AT, Steer RA, Brown GK. BDI-FastScreen for medical patients manual. San Antonio: The Psychological Corporation; 2000.
9. Benedict RHB, Fishman I, McClellan MM, Bakshi R, Weinstock-Guttman B. Validity of the beck depression inventory – fast screen in multiple sclerosis. Mult Scler. 2003;9:393–6.
10. Strober LB, Arnett PA. Depression in multiple sclerosis: the utility of common self-report instruments and development of a disease-specific measure. J Clin Exp Neuropsychol (in revision).
11. Honarmand K, Feinstein A. Validation of the Hospital Anxiety and Depression Scale for use with multiple sclerosis patients. Mult Scler. 2009;15:1518–24.
12. Benedict RHB, Wahlig E, Bakshi R, Fishman I, Munschauer F, Zivadinov R, Weinstock-Guttman B. Predicting quality of life in multiple sclerosis: accounting for physical disability, fatigue, cognition, mood disorder, personality, and behavior change. J Neurol Sci. 2005;231:29–34.
13. Bruce JM, Hancock LM, Arnett PA, Lynch S. Treatment adherence in multiple sclerosis: association with emotional status, personality, and cognition. J Behav Med. 2010;33:219–27.
14. Mohr DC, Goodkin DE. Treatment of depression in multiple sclerosis: review and meta-analysis. Clin Psychol Sci Practice. 1999;6(1):1–9.
15. Mohr DC, Hart SL, Julian L, Catledge C, Honos-Webb L, Vella L, Tasch ET. Telephone-administered psychotherapy for depression. Arch Gen Psychiatry. 2005;62:1007–14.
16. Mohr DC, Likosky W, Bertagnolli A, Goodkin DE, Van Der Wende J, Dwyer P, Dick LP. Telephone-administered cognitive-behavioral therapy for the treatment of depressive symptoms in multiple sclerosis. J Consult Clin Psychol. 2000;68(2):356–61.
17. Cooper CL, Hind D, Parry GD, Isaac CL, Dimairo M, O'Cathain A, Sharrack B. Computerised cognitive behavioural therapy for the treatment of depression in people with multiple sclerosis: external pilot trial. Trials. 2011;12:259.

18. Askey-Jones S, David AS, Silber E, Shawd P, Chalder T. Cognitive behaviour therapy for common mental disorders in people with multiple sclerosis: a bench marking study. Behav Res Ther. 2013;51:648–55.
19. Stanton BR, Barnes F, Silber E. Sleep and fatigue in multiple sclerosis. Mult Scler. 2006;12:481–6.
20. Rimer J, Dwan K, Lawlor DA, Greig CA, McMurdo M, Morley W, Mead GE. Exercise for depression. Cochrane Database Syst Rev. 2012;7.
21. Stroud NM, Minahan CL. The impact of regular physical activity on fatigue, depression and quality of life in persons with multiple sclerosis. Health Qual Life Outcomes. 2009;7:68.
22. Ensari I, Motl RW, McAuley E, Mullen SP, Feinstein A. Patterns and predictors of naturally occurring change in depressive symptoms over a 30-month period in multiple sclerosis. Mult Scler. 2014;20:602–9.
23. Jensen MP, Molton IR, Gertz KJ, Bombardier CH, Rosenberg DE. Physical activity and depression in middle and older-aged adults with multiple sclerosis. Disabil Health J. 2012;5:269–76.
24. Suh Y, Motl RW, Mohr DC. Physical activity, disability, and mood in the early stage of multiple sclerosis. Disabil Health J. 2010;3(2):93–8.
25. Snook EM, Motl RW. Effect of exercise training on walking mobility in multiple sclerosis: a meta-analysis. Neurorehabil Neural Repair. 2008;23(2):108–116.
26. Feinstein A, Rector N, Motl R. Exercising away the blues: can it help multiple sclerosis-related depression? Mult Scler. 2013;19:1815–9.
27. Ahmadi A, Arastoo AA, Nikbakht M, Zahednejad S, Rajabpour M. Comparison of the effects of 8 weeks aerobic and yoga training on ambulatory function, fatigue and mood status in MS patients. Iran Red Crescent Med J. 2013;15:449–54.
28. Briken S, Gold SM, Patra S, Vettorazzi E, Harbs D, Tallner A, Heesen C. Effects of exercise on fitness and cognition in progressive MS: a randomized controlled pilot trial. Mult Scler. 2014;20:382–90.
29. Bombardier CH, Ehde DM, Gibbons LE, Wadhwani R, Sullivan MD, Rosenberg DE. Telephone-based physical activity counseling for major depression in people with multiple sclerosis. J Consult Clin Psychol. 2013;81:89–99.
30. Burke BL, Arkowitz H, Menchola M. The efficacy of motivational interview: a meta-analysis of controlled clinical trials. J Consult Clin Psychol. 2003;71:843–61.
31. Paul L, Coulter EH, Miller L, McFadyen A, Dorfman J, Mattison PGG. Web-based physiotherapy for people moderately affected with multiple sclerosis; quantitative and qualitative data from a randomized controlled pilot study. Clin Rehabil. 2014;28:924–35.
32. Ashworth NL, Chad KE, Harrison EL, Reeder BA, Marshall SC. Home versus center based physical activity programs in older adults. Cochrane Database Syst Rev. 2005;25.
33. Romberg A, Virtanen A, Ruutiainen J. Long-term exercise improves functional impairment but not quality of life in multiple sclerosis. J Neurol. 2005;252:839–45.
34. Watson D. Intraindividual and interindividual analyses of positive and negative affect: their relation to health complaints, perceived stress, and daily activities. J Pers Soc Psychol. 1988;54:1020–30.
35. Kelsey KS, DeVellis BM, Begum M, Belton L, Hooten EG, Campbell MK. Positive affect, exercise, and self-reported health in blue collar women. Am J Health Behav. 2006;30:199–207.
36. Kratz AL, Ehde DM, Bombardier CH. Affective mediators of a physical activity intervention for depression in multiple sclerosis. Rehabil Psychol. 2014;59:57.
37. Suh Y, Motl RW, Mohr DC. Physical activity, social support, and depression: possible independent and indirect associations in persons with multiple sclerosis. Psychol Health Med. 2012;17(2):196–206.
38. MacAllister WS, Krupp LS. Multiple sclerosis-related fatigue. Phys Med Rehabil Clin N Am. 2005;16:483–502.

39. Roppolo M, Mulasso A, Gollin M, Bertolotto A, Ciairano S. The role of fatigue in the associations between exercise and psychological health in multiple sclerosis: direct and indirect effects. Ment Health Phys Act. 2013;6:87–94.
40. Asano M, Duquette P, Andersen R, Lapierre Y, Mayo NE. Exercise barriers and preferences among women and men with multiple sclerosis. Disabil Rehabil. 2013;35(5):353–61.
41. Morrison JD, Stuifbergen AK. Outcome expectations and physical activity in persons with longstanding multiple sclerosis. J Neurosci Nurs. 2014;46(3):171–9.

Part III
Cognitive Perspective

Chapter 13
Introduction to Social Cognition

Cécile Dulau

Abstract Social cognition (SC) encompasses the cognitive processes that underlie human relationships. Understanding others' belief, thoughts, intentions, and emotions allows adapting its own behavior. It has been separated in various domains such as theory of mind (ToM), empathy, emotion processing, social knowledge, and social perception. There is overlapping among those terms. Neural basis concerns essentially the prefrontal cortex, amygdala, temporal poles, and temporoparietal junction. Many ways to assess SC, essentially ToM tasks and facial emotion recognition tasks, have been studied in neurological and psychiatric disorders. The relationship between neurocognition and SC remains discussed, but many arguments suggest a link, especially with executive functions and episodic memory for ToM and attention for facial emotion recognition. Neuropsychological and pharmacological management of SC, mostly studied in traumatic brain injury or in schizophrenia, is at the very beginning.

Keywords Social cognition • Theory of mind • Mentalizing • Empathy • Emotion processing • Emotion • Prefrontal cortex • Amygdala

Abbreviations

FFA	Fusiform face area
fSTS	Facial part of the superior temporal sulcus
MS	Multiple sclerosis
SC	Social cognition
ToM	Theory of mind

C. Dulau, M.D. (✉)
Centre Hospitalier Universitaire Pellegrin, Bordeaux, France

Department of Neurology, Université de Bordeaux, Bordeaux 33076, France
e-mail: cecile.dulau@chu-bordeaux.fr

B. Brochet (ed.), *Neuropsychiatric Symptoms of Inflammatory Demyelinating Diseases*, Neuropsychiatric Symptoms of Neurological Disease,
DOI 10.1007/978-3-319-18464-7_13

Definition

Social cognition (SC) is described as the cognitive processes underlying personal relationships [1]. It is the capacity to recognize and interpret interpersonal cues that guide social behavior [2]. In some models, in particular in traumatic brain injury [3], SC encompasses "hot" processes, that is, emotional perception and empathy, and "cold" processes, which are the ability to infer the beliefs, feelings, and intentions of others (theory of mind, ToM) to understand their point of view (cognitive empathy) and what they mean when they communicate (pragmatic inference) [2]. A consensus-building meeting on SC in schizophrenia was held at the National Institute of Mental Health in March 2006. They reviewed recent publication research about SC in schizophrenia. SC would fit into the following five areas [4]: ToM, social perception, social knowledge, attribution bias, and emotional processing. ToM and emotional processing will be further developed. Tests about social perception assess one's ability to identify social roles, societal rules, and social context [4]. Social knowledge refers to awareness of the roles, rules, and goals that characterize social situations and guide social interactions. Unlike mental state attribution, attribution bias reflects how people typically infer the causes of particular positive and negative events. There is considerable overlap between the terms. For example, identifying emotions is clearly a component of emotional processing but is sometimes considered to be an aspect of ToM. Likewise, social knowledge overlaps with social perception [4].

Emotional Processing

Emotional processing refers broadly to perceiving and using emotions [4]. One influential model of emotional processing defines emotional intelligence as a set of four components, including identifying emotions, facilitating emotions, understanding emotions, and managing emotions [4].

Facial Emotion Recognition

Faces convey a wealth of social signals [5]. Recognizing emotions from facial expressions is essential for perceiving the intentions and dispositions of others. This can be considered a key skill for the understanding of relevant social information in everyday life [6]. Emotional facial recognition tends to change the observer's behavior (e.g., withdrawal when faced with an angry expression) [7]. As a consequence, adequate social interaction requires accurate recognition of the emotional facial expression of other individuals [8].

Primary emotions, as happiness, surprise, fear, anger, disgust, and sadness, need a primary emotional treatment, and complex emotions (like "preoccupied," "joker," etc.) involve mental state attribution [9]. Face perception is a complex task that involves the concerted action of different functional components [5]. It requires neural systems that connect such perception to motivation, emotion, and adaptive behavior [10].

Functional imaging studies allow a better understanding of neural basis of emotional facial recognition. The key brain structures that participate in the recognition of basic emotions are the occipitotemporal cortices, amygdala, prefrontal cortex, basal ganglia, and right parietal cortices [11]. After activating early visual areas, facial stimuli are selectively processed in a region of the fusiform gyrus, in the fusiform face area (FFA), and in the facial part of the superior temporal sulcus (fSTS) [5]. Generally, distinction is made between the FFA, which treats facial identity, and the fSTS which codes the changeable aspects of the face, like lip speech, gaze fixations, and emotional facial expressions [12]. The ventrolateral prefrontal cortex has been described to be important for the processing of facial expression tasks, and the anterior cingulate cortex generates the appropriate "motivational state" [13]. The amygdala plays an essential role in processing the emotional part of facial expressions. It is involved in extracting emotional stimuli from emotional external cues [14]. Associated with the amygdala, the orbitofrontal cortex is involved in the regulation of control to the social stimuli (via connections to motor structures, hypothalamus, and brainstem nuclei) [8]. The anterior insula is involved in creating a somatosensory feeling of the emotion, like the mirror-neuron system applied to emotions [15]. In addition to the insula, there is good evidence that recognition of facial emotions (like disgust) requires the integrity of the basal ganglia and parietal cortex (integrity of the somatosensory-related system). All these structures are interconnected and engaged in multiple processes [8].

Affective Prosody Recognition

Emotion-specific impairments are often not only restricted to the recognition of facial emotions; vocal expressions are generally also affected [5]. Affective prosody refers to the communication of emotion by variation in tone of voice and other acoustic parameters. Much of the research conducted to date has examined the relative contributions of the left and right hemispheres to the expression and comprehension of affective prosody [16]. Deficits in affective prosody recognition may make misunderstandings and poor communication more likely [17].

Emotional Awareness and Alexithymia

Lane and Schwartz [18] have divided emotional awareness into five separate levels of increasing emotional awareness, from no emotional awareness to blended emotional awareness. This model was created to provide an organizing framework for understanding individual differences in the experience and expression of emotion. It can be assessed by the Levels of Emotional Awareness Scale (LEAS) that exposes interpersonal situations and elicits descriptions of the emotional responses of self and others [19].

Alexithymia will be developed in Chap. 14.

Theory of Mind and Empathy

Premack and Woodruff have firstly described ToM in 1978 [20]. ToM is acquired in normal development around the age of 5 [21]. ToM refers both to cognitive processing "cognitive ToM" and to affective processing "affective ToM" [22]. "Cognitive ToM" refers to the ability of making inferences about mental states (intentions, dispositions, and beliefs) of other people [23]. "Affective ToM" is the ability to make affective inferences about what another person is thinking or feeling and requires empathy [22]. ToM is also called mental state attribution, mentalizing [4], or mind reading. Mind reading allows a person to explain and predict people's behavior on the basis of their mind and to recognize that another person's knowledge is different from our own [21]. Mind reading is possible by decoding nonverbal cues such as facial expression, eye gaze and body gestures, and complex abstract reasoning about verbal information enable mind reading [24]. As a metacognitive processing, ToM regroups mental representations that fit into each other, like nesting dolls [25]. Two levels of mentalization are described. The first order consists of mental representation of another's perspective. The second order consists of mental representation of another's mental representation of another's perspective [26] (see Table 13.1: first- and second-order false belief tasks).

Empathy is strongly related to ToM skills, because it describes the ability to have insight into emotional stages and feelings of others [39]. The basic level of empathy is characterized by emotion recognition, understanding the other person's feeling. The second level of empathy implies emotional state reasoning, which allows the empathizer to understand the other person's feeling and to predict the consequences of those feelings. That requires more cognitive effort than the basic level of empathy. This second level of empathy can be associated with "affective ToM" because it implies emotional state reasoning which enables the empathizers to understand others' feelings and predict their consequences [22].

Neural Basis of Theory of Mind and Empathy

What About Mirror Neurons?

Our ability to perceive the goals and intentions of others from nonverbal cues is often ascribed to mirror neurons. These neurons become active when animals observe an action as well as when they execute the same action [15]. In humans, a more extensive mirror system has been identified in the premotor cortex (Brodmann

Table 13.1 Examples of social cognition assessment from tests used in social cognition in MS studies [9, 24, 27–37]

Social cognition domain	Test	Brief description
Facial emotion recognition	Benton Facial Recognition Test (BFRT) [28]	A measure of facial discrimination of non-emotional stimuli
	Ekman and Friesen (1976) [29]	110 faces representing standardized poses of fundamental human emotions
	Face test Baron-Cohen [30]	20 faces of the same actress who portrayed primary and complex mental states
	Florida Affect Battery (FAB) [31]	Faces of women with primary emotions (happiness, sadness, anger, fear, or neutral), non-emotional facial identity task, facial emotion naming, facial emotion selection task, facial emotion matching
Affective prosody	Aprosodia battery [32]	Prosodic comprehension and prosodic discrimination of 12 randomized sentences with different intonations
Cognitive ToM	Reading the Mind in the Eyes Test [9]	Recognition of emotional state in 38 photos of sets of eyes with a forced choice between four descriptions. ToM task because it makes reference to global mental states ("joking," "preoccupied")
	First- and second-order false belief tasks [33]	Understanding the perspective of another The first-order false belief task tests subject's ability to infer that someone can have a mistaken belief that is different from their own true belief The second-order false belief tasks tests the ability to understand what someone else thinks about what another person thinks
	Happé's strange stories [34]	Strange stories-mental require an appreciation of a character's mental states (desires, thoughts) Strange stories-physical involving making a physical inference
Affective ToM/empathy	Empathy Quotient [35]	Assessing both affective ToM and emotional reactivity (self-administered questionnaire)
	Hogan Empathy Scale (HES) [36]	Assessing interpersonal empathy (self-administered questionnaire)
Both cognitive and affective ToM	Movie for the Assessment of Social Cognition (MASC) [37]	Short film with several questions referring to the actors' mental states
	Faux Pas Test [24]	Detecting social faux pas. Cognitive ToM because of decoding something said socially inappropriate and affective ToM because of decoding something said that might hurt

(continued)

Table 13.1 (continued)

Social cognition domain	Test	Brief description
Decision-making	Gambling task [27]	Simulating real-life decisions in terms of uncertainty, reward, and punishment. In this task, subjects have to make a long series of decisions, picking cards from four decks that are either advantageous or disadvantageous. Over a hundred card selections, healthy subjects typically learn to avoid disadvantageous decks. This behavioral process requires not only an adequate cognitive analysis of the situation but also an effective emotional regulation, because the subject must be able to avoid immediate reward to ensure a final favorable issue [38]

area 44) and parietal regions (anterior intraparietal sulcus) [23]. Some studies suggest a brain's mirror system also for emotions [15]. The brain's mirror system would be a first step for mentalizing [39].

Brain Regions Implicated in Theory of Mind and Empathy

Neural basis of ToM abilities has been widely investigated using advanced neuroimaging methods in healthy subjects and in clinical conditions showing social cognitive impairments, such as autism spectrum disorder and schizophrenia [40, 41].

Cumulative evidence from morphological and functional MRI studies suggests that neuronal processes of ToM and empathy recruit a widespread cerebral network between the ventromedial prefrontal cortex, frontotemporal cortex, and temporoparietal junction [39]. The medial prefrontal cortex has been hypothesized to play a critical role in cognitive ToM or mentalizing [1, 42].

Affective ToM or empathic responding requires the additional recruitment of the networks involved in emotional processing, such as the anterior cingulated cortex, limbic areas, and somatosensory and anterior insular cortices [43]. The right orbitofrontal cortex and inferior frontal cortex are hypothesized to be selectively important for emotional contagion [42].

Finally, ToM is dependent on the integrity of the dopaminergic and serotoninergic systems which are primarily engaged in the maintenance and application of represented mental states [44].

Decision-Making

Some consider that decision-making is a part of SC because altered decision-making is a known cause of functional impairment in the social environment. It is one of the constant challenges in daily life [45]. According to Damasio's somatic marker hypothesis [46], decision-making is primarily dependent on emotional reactivity induced by a particular event in the environment [38]. Somatic markers may play a role of incitement or constraining to decision-making. They are acquired during education in childhood. Prefrontal structures could be a link between decision-making and emotional statements. Decision-making capacity can be measured using the gambling tasks which simulate real-life decisions in terms of uncertainty, reward, and punishment [27] (see Table 13.1).

Social Cognition Impairment Repercussion

SC impairment involves repercussions in daily life functioning, interpersonal functioning with family and friends, and vocational achievement. Social cognitive abilities enable subjects to interact effectively with their social environment. Deficits in SC could lead to social misperceptions, resulting in inappropriate interpersonal reactions or social withdrawal [4].

Social Cognition in Psychiatric Illness as a Trait Component

How to understand SC? By understanding its primary impairment in several psychiatric illness [47], especially autism spectrum disorder, schizophrenia, or social communication disorder. In these pathologies, SC is the core problem [48]. DSM-IV [49] defines pervasive developmental disorders (PDDs) as an impaired development of reciprocal social interaction associated with impairments in either verbal or nonverbal communication skills. PDD includes Asperger's syndrome. DSM-V proposed to delete the term PDD and recommended a new diagnostic category: autism spectrum disorder. Social impairments in DSM-V for autism spectrum disorder must meet the criteria "persistent deficits in social communication and social interaction across contexts (…) manifesting by all 3 of the following: deficits in social-emotional reciprocity, deficits in nonverbal communicative behaviors used for social interaction and deficits in developing and maintaining relationships." DSM-V includes a new category: social communication disorder. It is defined by persistent difficulties in the social use of verbal and nonverbal communication. These deficits must result in functional limitations in effective communication, social participation, social relationships, academic achievement, or occupational performance [50].

In schizophrenia domain, SC associated with neurocognition has a potential role as putative endophenotype marker [51, 52] and predictor of functional outcome [53, 54].

Depression is associated with deficits in emotional expression recognition [55, 56] which may lead to inappropriate reaction to others' emotions thereby interfering with successful social interaction.

Social Cognition in Neurological Disorders

In neurological disorders, SC has been mostly investigated in traumatic brain injury [3]. Executive functions and SC impairments are common consequences of traumatic brain injury that are linked with poor functional outcome [57]. The notable case is Phineas Gage described by Harlow [58], a railway worker who survived an anterior skull-penetrating injury resulting in profound behavioral changes. Recent research demonstrated that he did not have a focal brain injury but a widespread damage in white matter pathways, showing the importance of fiber tract integrity in SC impairment [59].

SC seems to be impaired in both focal and diffuse brain lesions [42]. With the knowledge of the neural networks that underlie SC, Ibanez and Manes proposed the social context network model (SCNM), a fronto-insular-temporal network responsible for processing social contextual effects [60]. Assuming this, SC would be impaired in diseases in which the "SCNM" is disrupted. Indeed, the ability to recognize emotional facial expressions, ToM, and empathy can be impaired in several neurological disorders. SC might be impaired in traumatic brain injury [3], acquired brain lesions involving the VLPFC or amygdala [42, 61, 62], Parkinson's disease [63] Huntington's disease [64], behavioral variant of the frontotemporal dementia [60, 65] Alzheimer's disease [65], and chronic temporal epilepsy [66, 67].

SC and multiple sclerosis will be detailed in Chap. 15.

Distinctiveness of Social Cognition and Neurocognitive Factors

Distinctiveness of SC and neurocognitive factors remains a controversial subject. In a recent review of 20 studies of SC in schizophrenia [51], SC appears to be a separate cognitive domain, but there is a lack of studies of factor structure in large samples of patients. SC might be independent or dissociable from general intelligence [68]. In autism spectrum disorder such as Asperger's syndrome, intelligence efficiency appears to be normal, but ToM is impaired [9]. On the contrary, in Williams syndrome, where social abilities are preserved, general intelligence is impaired [11].

SC guides both automatic and volitional behaviors, being composed of a variety of cognitive, emotional, and motivational processes that modulate behavioral responses. Memory, decision-making, attention, motivation, and emotion are all prominently recruited when socially relevant stimuli elicit behavior [69].

SC supposes perception and other cognitive functions as language (especially access to the verbal lexicon), visuospatial exploration, memory, and executive function [70]. SC often involves a deep level of abstraction, inference, and counterfactual thinking recruiting executive functions [47]. Assuming this, overlapping between SC network and cognitive processing suggests a cognitive treatment of SC

[1], especially those involved in the cognitive elaboration of stimuli, such as the temporoparietal junctions and the medial prefrontal cortex [71]. Autobiographical memory and mentalizing processing share a large part of their neural basis [1]. Attentional control may be necessary to distinguish specific features of different facial expressions [72]. It has been suggested that appreciating another's mental state often requires inhibition of one's own perspective [73]. Other studies tend to demonstrate the role of executive functions in ToM, especially in a context of a dual task [25, 73]. A review about ToM and executive functions [74] suggests that both appear tightly associated, but no elementary executive process could be specifically associated with ToM performances. Moreover, developmental studies suggest a functional dependency between the development of executive functions and social interaction [75]. In a similar way, ToM acquirement might be associated with episodic memory development [76].

How to Assess Social Cognition

SC is a multidimensional construct that can be assessed by various methodologies. The available tools evaluate mainly ToM and facial emotion recognition and were first studied in autism spectrum disorder [77] or schizophrenia studies [52]. A French team attempted to develop a protocol for SC assessment in adults encompassing the two aspects of SC, "cognitive" and "affective" SC. It regroups tests assessing recognition of facial emotions, ToM, emotional fluency, emotional awareness, and alexithymia [78]. Examples of non-exhaustive SC domains with their tests available in the English language and their description are presented in Table 13.1. For greater clarity and simplification, only SC tests used in the domain of multiple sclerosis are presented.

Rehabilitation of Social Cognitive Impairment

Neuropsychological Rehabilitation

Psychosocial rehabilitation is not a clear and validated approach. Most of this work has been conducted in the context of autism spectrum disorder, schizophrenia, and traumatic brain injury. An important issue is whether a specific training in emotion perception can enhance residual function in emotion recognition networks or develop other, perhaps more conscious strategies to be applied to this problem [57].

Rehabilitation of SC impairments is often associated with rehabilitation of executive functions and working memory. Two main approaches to improving SC have been explored: retraining social cue perception and social skills training [57]. Retraining social cue perception consists of training in game-like tasks about static visual, dynamic auditory-visual, and complex social emotions. Social skills training consists of role-play in different themes such as agreeing, disagreeing, and listening [79]. Basic perceptual decoding of emotional cues and social skills training can respond to training [80], but the generalization of this gain to everyday situation and functional outcome needs more well-designed and powered studies. Social support therapy and music therapy seemed to have some success to improve empathy in a small randomized trial [81].

Pharmacological Treatments

In schizophrenia, antipsychotic drugs seem not to have effects on CS [82], especially on facial emotion recognition [83]. They have various effects on ToM tasks [84]. Recent works may have found a beneficial effect of oxytocin in SC impairment in particular in autism spectrum disorder [85], in frontotemporal dementia [86], in schizophrenia [87], and in patients with amygdala lesions [88].

Perspectives

More studies are necessary to investigate the potential beneficial effect of pharmaceutical treatments on SC. Focused neuromodulatory treatments, such as transcranial magnetic stimulation or transcranial direct current stimulation, may be useful, with the knowledge of the neural networks that underlie affective perspective-taking [42].

References

1. Frith CD, Frith U. Social cognition in humans. Curr Biol. 2007;17:R724–32.
2. Frith CD, Frith U. Mechanisms of social cognition. Annu Rev Psychol. 2012;63:287–313.
3. McDonald S. Impairments in social cognition following severe traumatic brain injury. J Int Neuropsychol Soc. 2013;19:231–46.
4. Green MF, Penn DL, Bentall R, Carpenter WT, Gaebel W, Gur RC, et al. Social cognition in schizophrenia: an NIMH workshop on definitions, assessment, and research opportunities. Schizophr Bull. 2008;34:1211–20.
5. Calder AJ, Young AW. Understanding the recognition of facial identity and facial expression. Nat Rev Neurosci. 2005;6:641–51.
6. Brothers L. The social brain – a project for integrating primate behavior and neurophysiology in a new domain. Concepts Neurosci. 1990;1:27–51.
7. Van Kleef GA. How emotions regulate social life. The emotions as social information (EASI) model. Curr Dir Psychol Sci. 2009;18:184–8.
8. Adolphs R. Neural systems for recognizing emotion. Curr Opin Neurobiol. 2002;12:169–77.
9. Baron-Cohen S, Wheelwright S, Hill J, Raste Y, Plumb I. The "reading the mind in the eyes" test revised version: a study with normal adults, and adults with Asperger syndrome or high-functioning autism. J Child Psychol Psychiatry. 2001;42:241–51.

10. Adolphs R. The neurobiology of social cognition. Curr Opin Neurobiol. 2001;11:231–9.
11. Adolphs R. Cognitive neuroscience: cognitive neuroscience of human social behaviour. Nat Rev Neurosci. 2003;4:165–78.
12. Allison T, Puce A, McCarthy G. Social perception from visual cues-role of the STS region. Trends Cogn Sci. 2000;4:267–78.
13. Phillips ML, Drevets WC, Rauch SL, Lane R. Neurobiology of emotion perception I: the neural basis of normal emotion perception. Biol Psychiatry. 2003;54:504–14.
14. Vuilleumier P, Pourtois G. Distributed and interactive brain mechanisms during emotion face perception – evidence from functional neuroimaging. Neuropsychologia. 2007;45:174–94.
15. Rizzolatti G, Craighero L. The mirror-neuron system. Annu Rev Neurosci. 2004;27:169–92.
16. Beatty WW, Orbelo DM, Sorocco KH, Ross ED. Comprehension of affective prosody in multiple sclerosis. Mult Scler. 2003;9:148–53.
17. Kraemer M, Herold M, Uekermann J, Kis B, Daum I, Wiltfang J, Berlit P, Diehl RR, Abdel-Hamid M. Perception of affective prosody in patients at an early stage of relapsing-remitting multiple sclerosis. J Neuropsychol. 2013;1:91–106.
18. Lane RD, Schwartz GE. Levels of emotional awareness – a cognitive-developmental theory and its application to psychopathology. Am J Psychiatry. 1987;144:133–43.
19. Lane RD, Quilan DM, Schwartz GE, Walker PA, Zitlin SB. The levels of emotional awareness scale – a cognitive-developmental measure of emotion. J Pers Assess. 1990;52(1–2):124–34.
20. Premack D, Woodruff G. Does the chimpanzee have a theory of mind. Behav Brain Sci. 1978;1:515–26.
21. Frith C, Frith U. Theory of mind. Curr Biol. 2005;15:R644–5.
22. Shamay-Tsoory SG, Aharon-Peretz J. Dissociable prefrontal networks for cognitive and affective theory of mind: a lesion study. Neuropsychologia. 2007;45:3054–67.
23. Baron-Cohen S. Theory of mind and autism – a review. Int Rev Res Ment Ret. 2001;23:169–84.
24. Stone VE, Baron-Cohen S, Knight RT. Theory of mind. Autism Sci Ment Health. 2013;226.
25. Duval C, Piolino P, Bejanin A, Laisney M, Eustache F, Desgranges B. La théorie de l'esprit: aspects conceptuels, évaluation et effets de l'âge. Rev Neuropsychol. 2011;3:41–51.
26. Miller SA. Children's understanding of second-order mental states. Psychol Bull. 2009;135:749–73.
27. Bechara A, Damasio AR, Damasio H, Anderson SW. Insensitivity to future consequences following damage to human prefrontal cortex. Cognition. 1994;50:7–15.
28. Benton AL, Hamsher K, Varney NR, Spreen O. Contributions to neuropsychological assessment. New York: Oxford University Press; 1983.
29. Ekman P, Friesen W. Pictures of facial affect. Palo Alto: Consulting Psychologist Press; 1976.
30. Baron-Cohen S, Wheelwright S, Jolliffe T. Is there a "language of the eyes"? Evidence from normal adults, and adults with autism or Asperger syndrome. Vis Cogn. 1997;4:311–31.
31. Bowers D, Blonder LX, Heilman KM. The Florida affect battery, manual (revised). Gainesville: Center for Neuropsychological Studies, University of Florida; 2001.
32. Ross ED, Thompson R, Yenkosky J. Lateralization of affective prosody in brain and callosal integration of hemispheric language functions. Brain Lang. 1997;56:27–54.
33. Perner J, Wimmer H. "John thinks that Mary thinks that…" attribution of 2nd-order beliefs by 5-year-old to 10-year-old children. J Exp Child Psychol. 1985;39:437–71.
34. Happé FGE. An advanced test of theory of mind: understanding of story characters' thoughts and feelings by able autistic, mentally handicapped and normal children and adults. J Autism Dev Disord. 1994;24:129–54.
35. Baron-Cohen S, Wheelwright S. The empathy quotient: an investigation of adults with Asperger syndrome or high functioning autism, and normal sex differences. J Autism Dev Disord. 2004;2:163–75.
36. Hogan R. Development of an empathy scale. J Consult Clin Psychol. 1969;33:307–16.
37. Dziobek I, Fleck S, Kalbe E, Rogers K, Hassenstab J, Brand M, et al. Introducing MASC: a movie for the assessment of social cognition. J Autism Dev Disord. 2006;36:623–36.

38. Kleeberg J, Bruggimann L, Annoni J-M, van Melle G, Bogousslavsky J, Schluep M. Altered decision-making in multiple sclerosis: a sign of impaired emotional reactivity? Ann Neurol. 2004;56:787–95.
39. Frith CD, Frith U. The neural basis of mentalizing. Neuron. 2006;50:531–4.
40. Schulte-Ruther M, Greimel E, Markowitsch HJ, Kamp-Becker I, Remschmidt H, et al. Dysfunctions in brain networks supporting empathy – an fMRI study in adults with autism spectrum disorders. Soc Neurosci. 2011;6:1–21.
41. Herold R, Feldmann A, Simon M, Tenyi T, Kover F, et al. Regional gray matter reduction and theory of mind deficit in the early phase of schizophrenia – a voxel-based morphometric study. Acta Psychiatr Scand. 2009;119:199–208.
42. Hillis AE. Inability to empathize: brain lesions that disrupt sharing and understanding another's emotions. Brain. 2014;137:981–97.
43. Völlm BA, Taylor ANW, Richardson P, Corcoran R, Stirling J, McKie S, et al. Neuronal correlates of theory of mind and empathy: a functional magnetic resonance imaging study in a nonverbal task. Neuroimage. 2006;29:90–8.
44. Abu-Akel A, Shamay-Tsoory S. Neuroanatomical and neurochemical bases of theory of mind. Neuropsychologia. 2011;49:2971–84.
45. Bar-On R, Tranel D, Denburg NL, Bechara A. Exploring the neurological substrate of emotional and social intelligence. Brain. 2003;126:1790–800.
46. Damasio AR. The somatic marker hypothesis and the possible functions of the prefrontal cortex. Philos Trans R Soc Lond. 1996;351:1413–20.
47. Kennedy DP, Adolphs R. The social brain in psychiatric and neurological disorders. Trends Cogn Sci. 2012;16:559–72.
48. Korkmaz B. Theory of mind and neurodevelopmental disorders of childhood. Pediatr Res. 2011;69:101R–8.
49. American Psychiatric Association. Diagnostic and statistical manual of mental disorders. 4th ed., text rev. Washington, DC: Authors; 2000.
50. American Psychiatric Association. Diagnostic and statistical manual of mental disorders. 5th ed. Arlington: American Psychiatric Publishing; 2013.
51. Mehta UM, Thirthalli J, Subbakrishna DK, Gangadhar BN, Eack SM, Keshavan MS. Social and neuro-cognition as distinct cognitive factors in schizophrenia: a systematic review. Schizophr Res. 2013;148:3–11.
52. Savla GN, Vella L, Armstrong CC, Penn DL, Twamley EW. Deficits in domains of social cognition in schizophrenia – a meta-analysis of the empirical evidence. Schizophr Bull. 2012. doi:http://dx.doi.org/10.1093/schbul/sbs080.
53. Kalkstein S, Hurford I, Gur RC. Neurocognition in schizophrenia. Curr Top Behav Neurosci. 2012;4:373–90.
54. Fett AK, Viechtbauer W, Dominguez MD, Penn DL, van Os J, Krabbendam L. The relationship between neurocognition and social cognition with functional outcomes in schizophrenia – a meta-analysis. Neurosci Biobehav Rev. 2011;35(3):573–88.
55. Persad SM, Polivy J. Differences between depressed and nondepressed individuals in the recognition of and response to facial emotional cues. J Abnorm Psychol. 1993;102:358–68.
56. Asthana HS, Mandal MK, Khurana H, Haque-Nizamie S. Visuospatial and affect recognition deficit in depression. J Affect Disord. 1998;48:57–62.
57. Manly T, Murphy FC. Rehabilitation of executive function and social cognition impairments after brain injury. Curr Opin Neurol. 2012;25:656–61.
58. Harlow JM. Passage of an iron rod through the head. Boston Med Surg J. 1848;39:389–93.
59. Van Horn JD, Irimia A, Torgerson CM, Chambers MC, Kikinis R, Toga AW. Mapping connectivity damage in the case of Phineas gage. Sporns O, éditeur. PLoS One. 2012;7, e37454.
60. Ibañez A, Manes F. Contextual social cognition and the behavioral variant of frontotemporal dementia. Neurology. 2012;78:1354–62.
61. Stone VE, Baron-Cohen S, Calder A, Keane J, Young A. Acquired theory of mind impairments in individuals with bilateral amygdala lesions. Neuropsychologia. 2003;41:209–20.

62. Happé F, Brownell H, Winner E. Acquired theory of mind' impairments following stroke. Cognition. 1999;70:211–40.
63. Sprengelmeyer R, Young AW, Mahn K, Schroeder U, Woitalla D, Buttner T, et al. Facial expression recognition in people with medicated and unmedicated Parkinson's disease. Neuropsychologia. 2003;41(8):1047–57.
64. Sprengelmeyer R, Schroeder U, Young AW, Epplen JT. Disgust in pre-clinical Huntington's disease – a longitudinal study. Neuropsychologia. 2006;44(4):518–33.
65. Gregory C, Lough S, Stone V, Erzinclioglu S, Martin L, Baron-Cohen S, Hodges JR. Theory of mind in patients with frontal variant frontotemporal dementia and Alzheimer's disease – theoretical and practical implications. Brain. 2002;125:752–64.
66. Broicher SD, Kuchukhidze G, Grunwald T, Krämer G, Kurthen M, Jokeit H. "Tell me how do I feel" – emotion recognition and theory of mind in symptomatic mesial temporal lobe epilepsy. Neuropsychologia. 2012;50:118–28.
67. Schacher M, Winkler R, Grunwald T, Kraemer G, Kurthen M, Reed V, et al. Mesial temporal lobe epilepsy impairs advanced social cognition. Epilepsia. 2006;47:2141–6.
68. Shamay-Tsoory SG, Tomer R, Aharon-Peretz J. The neuroanatomical basis of understanding sarcasm and its relationship to social cognition. Neuropsychology. 2005;19:288–300.
69. Adolphs R. The social brain: neural basis of social knowledge. Annu Rev Psychol. 2009;60:693–716.
70. Merceron K, Prouteau A. Évaluation de la cognition sociale en langue francaise chez l'adulte : outils disponibles et recommandations de bonne pratique clinique. Evol Psychiatr. 2013;78:53–70.
71. Van Overwalle F. Social cognition and the brain – a meta-analysis. Hum Brain Mapp. 2009;30:829–58.
72. Palermo R, Rhodes G. Are you always on my mind? A review of how face perception and attention interact. Neuropsychologia. 2007;45:75–92.
73. Bull R, Phillips LH, Conway CA. The role of control functions in mentalizing – dual-task studies of theory of mind and executive function. Cognition. 2008;107:663–72.
74. Aboulafia-Brakha T, Christe B, Martory MD, Annoni JM. Theory of mind tasks and executive functions: a systematic review of group studies in neurology. J Neuropsychol. 2011;1:39–55.
75. Moriguchi Y. The early development of executive function and its relation to social interaction: a brief review. Front Psychol. 2014;5:1–6.
76. Perner J, Kloo D, Gornik E. Episodic memory development – theory of mind is part of re-experiencing experienced events. Infant Child Dev. 2007;16:471–90.
77. Baron-Cohen S, Leslie AM, Frith U. Does the autistic child have a theory of mind? Cognition. 1985;1:37–46.
78. Etchepare A, Merceron K, Amieva H, Cady F, Roux S, Prouteau A. Évaluer la cognition sociale chez l'adulte – validation préliminaire du Protocole d'évaluation de la cognition sociale de Bordeaux (PECS-B). Rev Neuropsychol. 2014;2:138–49.
79. Driscoll DM, Dal Monte O, Grafman J. A need for improved training interventions for the remediation of impairments in social functioning following brain injury. J Neurotrauma. 2011;28:319–26.
80. Bornhofen C, McDonald S. Comparing strategies for treating emotion perception deficits in traumatic brain injury. J Head Trauma Rehabil. 2008;23:103–15.
81. Eslinger PJ. Neurological and neuropsychological bases of empathy. Eur Neurol. 1998;39:193–9.
82. Sergi MJ, Green MF, Braff DL, et al. Social cognition and neurocognition – effects of risperidone, olanzapine, and haloperidol. Am J Psychiatry. 2007;164:1585–92.
83. Hempel RJ, Dekker JA, van Beveren NJM, Tulen JHM, Hengeveld MW. The effect of antipsychotic medication on facial affect recognition in schizophrenia – a review. Psychiatry Res. 2010;178:1–9.
84. Savina I, Beninger RJ. Schizophrenic patients treated with clozapine or olanzapine perform better on theory of mind tasks than those treated with risperidone or typical antipsychotic medications. Shizophr Res. 2007;94:128–38.

85. Andari E, Duhamel J-R, Zalla T, Herbrecht E, Leboyer M, Sirigu A. Promoting social behavior with oxytocin in high-functioning autism spectrum disorders. Proc Natl Acad Sci. 2010;107:4389–94.
86. Jesso S, Morlog D, Ross S, Pell MD, Pasternak SH, Mitchell DG, et al. The effects of oxytocin on social cognition and behaviour in fronto-temporal dementia. Brain. 2011;134:2493–501.
87. Fischer-Shofty M, Brüne M, Ebert A, Shefet D, Levkovitz Y, Shamay-Tsoory SG. Improving social perception in schizophrenia – the role of oxytocin. Schizophr Res. 2013;146:357–62.
88. Hurlemann R, Patin A, Onur OA, Cohen MX, Baumgartner T, Metzler S, et al. Oxytocin enhances amygdala-dependent, socially reinforced learning and emotional empathy in humans. J Neurosci. 2010;30:4999–5007.

Chapter 14
Psychopathology of Alexithymia and Multiple Sclerosis

Khadija Chahraoui, Emmanuelle Dieu, and Thibault Moreau

Abstract The notion of alexithymia was introduced by Sifneos (Psychother Psychosom 22:255–62, 1973) to define a set of affective and cognitive characteristics observed in patients with psychosomatic diseases. Alexithymia appears to be a multidimensional and transnosographic concept ranging from normal to pathological, and it is important to regard alexithymia as a disorder of emotional regulation, which can be found in different populations and not only in somatic diseases.

We present here the many theories of alexithymia (cognitive, neuropsychological, cultural, psychological (attachment theory), and psychoanalytical conceptions) and some evaluation tools to assess alexithymia (quantitative methods: TAS, BIQ) or qualitative methods (Rorschach, analysis of discourse). Alexithymia has been frequently observed in subjects with MS, reflecting a form of emotional disturbance related to the traumatic aspects of the disease. But it may also constitute a form of defense by the freezing or the denial of emotions, which allows patients to adapt psychologically to the situation by reducing distress. It is necessary to take these problems of emotional regulation into account and above all to understand what meaning the patient gives to these difficulties.

K. Chahraoui, Ph.D.
Laboratoire de Psychopathologie et de Psychologie Médicale,
Université de Bourgogne, Esplanade Erasme, Dijon 21000, France
e-mail: Khadija.Chahraoui@u-bourgogne.fr

E. Dieu, Ph.D.
Laboratoire De Psychopathologie Et Psychologie Médicale,
Université De Bourgogne, Dijon, France

Centre Hospitalier Universitaire de Dijon, Service De Psychiatrie
Adulte Et Addictologie – Service De Neurologie,
Boulevard Jeanne D'arc, Dijon 21000, France
e-mail: emmanuelle.dieu@wanadoo.fr

T. Moreau, M.D., Ph.D. (✉)
Department of Neurology, Centre of Epidemiology of the Populations EA 4184,
University Hospital of Dijon, 14 rue Gaffarel, 21000 Dijon, France
e-mail: thibault.moreau@chu-dijon.fr

B. Brochet (ed.), *Neuropsychiatric Symptoms of Inflammatory Demyelinating Diseases*, Neuropsychiatric Symptoms of Neurological Disease,
DOI 10.1007/978-3-319-18464-7_14

Keywords Alexithymia • Multiple sclerosis • Emotional regulation • Defense mechanisms • Psychic trauma

The Concept of Alexithymia

Definition of the Notion of Alexithymia

The notion of alexithymia is a neologism that means "the absence of words to express ones emotions." This psychological term was introduced by Sifneos [1] to define a set of affective and cognitive characteristics observed in patients with psychosomatic diseases. Alexithymia was characterized according to several dimensions in these patients: firstly, the difficulty in identifying and describing feelings to others; secondly, restricted imaginative processes; thirdly, a propensity to act in order to avoid resolving conflicts; and fourthly, a detailed description of facts, events, and physical symptoms [2, 3]. For Sifneos [1], alexithymic subjects tend to use acts to resolve conflicts. They show a pronounced difficulty to recognize and describe their own sentiments and to distinguish between emotional states and physical sensations. He considered alexithymia an inability to associate visual images, fantasies, and thoughts with specific emotional state [4].

Following on from these early works, the concept of alexithymia developed considerably in most western countries, and the hypothesis of a relationship between impaired expression of emotion and the onset of psychosomatic disorders was shared by a large number of clinicians.

More recent research has shown that alexithymia was not specific to somatic disorders but could also occur in other diseases. It was subsequently observed also among patients with a variety of psychiatric disorders that involved disturbances in emotional regulation. Alexithymia was thus found to be a mode of psychological functioning in various disorders like addiction [5], eating disorders [6, 7], and post-traumatic stress [8] and in the population at large [9]. Alexithymia has thus progressively become a personality trait that constitutes a factor of vulnerability in a certain number of psychiatric or somatic diseases [10].

Studies have also focused on two types of alexithymia [11–14]; one seems to be primary and dispositional; it seems to condition an inappropriate reaction to stress and could thus increase the probability of psychosomatic functioning modes occurring [1]. The other seems to be secondary and could be perceived rather as a consequence of stress and thus be considered a form of reaction that inhibits adaptation to situations of stress. In this version, alexithymia is not only a factor of vulnerability but also a defense strategy to cope with unpleasant emotions [11].

Alexithymia and Emotion

Alexithymia: A Disorder of Emotional Regulation

Alexithymia appears to be a multidimensional and transnosographic concept ranging from normal to pathological [10]. It is important to regard alexithymia above all as a disorder of emotional regulation, which can be found in different populations and not only in somatic diseases. Emotions play a central role in an individual's psychic equilibrium and in the development and organization of psychic functioning.

Emotion involves several levels of reality [15]: (1) neurophysiological processes (neuroendocrine activation and autonomous nervous system), (2) motor or expression behavior processes (facial expressions, changes in posture or tone of voice), and (3) cognition-experience and interpersonal processes (subjective conscience of emotional states that motivate the subject and the verbal expression of this conscience).

Psychologists have analyzed this third dimension in particular, and many studies have shown that emotion has an impact at different levels, in attention and perception processes, in decision-making, and in the interpersonal relationships [15, 16]. Emotion is no longer regarded as uniquely negative but also as an adaptation process that plays a major role in adaptation and the regulation of interpersonal relationships.

Alexithymia, Emotion, and Cognitive Theories

The cognitive theories of alexithymia are generally based on the postulate that emotions are determined by cognitive phenomena, which implies that the absence of the individual representation of emotions leads to suppression of their expression.

Several authors insist more specifically on the links between alexithymia and impaired emotional conscience [17, 18]. This includes several levels and stages of development and corresponds to the ability to represent emotional experience and to the verbal representations used to express it. Alexithymia could be situated at the initial stage of affective development, at which emotions have an essentially somatic expression (sensorimotor stage) [17, 19].

Neuropsychological Data of Alexithymia

According to neuropsychological data, alexithymia could be understood as a functionally or structurally impaired connection between the cerebral hemispheres. Alexithymia could thus be related to the impossibility to transfer information between the left hemisphere (analytical) and the right hemisphere (emotional), but also to functional limitations of the right hemisphere [20]. One of the first studies on this hypothesis [21] showed that patients who had undergone section of the corpus callosum and the anterior commissure in a context of refractory epilepsy found it difficult to name the sentiments and emotions felt in painful situations and developed a mode of discourse that focused on facts and concrete situations. Many subsequent studies showed that alexithymia could be due to the impaired transmission of emotional information, notably interoceptive information, to the anterior cingulate cortex [17]. Moreover, the results of Berthoz et al. [17] are coherent with Sifneos' hypothesis that the connection between the limbic system and the neocortex was impaired and, as shown by MRI, that the limbic and paralimbic regions seem to play a role in the emotional response [17]. Today, most of these studies have been combined with cognitive approaches because the definition of alexithymia cannot be reduced to the specificity of biological mechanisms and the associated phenomena. In order to understand the alexithymic response, it is necessary to include the notion of perception and the representation of emotion and thus cognitive aspects [17, 18].

Emotional Regulation and Attachment Theory

Attachment theory [22] particularly insists on the role of interpersonal relationships in emotional regulation. Attachment could also regulate emotions in relationships [23–25] by exercising an adaptive role. People may adopt different strategies for emotional regulation depending on their attachment profile (secure, anxious–preoccupied, dismissive–avoidant, fearful–avoidant) [23, 26]. Secure attachment could thus represent an internal resource for individuals that would allow them to evaluate certain stressful experiences as positive and to better adjust, whereas insecure attachment may lead to far less effective adjustment strategies. Several studies have reported an association between insecure attachment and alexithymia [27].

The style of attachment is a major individual variable as it interacts with the emotional regulation, adjustment strategies, and the search for social support [28]. Secure subjects manage to cope with and to recognize their attachment needs in a supple and coherent manner and thus to regulate their emotions. Preoccupied subjects tend to avoid their emotions and to bypass their attachment needs to reduce conflict with figures of attachment and their emotional distress [26]; anxious-avoidant insecure subjects are invaded by their memories and their emotions with regard to attachment; they overactivate their attachment system [29] to try to find secure attachment. This creates excessive focalization on attachment and increases emotional distress.

Cultural Conceptions

The expression of emotions is not only the product of personality or of a biological dysfunction, but it is also the product of a person interacting with another one in a given context. In this sense, it is also a cultural manifestation since culture defines ways to behave and of ways to feel depending on the values proposed in a given society. It is not possible to assess emotions without referring to the way in which the subject was shaped in a given culture, depending on his/her education and experiences in his/her peer group [30].

Certain currents believe that emotions are universal. Izard, for example, described several basic emotions (*joy, sadness, surprise, disgust, anger, culpability, fear, timidity, mistrust, interest, attention*). These emotions were invariable and had a biological foundation. Culture only affected the way in which these emotions could be coded [31]. In contrast, the American culturalist current regard emotions as specific to a given culture and closely tied to specific aspects of upbringing in each society. The work of Margaret Mead [32] showed, for example, how children in Samoa developed insensitivity and affective distancing associated with the methods of upbringing. These studies were nonetheless severely criticized, and the universal approach to emotions was predominant with the idea that what differentiated emotions from one culture to another was more due to differences in modes of coding [30]. Depending on the context, the culture, and the individual, emotions may be expressed differently in the language, the words, and the metaphors. The expression of emotions is dependent on education and upbringing and can be encouraged to varying degrees depending on the closeness of ties, the period, and the place. Emotions can be strongly codified and structured, for example, during funeral ceremonies in certain societies with the interdiction of weeping, for instance, or with the weeping being provided by professional weepers [33].

In our clinical observations of alexithymic subjects, we often found a family upbringing marked by the prohibition of emotional expression and the denial of emotions in situations of conflict.

It is without doubt just as important to regard the medical environment as a specific sociocultural context in which the expression of emotions may have a particular meaning; for example, the difficulty of patients to express their emotions may also be considered a way to protect their psychic intimacy when their body is naked and exposed to the medical world.

Psychoanalytical Conceptions

Alexithymia is a notion close to the operative thought concept developed in the field of psychoanalysis [34] which refers to the inability to mentalize conflicts with a risk of shifting these to the somatic level. According to Corcos et al. [35], alexithymia can be considered a component of operative thought in that it shows an inability to gain access to one's inner world and a cognitive style characterized by fact-focused thinking.

According to psychoanalysts, alexithymia is related to early affective development disorders and in particular to affective deficits in relationships with an absent and non-interiorized object of love. These absences and early affective deficits hamper the development and the construction of the self with a profound affective disinvestment of the maternal object [35]. For psychoanalysts [13], the mother (or the parent who provides care) plays an early role in the infant's capacity to identify, to express, and to relieve its emotional experiences. This is what allows the child to grasp its own psychic and corporal reality, and the maternal function has a mediating role in this sense. The deficit of care and of early holding and handling could increase vulnerability to the development of somatic difficulties [35] as shown by Spitz in babies deprived of relationships with their mothers and brought up in a nursery [36]. In the absence of the early communication of emotions, a certain number of affective and cognitive models in a relationship cannot develop [35].

These subjects not only find it difficult to express emotions, but also invest too much in action to the detriment of thinking. This is an overinvestment in external reality to the detriment of interior life.

For psychoanalysts, alexithymia is not only a personality trait, but also a defense mechanism that could be related to the individual's early development. These mechanisms, which come into being early, protect the infant from loss and absence. In this context, alexithymia may be a mode of response every time a factor of stress, whether internal or external, arises. It may thus be secondary to psychic trauma that violates the psychic apparatus and overwhelms the capacities of defense and mentalization. Subsequently, the subject could thus temporarily resort to defense mechanisms that involve overinvestment in the sensorimotor-type world. Early affective deficits will remain etched in the corporal schema and may come to the fore in cases of major trauma [35].

What must be remembered from the psychoanalytical approach is that alexithymia is above all a mode of response to pain, which cannot be expressed otherwise. It expresses frozen emotions, a major defense against distress.

Assessment of Alexithymia

Since Sifneos' initial definition, many evaluation tools to measure alexithymia have been proposed. There are internationally recognized validated questionnaires to assess alexithymic functioning, such as the BIQ (Beth Israel Questionnaire of Sifneos) or the Toronto Alexithymia Scale (TAS). There has also been qualitative research on the specificities of the alexithymic discourse. We present here a few examples of these types of evaluation.

Toronto Alexithymia Scale (TAS)

Internationally, the Toronto Alexithymia Scale (TAS) is the most widely used scale to measure alexithymia. The TAS-20 is a self-report scale that comprises 20 items. TAS-20 has three subscales [37, 38]:

- Difficulty identifying feelings (seven items: 1 + 3 + 6 + 7 + 9 + 13 + 14)
- Difficulty describing feelings (five items: 2 + 4 + 11 + 12 + 17)
- Externally oriented thinking (eight items: 5, 8, 10, 15, 16, 18, 19, 20)

Items are rated using a 5-point Likert scale whereby 1 = strongly disagree and 5 = strongly agree. There are five items that are negatively keyed (items 4, 5, 10, 18, and 19). The total alexithymia score is the sum of responses to all 20 items, while the score for each subscale factor is the sum of the responses to that subscale. The TAS-20 uses cutoff scoring as follows: 20–50, non-alexithymic patients; 51–60, borderline alexithymic patients; and 61–100, alexithymic patients.

Research demonstrates good internal consistency, test-retest reliability, and validity. In addition, it has been found to be stable and replicable across clinical and nonclinical populations [37, 38].

Alexithymia and Rorschach

Certain authors [12] studied the specificity of alexithymic responses to the Rorschach test by developing several hypotheses. Alexithymia reflects a major impairment in the ability to regress in an adaptive manner, which is a fundamental element in problem resolution, creativity, and effective adaptation. In so far as alexithymic disorders constitute a perturbation of the relationship with the physical, social, and emotional environment, they function as risk factors in situations of environmental stress and increase the probability of psychosomatic adaptation modes occurring. Alexithymic indexes on the Rorschach test show impoverishment or absence of human movements and human content, impoverishment of color use, a lack of originality, and rigid control. These hypotheses are reflected by a certain number of indicators revealed by analysis of the Rorschach test: (1) *fantasies*, few responses in general (*R*), no human movement (*M*), or human movement (*K*); (2) *affects*, few color responses (*Sum C*) (affect), few primary-form color responses (few *FC*); (3) *cognition*, concrete thoughts (few *blends*), stereotypic ideation (high *lambda*); and (4) *adaptive resources*, few representations and limited internal resources (low *EA*).

Analysis of Discourse

Researchers [39–42] have also investigated the alexithymic discourse and have shown that alexithymics have certain language characteristics. In particular, more passive forms, fewer adverbs [39], more frequent use of the present, more infinitive forms, a greater frequency of the verb "to be" than the verb "to have," more description and focusing on events, a short discourse, less frequent use of "I" and more frequent use of "one," [40] higher occurrence of factive verbs than stative verbs [39, 42] and of words less loaded with affectivity, fewer adjectives, more auxiliary verbs, and more incomplete phrases [40]. As underlined by Donabedian [43] in alexithymic subjects, action words seem to be more common. Such words are used to describe an event, in which the affect is suggested but is not linked to psychic work given the lack of construction of representations.

Alexithymia and SEP

Prevalence of Alexithymia in MS

The prevalence of alexithymia in the MS population is estimated at between 40 and 50 % [44, 45]. However, studies seem to show of differences between countries. An Italian study that used the North American cutoff value for determining the presence of alexithymia (i.e., >60) observed a prevalence of 13.8 % in a sample of 58 patients [46]. A study conducted in France found a rate of 23.2 % among a sample of 115 patients [47], and another [48] found a prevalence of around 30 % of alexithymic subjects in a sample of 61 patients with MS using the international TAS-20 cutoff values. The discrepancies could be explained by the different clinical characteristics of the different samples (e.g., lower level of handicap and higher proportion of male patients in the Italian study) or by cultural differences that remain to be elucidated.

In patients with MS, alexithymia mainly manifests itself as difficulty identifying and describing their emotions, a paucity of fantasies (e.g., lack of daydreams or dreams), and a discourse centered on facts and symptoms [44]. Alexithymia may play a role in the development and the severity of depression [46, 47]. Indeed, MS ultimately leads to significant limitations and loss of autonomy due to the evolving nature of the disease, and these changes can require considerable and repeated social, professional, and familial adjustments. The profound changes and progressive loss of autonomy can create negative emotions and painful psychological experiences. Therefore, the patient affected by MS must work through a period of "grief" in order to be able to assimilate these losses in psychological terms. Alexithymia could therefore be a major factor of vulnerability in this respect, in that it contributes to the inhibition of emotional expression and the capacity of mentalization of the psychic trauma associated with the disease and its course. Alexithymia may also represent a key psychological factor that hampers true emotional and cognitive integration of the changes related to the disease.

Evolution of Alexithymia in MS

Few studies have evaluated the evolution of alexithymia in multiple sclerosis. In a recent study [48], we investigated the evolution of alexithymia and its relationship with anxiety and depression in patients with MS over a period of 5 years. Improved knowledge of the evolution of emotional disturbances over time could help us to better understand how MS patients adapt to their handicap in psychological terms over time.

In this research [48], we found a relationship between alexithymia and both anxiety and depression and at both time points of the study. The rates of anxiety and depression were consistently high, with around 40 % of MS patients' distress from anxiety problems, and this finding was stable over time. Conversely, the rate of depression tended to decrease between the two evaluations, falling from 40 to 26 %. Multivariate logistic regression showed that alexithymia seems to be more strongly associated with anxiety. These results underline the similar manner in which alexithymia and anxiety are mediated, as well as the stability of these disorders over time. The great emotional difficulty experienced by patients with MS has consistently been reported in the literature [49, 50], and the persistence of emotional disturbances over time has previously been highlighted by other authors [51, 52]. It is possible that this persistence arises from a permanent incapacity of these patients to cope with the disease, particularly since prognosis is very uncertain in terms of progression of handicap [53]. The unpredictable nature of the progression of MS appears to be a central component in understanding the persistence of the emotional problems. Indeed, MS is characterized by the occurrence of attacks (relapses) of worsening neurological function that are highly unpredictable, and the patient cannot anticipate either the occurrence of an attack or the type or intensity of symptoms. Predicting disease progression is therefore a dimension of MS that is extremely challenging. For example, some patients may experience several relapses in the same year, whereas others may go 10 years without an attack. Furthermore, while some attacks can have more or less severe effects that may partially or totally recede, others may herald a functional deficit, such as impaired motor function that can remain and become permanent. While the relapsing-remitting form of MS highlights in particular the uncertainty experienced by patients with MS, the progressive form, with its slowly but constantly worsening neurological function, also leaves patients feeling insecure and uncertain about their future, once the symptoms or irreversible deficits begin to appear. The long-term course of MS is characterized by ambiguity, which can lead to a real fear of what lies ahead and a permanent state of anxiety. Patients are in a constant state of worry; when they are well, they worry that they may soon suffer a relapse, and when they do suffer an attack, they worry that it may be the beginning of a rapid decline or the harbinger of further deficits. Furthermore, MS attacks can be experienced as a veritable traumatic event by patients, as onset is often sudden and unexpected, which can be complex and painful to cope with and accept. Many patients find it hard to let go of their hopes of living normally and to accept the physical constraints imposed on them by their disease. The wide gap between what they are physically able to do and what they would like to be able to do is often hard to accept.

Another noteworthy point in this study [48] is that only two dimensions of alexithymia, namely, difficulty describing and difficulty identifying feelings, correlated with anxiety and depression, whereas the third component of alexithymia (externally oriented thinking (EOT)) was independent of both these disorders. We also observed that this latter factor was the only one to evolve over time, with a significant fall in this dimension observed at 5 years. It is also the only factor that correlated specifically with the number of MS relapses. Given that EOT did not correlate with either anxiety or depression, it is possible that it may be a form of defensive strategy for coping with the traumatic experience of MS relapses. Accordingly, by orienting their preoccupations and thoughts externally, the patient is able to avoid facing up to their interior feelings and, more particularly, the anxiety arising from the traumatic nature of the course of the disease. We could even go so far as to hypothesize that EOT may represent a form of avoidance and denial of reality employed by the patient to protect themselves against excessively distressing feelings.

The fact that the effect of this factor decreases over time could suggest more successful adaptation to the disease, in so far as the patient has less need to use this defensive strategy. This is in line with the reduction in depression over time, which may also indicate better adjustment to the disease after a number of years. These findings are in line with the study by Chwastiak et al. [54] who reported that depressive symptoms decreased in the long term after diagnosis. The question arises, therefore, as to whether the reduction in depression over time can be explained by better adjustment to the different disease-related handicaps and by improved coping strategies that allow the patient to adapt better, thereby reducing depression. However, these results should be interpreted in the context of the clinical characteristics of our study population, where the level of handicap was moderate overall (3.83 ± 2.36), meaning that these patients were able to maintain active social and professional lives. Indeed, 56 % of the patients in our study were still professionally active, which could be considered to protect them to a certain extent by maintaining a network of social contact and support, whose positive effects on MS have previously been documented [47].

Anxiety, Psychic Trauma, and Alexithymia: Patients' Accounts

Announcement of the Diagnosis, Psychic Trauma, and Alexithymia

The moment the diagnosis is announced is a strong subjective experience for patients who are already marked by not only physical but also psychological frailty as they are hearing the announcement of their vulnerability, their helplessness, and their finiteness, in an outlook that seems to be sealed. Announcement of the diagnosis is not an epiphenomenon; it needs to be historicized in a dynamic perspective with regard not only to the pre-symptoms, which had already caused distress: upheaval of the identity, the absence of self-determination, and the problem of anxiety, but also their own trajectory, their story, their representations, and their desires.

Announcement of the diagnosis is a subjective experience, faced alone, and the moment of a face-to-face encounter with the clinician. Though certain authors contest the notion of trauma at the announcement of the diagnosis of a chronic disease, preferring the notions of initial shock and of grief [55], it seems that the psychic bewilderment, the rupture between before and after, the repeated reference in the narration, and the crystallizations reflect true psychic trauma.

Beyond the association with the subjective experience of the patients encountered, the notion of trauma takes into account and recognizes the patient's distress and makes it possible, paradoxically, to think this moment of frozen emotions dynamically. It can be considered the crucible of alexithymic symptoms.

Announcement of the diagnosis of multiple sclerosis is always a moment of emotional shock that perturbs the patient indefinitely. Even several years after the diagnosis, patients first talk about this moment, which marked them deeply. First of all, there is an intellectual dimension, in which they express their incomprehension "I didn't understand," incredulity "so, it's not true; is it true or not; I didn't really believe it; I said to myself, it's not possible," ignorance "it didn't mean anything to me," and refusal to be concerned and non-acceptation. Concerning the more emotional dimension, patients recount, relive the effects of bewilderment, of surprise: "I had the feeling I was rooted to the spot, flabbergasted; I was neither angry nor happy; I didn't say much"; "All the same, it was a terrible shock"; "It was like a slap in the face." As for the concrete dimension, patients talk about their anguish, fear, and desperation: "the ground seemed to fall away beneath my feet."

Anxiety disorders are particularly prevalent in this period surrounding this announcement. In their study, Giordano et al. [53] reported levels of anxiety that were higher than levels of depression around the announcement of the diagnosis: more than 40 % of patients (in a population of 120 persons) suffered from anxiety, which persisted for 6 months after the announcement (36 %), and this was particularly the case in women.

The expression of emotions in this period is limited, which reinforces the hypothesis of feelings of traumatic bewilderment with a feeling of violation, which at the moment in question cannot be put into words. There is sometimes a feeling of "dissociation": "It isn't me; it wasn't me," which reflects the notion of loss of identity.

In addition, subjects often feel the need to speak for a long time and in detail about the period leading up to discovery of the disease. The symptoms, generally speaking, are described in great detail with reliving of the pain that accompanied them, the handicaps, which were sometimes extremely disabling or even the massive all-consuming fatigue. The recovery after the flares sometimes took on a miraculous aspect, which, at certain moments, reinforced the notion of disease denial: "When I lost my balance, even though I was feeling well, (…), that was really horrible, it was too much, everything seemed to be spinning around, I couldn't even open my eyes…"; "One side was paralyzed, but it wasn't violent (…), I could feel the pain"; "I had optic neuritis, (…) I went to accident and emergency (…), I had a terrible pain in my eye and I didn't know where it was coming from…."

Certain patients talk about the impossibility to understand the onset of unusual symptoms or being invaded by fatigue, which seemed foreign, foreign to their selfhood. Anxiety is often a dominant feature, notably the fear of being affected by an incurable disease, like cancer, with a deadly outcome. "Me, I didn't know anything about the disease, I'd heard about it in passing, and I immediately thought about incurable diseases like cancer and all of that; it's true, when you don't know…."

The course of the symptoms is sometimes marked by long-lasting medical involvement due to the need to carry out a certain number of examinations. Patients express the extent to which this aggravates their anxiety and distress marked by doubt, fear, and speculation.

These extracts from interviews show to what degree the context of the announcement of the diagnosis may through the traumatic impact contribute to the psychic bewilderment, which hinders any elaboration and mentalization. The impossibility to express ones emotions and the emotional freeze that can follow on from this traumatic event can be considered a consequence of traumatic bewilderment. It is for this reason that clinicians must be particularly attentive to the way in which the patient may experience this moment, and they must foster the work of narration to allow the patient to integrate this major event of his/her existence.

Anxiety, Alexithymia, and Unpredictability of the Disease

Multiple sclerosis stands out because of its course, which varies from patient to patient, thus leaving even the medical world in a state of uncertainty, as doctors are unable to predict the course of the disease in any given patient. Despite advances in research, the etiology of the disease is still unknown, and there is no cure.

The most notable feature of this chronic disease is its unpredictability. The onset of disorders or handicap is thus unpredictable, and this puts patients in a situation of uncertainty with regard to their future prospects. Subjected to a lack of control and anticipation, they are forced to make constant adjustments. This situation can generate stress and anxiety, above and beyond the loss of physical abilities and identity, which can manifest themselves as depression and grief. The unpredictability of the disease may lead to alexithymic-type modes of adaptation [48].

This experience of not knowing accompanies the traumatic violation of the body and mind induced by the disease and contributes to the confusion of internal references: patients thus need to reconstruct a life project and find meaning by returning to their own life story [56, 57].

MS and its unpredictability "contaminate" among other things the family with processes and a context of psychological distress similar to those in marital discord [58]. The quality of intra-family support, together with the components of personality, is an essential factor in adaptation [59]. The perception of good social support and quality conjugal relationships [60] allows patients to cope with what they cannot control and to continue investing in life despite the uncertain outlook.

Patients cannot anticipate the onset of a flare, the nature of the symptoms, and their intensity. Prediction is extremely complicated: several flares may occur in the same year, or flares may occur 10 years apart. In addition, certain flares are followed by partial and sometimes total remission, while others announce a "functional deficit," a permanent motor or skeletal disorder.

Though the relapsing-remitting form particularly accentuates the uncertain nature of the disease, the progressive form, with its slow but relentless deterioration, also leaves patients in a situation of uncertainty about their future and about the moment when irreversible symptoms or disorders will occur.

Very often, fear of the "wheelchair" comes to the fore when patients talk about their illness, even though their trajectories remain similar for certain points and singular for others and are accompanied by adaptive modes necessary to carry on with life.

The long-term evolution of the disease and the inherent feeling of uncertainty lead to a true fear of the future, a state of permanent anxiety and in particular difficulty to adjust in the long term. These are principally problems of psychological adaptation, which are paramount in this context of uncertainty, and a source of intense distress as illustrated by these extracts: "Um … let's say as I was saying to my colleague this afternoon, in fact it's worrying if there's a flare or not a flare,… I say to myself, OK, well, if it gets worse, it may get worse quicker, and then when everything's OK it's still worrying because I say to myself – er – I've planned something, but maybe I planned too much. Maybe I'll have a flare. Maybe I'm going too far, and so I don't do anything. I'm not alive any more. So, at the end of the day, if I'm feeling OK I'm worried and when I'm bad I'm worried: so at the end of the day there aren't many times when I can completely unwind, and tell myself I'm not bothered and that everything's fine and live normally…."

Flares are very particular moments for patients, who may experience them as traumatic, violent, and brutal. These periods of flares lead patients to adopt a complex and painful form of adaptive management with difficulty to accept and come to terms with the grief due to the limitations imposed by their bodies: "yes there, yes, I've got it… it's as if suddenly, though I was living almost normally apart from the fact that I stopped driving for a while, suddenly, I realized that I wasn't going to be my usual self and that was er not a shock but a refusal to say no, I don't want it, I don't want it and to realize that there was a huge difference between what you want and what you get. Because you want to be normal and you can't manage it and your body it doesn't do what you want it to do and it's this difficulty to cope with your own body that was difficult. There's a sort of resignation. That's the word I can't get into my mind (laughter) that's the word I was looking for."

The traumatic and violent effect of flares, which sometimes come on by complete surprise, leads patients to search for meaning and an explanation for what is happening to them. The dimension of culpability is often present as shown in this extract : "Um, yes, yes, I think yes; I don't know, um, the last time when I had these famous pins and needles, um , it was nights when I wasn't sleeping at all well anyway, so um (laughter), as well as that as it was my mother's birthday in 8 days, and at the beginning I said to myself damn if it carries on like that, I didn't go for a walk, I was standing it was OK, but I said to myself, um I can't see myself going to work in a …, a wheelchair. It's true that I haven't set a time, but it's true it can happen very, very fast, at the same time, um the same thing has happened other times, why did it happen at that particular time, and then I said to myself, yes, there may be a reason. In fact, I didn't at all fancy going to my parents' place to celebrate my mother's birthday, and I said to myself, I expect that I'll feel much better on Sunday evening and Monday."

Given its ineluctable and unpredictable evolution, multiple sclerosis places the patient in a situation of "fundamental" uncertainty mixed with anguish. This can be latent or more paroxystic as is the case during a flare, which can be a true crisis with an uncertain outcome. Though the notion of uncertainty is related to the notions of fragility and precariousness, it is also related to the notion of worry, "agitation, and disorder" in which the mobilization of the patient plays a full part. Of course, these internal states are accompanied by psychic distress and may "spread" to the patient's entourage. They also represent a state of tension or even conflict from which patients may be able to draw certain adjustments which are sometimes favorable for their well-being.

Conclusion

Alexithymia can be understood as a difficulty with emotional regulation in the face of numerous disease-related changes and stress. MS is a major handicap with an uncertain course, and in patients, it engenders many of upheavals not only in physical and psychic domains, but also with regard to family, conjugal, social, and professional life. Alexithymia may reflect a form of emotional disturbance related to the traumatic aspects of the disease. It may also constitute a form of defense by the freezing or the denial of emotions, which allows patients to adapt psychologically to the situation by reducing distress. It is necessary to take these problems of emotional regulation into account and above all to understand what meaning the patient gives to these difficulties. Appropriate management throughout the medical process from the announcement of the diagnosis onward is essential to enable the patient to come to terms with the disease-related upheavals. Psychological support that takes these emotional difficulties into account and focuses on the expression and verbalization of emotions and giving them meaning through narration seems to us to be a key element in the management of these patients.

References

1. Sifneos PE. The prevalence of 'alexithymic' characteristics in psychosomatic patients. Psychother Psychosom. 1973;22:255–62.
2. Taylor GJ. Alexithymia: concept, measurement, and implications for treatment. Am J Psychiatry. 1984;141:725–32.
3. Nemiah JC, Freyberger H, Sifneos PE. Alexithymia, theoretical considerations. Psychother Psychosom. 1976;28:199–206.

4. Sifneos PE. Alexithymia: past and present. Am J Psychiatry. 1996;153–7:137–42.
5. Haviland MG, Hendryx MX, Shaw DG, Henry JP. Alexithymia in women and men hospitalized for psychoactive substance dependence. Compr Psychiatry. 1994;35(2):124–8.
6. Bourke MP, Taylor GJ, Parker JD, Bagby RM. Alexithymia in women with anorexia nervosa. A preliminary investigation. Br J Psychiatry. 1992;161:240–3.
7. Guilbaud O, Loas G, Corcos M, et al. Alexithymia in addictive behaviors and in healthy subjects: results of a study in French speaking subjects. Ann Med Psychol. 2002;160:77–85.
8. Hyer LA, Woods MG, Summers MN, et al. Alexithymia among Vietnam veterans with posttraumatic stress disorder. J Clin Psychiatry. 1990;51–6:243–7.
9. Loas G, Fremaux D, Otmani O, Verrier A. Prévalence de l'alexithymie en population générale chez 183 sujets « tout venant » et chez 263 étudiants. Ann Med Psychol. 1995;153(5):355–7.
10. Corcos M, Speranza M. Psychopathologie de l'alexithymie. Paris: Dunod; 2003.
11. Parker JD, Taylor GJ, Bagby RM. Alexithymia: relationship with ego defense and coping styles. Compr Psychiatry. 1998;39:91–8.
12. Acklin MW. Alexithymia: somatization and the rorschach response process. Rorschachiana. 1991;17:180–7.
13. Krystal H. Integration and self healing. Affect, trauma, alexithymia. Hillsdale: Analytic Press; 1988.
14. Pedinielli JL, Rouan G. Concept d'alexithymie et son intérêt en psychosomatique. Encycl Med Chir. 1998;37-400-D-20: 3p. Paris: Elsevier Psychiatrie.
15. Speranza M, Atger F. Approche développementale de l'alexithymie. In: Corcos M, Speranza M, editors. Psychopathologie de l'alexithymie. Paris: Dunod; 2003.
16. Luminet O. Psychologie des émotions. Nouvelles perspectives pour la cognition, la personnalité et la santé. Bruxelles: De Boeck; 2013.
17. Berthoz S, Martinot JL. Etudes neuropsychologiques et études en imagerie cérébrale dans l'alexithymie. In: Corcos M, Speranza M, editors. Psychopathologie de l'alexithymie. Paris: Dunod; 2003.
18. Bydlowsky S, Berthoz S, Corcos M, Consoli SM. Conscience émotionnelle et alexithymie: deux notions distinctes. In: Corcos M, Speranza M, editors. Psychopathologie de l'alexithymie. Paris: Dunod; 2003.
19. Lane RD, Schwartz GE. Levels of emotional awareness, a cognitive developmental theory and its application to psychopathology. Am J Psychiatry. 1987;144:133–43.
20. Montreuil M, Lyon-Caen O. Thymic troubles and relation between alexithymia and interhemispheric dysfunction in multiple sclerosis. Rev Neuropsychol. 1993;3:287–302.
21. Hoppe KD, Bogen JE. Alexithymia in twelve commissurotomized patients. Psychother Psychosom. 1977;28:148–55.
22. Bowlby J. La perte, tristesse et séparation. Paris: PUF; 1984. Traduction française de WEIL, 1969.
23. Guedeney N, Guedeney A. L'attachement : approche théorique : du bébé à la personne âgée. Paris: Masson; 2009.
24. Griffin D, Bartholomew K. Models of the self and other: fundamental dimensions underlying measures of adult attachment. J Pers Soc Psychol. 1994;67:430–45.
25. Pierrehumbert B, Bader M, Miljkovitch R, Mazet P, Amarn P, Halfon O. Strategies of emotion regulation in adolescents and young adults with substance dependence or eating disorders. Clin Psychol Psychother. 2002;9:384–94.
26. Miljkovitch R. L'attachement au cours de la vie. Paris: PUF; 2001.
27. Loas G. L'alexithymie. Ann Méd-Psychol. 2000;168. 2010;712–71.
28. Attale C, Consoli SM. Intérêt du concept d'attachement en médecine somatique. Presse Med. 2005;34:42–8.
29. Main M. Crosscultural studies of attachment organization: recent studies, changing methodologies, and the concept of conditional strategies. Hum Dev. 1990;33:48–61.
30. Heidenreich F, Revah-Lévy A. A propos des émotions : regards transculturels. In: Corcos M, Speranza M, editors. Psychopathologie de l'alexithymie. Paris: Dunod; 2003.
31. Izard CE. Innate and universal facial expressions: evidence from developmental and cross cultural research. Psychol Bull. 1994;115(2):288–99.

32. Mead M (1928–1935). Moeurs et sexualité en Océanie. Paris: Pocket; 1993.
33. Thomas LV. Rites de mort. Pour la paix des vivants. Paris: Fayard; 1985.
34. Marty P, M'Uzan M. De la pensée Opératoire. Rev Fr Psychanal. 1963;27:345–56.
35. Corcos M, Guilbaud O, Speranza M. Approche psychanalytique de l'alexithymie. In: Corcos M, Speranza M, editors. Psychopathologie de l'alexithymie. Paris: Dunod; 2003.
36. Spitz R. De la naissance à la parole. La première année de la vie. Paris: P.U.F; 1968.
37. Bagby RM, Parker JDA, Taylor GJ. The twenty-item Toronto Alexithymia Scale-I. Item selection and cross-validation of the factor structure. J Psychosom Res. 1994;38:23–32.
38. Parker JD, Graeme J, Taylorb R, Bagby M. The 20-Item Toronto Alexithymia Scale III. Reliability and factorial validity in a community population. J Psychosom Res. 2003;55:269–75.
39. Montreuil M, Benisty S, Lyon-Caen O. Alexithymie: étude des marqueurs langagiers en fonction de l'implication émotionnelle. Communication à la société française de Psychologie. Actes du Colloque, Université Paris VIII. Saint-Denis; 1999.
40. Pedinielli JL, De Bonis M, Somogyi M, et al. Alexithymie et récit de la maladie: contribution de la statistique textuelle à l'analyse des conduites langagières en psychopathologie. Rev Psychol Appl. 1989;39(1):51–67.
41. Taylor GJ. Psychosomatic medicine and contemporary psychoanalysis. Madison: International Universities Press; 1988.
42. Chahraoui K, et al. Alexithymie et traumatisme psychique. l'Encéphale. 2000;5:15–21.
43. Donabedian D. La fonction économique du langage, le mot action. Revue française de psychanalyse. 1998;3:810–819.
44. Montreuil M, Petropoulou H. Emotional disturbances in neurology and psychiatry: mood and emotions in multiple sclerosis. Neuropsy News. 2003;2:91–6.
45. Chahraoui K, Pinoit JM, Viegas N, Adnet J, Bonin B, Moreau T. Alexithymia and links with depression and anxiety in multiple sclerosis. Rev Neurol (Paris). 2008;164:242–5.
46. Bodini B, Mandarelli G, Tomassini V, et al. Alexithymia in multiple sclerosis: relationship with fatigue and depression. Acta Neurol Scand. 2008;118:18–23.
47. Gay MC, Vrignaud P, Garitte C, Meunier C. Predictors of depression in multiple sclerosis patients. Acta Neurol Scand. 2010;121:161–70.
48. Chahraoui K, Duchene C, Rollot F, Bonin B, Moreau T. Longitudinal study of alexithymia and multiple sclerosis. Brain Behav. 2014;4(1):75–82.
49. Dahl OP, Stordal E, Lydersen S, Midgard R. Anxiety and depression in multiple sclerosis. A comparative population-based study in Nord-Trondelag County, Norway. Mult Scler. 2009;15:1495–501.
50. Feinstein A, O'Connor P, GRAY T, Feinstein K. The effects of anxiety on psychiatric morbidity in patients with multiple sclerosis. Mult Scler. 1999;5:323–6.
51. Arnett PA, Randolph JJ. Longitudinal course of depression symptoms in multiple sclerosis. J Neurol Neurosurg Psychiatry. 2006;77:606–10.
52. Beal CC, Stuifbergen AK, Brown A. Depression in multiple sclerosis: a longitudinal analysis. Arch Psychiatr Nurs. 2007;21:181–91.
53. Giordano A, Granella F, Lugaresi A, et al. Anxiety and depression in multiple sclerosis patients around diagnosis. J Neurol Sci. 2011;307:86–91.
54. Chwastiak L, Ehde DM, Gibbons LE, Sullivan M, Bowen JD, Kraft GH. Depressive symptoms and severity of illness in multiple sclerosis: epidemiologic study of a large community sample. Am J Psychiatry. 2002;159:1862–8.
55. Giraudet JS. Annonce du diagnostic de maladie chronique à un patient. Synoviale. 2006;151:8–13.
56. Pakenham KL. Making sense of multiple sclerosis. Rehabil Psychol. 2007;52(4):380–9.
57. Russel CS, White MB, White CP. Why me? Why now? Why multiple sclerosis? Making meaning and perceived quality of life in a midwestern sample of patients with multiple sclerosis. Fam Syst Health. 2006;24–1:65–81.

58. McPheters JK, Sandber JG. The relationship among couples, relationship quality, physical functioning, and depression in multiple sclerosis patients and partners. Fam Syst Health. 2010;28–1:48–68.
59. Montreuil M. La prise en charge psychologique de la SEP. 2001. http://www.sclerose-en-plaques.apf.asso.fr
60. Dennison L, Moss-Morris R, Chalder T. A review of psychological correlates of adjustment in patients with multiple sclerosis. Clin Psychol Rev. 2009;29:141–53.

Chapter 15
Social Cognition and Multiple Sclerosis

Cécile Dulau

Abstract Social cognition (SC) impairment can contribute to everyday life difficulties in multiple sclerosis (MS). Studies tend to show that facial emotion recognition and generally emotion processing are impaired in MS. Only one study failed to show impairment in facial emotion recognition, but MS patients had a slow emotional processing. Deficits in cognitive and affective inferences about mental states of others have also been demonstrated in various heterogeneous groups and with various methodologies used to assess theory of mind. The relationship of cognitive impairment with SC impairment is still discussed. Results from functional imaging studies in MS suggested that there is brain dysfunction in areas previously demonstrated as key regions for emotional processing (ventrolateral prefrontal cortex, left anterior insula, limbic area). Results concerning morphological MRI studies, brain lesion volumes, and cortical thinning are still very preliminary, but one study suggested that cortical thinning in some regions of the brain correlated with emotion recognition test and mentalizing tasks. Results concerning correlation of SC with brain lesions are conflicting. Morphological and functional MRI studies are still limited in this area, and more studies are needed for a better understanding of these impairments.

Keywords Multiple sclerosis • Social cognition • Emotion recognition • Theory of mind (ToM) • Empathy • Functional MRI (fMRI) • Prefrontal cortex • Amygdala

Abbreviations

MS	multiple sclerosis
ToM	theory of mind
HCS	healthy control subjects
fMRI	functional MRI
VLPLC	ventrolateral prefrontal cortex

C. Dulau, MD (✉)
Centre Hospitalier Universitaire Pellegrin, Bordeaux, France

Department of Neurology, Université de Bordeaux, Bordeaux 33076, France
e-mail: cecile.dulau@chu-bordeaux.fr

B. Brochet (ed.), *Neuropsychiatric Symptoms of Inflammatory Demyelinating Diseases*, Neuropsychiatric Symptoms of Neurological Disease,
DOI 10.1007/978-3-319-18464-7_15

Introduction

In addition to the physical, cognitive, and psychological symptoms experienced by individuals with multiple sclerosis (MS), there is new research that indicates that some individuals with MS have difficulties in social cognition (SC). Studies about this topic in MS mainly investigated emotion recognition (especially facial emotion recognition) or theory of mind (ToM) in its two components, cognitive and affective ToM.

As fiber tracts and brain regions implicated in emotional recognition or ToM tasks may overlap with demyelination areas [1], morphological and functional magnetic resonance imaging (MRI, fMRI) correlation studies have been performed [2–5]. In particular fMRI studies investigated brain activation during recognition of emotional face expressions in MS [2–4].

Emotion Recognition and MS

Facial Emotion Recognition

Emotion recognition has been investigated in MS in a few studies [5–12] Authors used various methodologies to assess recognition of emotions: five studies [6, 7, 9, 11, 12] used the Ekman Faces Test [13], one [10] used the Baron-Cohen Faces Test [14], one used a computerized task [6], and another one [8] used the Florida Affect Battery [15]. In 1989, the first study of facial affect recognition in MS [11] showed that MS patients had significantly lower scores than healthy controls on the facial recognition test. However, the authors also found impairment in face identification so they argued that these results could be due to a general afferent visual defect. Another study did not found impairment in the facial identification task suggesting no visuoperceptual deficits in this task [9]. In a study performed in patients with a clinically isolated syndrome suggestive of MS, the authors did not found any emotion recognition impairment although patients had decreased reaction times regarding emotion recognition tests compared to healthy controls suggesting a slowing in facial emotional processing [5]. In another study [6], no overall group differences in facial affect recognition was identified, but specific difficulties were observed in decoding two facial emotions: anger and fear. The deficits in the emotion recognition task were related to cognitive domains, especially information processing speed.

The other five studies [7–10, 12] consistently showed significant differences between MS patients and healthy subjects for recognition of facial emotions. Among those studies, authors found correlations between cognitive domains such as information processing speed [6, 12], working memory [8, 12], and sustained attention [12]. However, facial recognition test correlated only with psychological and social aspects of quality of life in one study but not with neurocognitive tests [9]. Depression and fatigue were correlated with facial emotion recognition only in one study [8].

Affective Prosody Recognition

Two studies found impairment for identifying emotional states from prosodic cues in MS patients in comparison with healthy subjects [16, 17] However, the first study [16] used only primary emotional states that did not give any information on more subtle discriminations (such as sarcasm and sincerity), important in social functioning. The second study [17] investigated facial affect recognition in addition to affective prosody recognition and showed that both were impaired, at an early stage of the disease in relapse-remitting MS patients. Finally, both studies did not find any correlation with neurocognitive factors.

Theory of Mind and Empathy

So far, eight studies using ToM and empathy tests in MS adult patients have been published.

These studies suggested that MS patients may have difficulties in cognitive ToM tests or in attribution of intentions or thoughts, such as false-belief tasks [7, 18] and eyes test [6, 10]. Difficulties in strange stories tasks were found only in cognitively impaired MS patients [19]. Empathy has been studied in two studies [20] which found significant lower scores in MS patients. Some studies used tests assessing both cognitive and affective ToM, such as the faux pas test. The five studies using the latter test [7, 10, 18, 19, 21] found consistent scores significantly lower in MS patients than in controls, but with some important differences: it concerned only cognitively impaired MS patients in one study [19], patients with mild disability in another study [10], and only the cognitive ToM part of the faux pas in a third study [21].

Video tests reflect better everyday life and the complexity of real-world social interactions than classical tests. The Movie for the Assessment of Social Cognition (MASC) [22] and the Conversations and Insinuations Video (C&I) Task [19] have been used in MS. One study using the C&I test [19] found low scores only in cognitively impaired MS patients. Two studies [18, 23] using MASC found that identification of thoughts and intentions and emotion identification were both significantly impaired in MS [23].

It still remains unclear whether ToM impairment in MS is secondary to neurocognitive impairment or may occur independently. Less than half of the studies concerning ToM and empathy in MS [7, 10, 21] failed to demonstrate an association, but neuropsychological measures were different than in other studies, with only IQ or executive testing. The remaining studies found some correlations between ToM tasks and neurocognitive performances. For example, one study found a correlation between ToM impairment and performances on processing speed, memory, and executive function tasks, but ToM impairment remained significant in MS patients without neuropsychological deficits as compared to healthy subjects [23]. A second study found a correlation of ToM scores only with the Stroop test but not with working memory and set-shifting tests [18]. It has been also reported a relationship between empathy and executive functions in one study [20]. The last study found more important correlations between ToM tasks and impairment in several cognitive domains including attention, memory, and working memory [21]. The authors concluded: "MS, in the absence of cognitive deficits, cannot be considered to be a contributing factor to ToM deficits. The contributing factor is the cognitive deficits, not the MS per se."

Other Studies About Social Cognition in MS

Social Cognition in Pediatric-Onset MS Patients

MS also affects children and adolescents with an estimated frequency of 0.5 to 1.0 per 100,000 [24]. One study [25] tried to identify whether these youngest patients with MS are also at risk for SC impairment. Twenty-eight patients (mean age 16 years) were compared to 32 healthy controls using ToM tasks and a speed processing test (Symbol Digit Modalities Test, SDMT). All three scores of ToM tests were lower in MS patients than in healthy children, and group differences remained after controlling for SDMT z-scores.

Altered Decision-Making

By using the Gambling Task [26], a study showed a slower learning in avoiding disadvantageous choices than healthy subjects [27]. Patients were also tested by executive tests, a Dysexecutive Questionnaire and a scale of anxiety and depression. This slower learning was associated with impaired emotional reactivity, measured by a skin conductance response. Authors showed no correlations with executive tests. A correlation was found between anxiety and altered decision-making. They concluded that altered decision-making in MS probably depends on impaired emotional reactivity.

In summary, studies tend to show that emotion processing and ToM are impaired in MS. Only one study [5] failed to show any impairment in facial emotion recognition, but MS patients were at a very early stage and even had a slow emotional processing. Deficits in cognitive and affective inferences about mental states of others have also been demonstrated in heterogeneous groups with various methodologies. However, significant difference remains when "real-life testing" is used [18, 23] or when a study uses the largest sample of patients [7]. The relationship of cognitive impairment with SC processing is still discussed, but studies that assessed large cognitive batteries [19, 20] or sensitive neuropsychological measures to cognitive impairment in MS [8, 23] showed an association with speed processing in facial emotion recognition and with memory and executive functions in ToM tasks. Table 15.1 shows a review of the literature about SC in MS.

MRI Studies

Functional MRI Studies

Only a few functional MRI (fMRI) studies have investigated recognition of emotion in MS and none of ToM so far to our knowledge.

The first study [2] compared 11 MS patients found impaired for emotion recognition in a previous study [8] with 11 MS unimpaired and 11 healthy controls, by using lesion mapping on morphological MRI and fMRI during a facial affect matching task. The MS phenotypes were heterogeneous and unbalanced between groups (7 relapsing-remitting and 4 secondary progressive in the impaired MS group and 10 relapsing-remitting and 1 primary progressive in the unimpaired MS group). The impaired MS group showed decreased activation in the *left anterior insula* and in the *ventrolateral prefrontal cortex (VLPFC)* as compared to the unimpaired MS group and the healthy subjects. The decreased activation correlated with the percentage of correctly recognized unpleasant facial expressions. They also found a correlation between decreased emotion recognition and the presence of lesions in the left temporal white matter. The total lesion volume did not differ significantly between the groups.

In another study [3] 12 relapsing-remitting MS patients were compared to 12 healthy subjects using lesion mapping and fMRI during an emotion recognition task. MS patients were cognitively unimpaired and did not receive disease-modifying therapies. No impairment was found in emotional stimuli processing. Although no difference was found for structural measures, MS patients displayed even so significantly greater responses within the VLPFC compared to HC during the task. MS patients also showed a reduced functional connectivity between *two prefrontal areas* and the *amygdala*. Authors hypothesized that demyelination alter the axonal conduction between these areas. An enhanced regional response of task-dependent regions (i.e., the VLPFC) may represent a compensatory mechanism aimed to adapt the manifestation of emotional symptoms in MS.

Another study [4] compared 15 mildly disabled relapsing-remitting MS patients to 15 healthy subjects using fMRI during a behavioral testing for emotion recognition and lesion mapping. Patients showed no differences in cognitive testing. As the previous study, MS patients were not impaired in emotional recognition task. They reported an increased activation in the *posterior cingulate cortex (CC)* and *precuneus* in MS patients compared to healthy subjects for emotional faces (anger and disgust) and an increased activation of the occipital fusiform gyri and the anterior CC for neutral faces. No difference was found in structural measures (brain volume, lesion load).

Table 15.1 Literature review about social cognition in multiple sclerosis [5–12, 16–21, 23, 25, 27]

Study	Samples and MS types	Social cognition tests	Neurocognitive tests	Results	Correlation with cognitive impairment or other
Roca et al. [21]	*18 RR-MS* mild disability 16 HCS	Faux pas test	Executive functions	Significant impairment in cognitive ToM in faux pas test (ability to infer other's intention)	*NO* No correlation with executive dysfunction, depression, or fatigue
Charvet et al. [25]	*28 pediatric-onset MS* mean age 16 years EDSS 1-4 32 HCS	TOM tasks : Reading the mind in the eyes test, faux pas test False-belief task	SDMT (Symbol Digit Modalities Test)	Lower performance in all three ToM tasks	*NO* Group differences remained after controlling for SDMT z-scores
Berneiser et al. [8]	*61 MS* 47 RR; 3 PP; 11SP 53 HCS	FAB Florida Affect Battery	Paced Auditory Serial Addition Test PASAT Beck Depression Inventory BDI MS-specific fatigue scale : MS-FS	Poor performance in all subtests that required emotion recognition	*YES* Correlations with depression and fatigue in multivariate approach with fatigue depression cognition in univariate approach

Pottgen et al. [23]	*45 MS*	MASC Movie for Assessment of Social Cognition	SDMT	Significantly impaired ToM (impairment in emotion identification more than identification of thoughts or intentions)	*YES*
	31 RR; 8 SP; 13 PP		Multiple Choice Vocabulary Intelligence Test		Correlation with processing speed memory executive functions results remained significant in MS patients without neuropsychological deficits
	45 HCS		Verbal Learning and Memory Test		
			Executive functions		
Kraemer et al. [17]	*25 RR-MS* at an early stage < 2 years	"Tübingen affect prosody"	Short battery with attention and memory tests	Poor performance in affective prosody, except for the emotion "angry"	*NO*
	EDSS ≤ 2				No correlation with the neuropsychological tests
	25 HCS				
Kraemer et al. [18]	*25 RR-MS*	MASC Movie for Assessment of Social Cognition	Stroop test	Significant (impairment in attribution of emotions, intentions, or thoughts); empathy quotient significantly worse in MS patients	*YES*
	Early stage < 2 years	False-belief task	Span (WAIS III)		Only with Stroop test
	EDSS ≤ 2	Faux pas test	Trail Making Test		
	25 HCS	Empathy quotient Baron-Cohen			

(continued)

Table 15.1 (continued)

Study	Samples and MS types	Social cognition tests	Neurocognitive tests	Results	Correlation with cognitive impairment or other
Prochnow et al. [12]	*35 MS*	Benton Facial Recognition Test (BFRT)	Faces Symbol Test (FST) (information processing speed, working memory, sustained attention)	Normal neutral face recognition (BRFT)	*YES*
	29 SP; 5RR; 1 PP	Recognition of neutral faces	BDI	Poor performance in the PCFAE as well as in the Ekman-60-Faces Test : abnormalities of affect recognition in faces (fear, surprise, anger, and sadness)	MS patients scored under average in the FST that measures different cognitive domains such as information processing speed, working memory, and sustained attention. The FST scores significantly correlated with the PCFAE score
	61 HCs	Perceptual Competence of Facial Affect Recognition (PCFAE, computerized timed test not published)			
		Ekman-60-Faces Test			
Henry et al. [7]	*64 RR-MS*	False-belief tasks	IQ (WAIS -R/7)	Significant lower scores in all 3 tests in MS patients	*NO*
	30 HCS	Faux pas test (2 stories)	Executive functions: similarity test (WAIS-R), Brixton Spatial Anticipation Test (1997)		
		Ekman Faces Test 2002			

Phillips et al. [9]	*32 MS*	Faces test (Ekman and Benton)	Verbal fluency	Significant results in both static and dynamic facial emotion recognition	*NO*
	27 RR; 2PP; 3 SP	Emotion perception video (Sullivan)	Screening Examination for Cognitive Impairment (SEFCI)	No impairment in identity recognition	But correlation with psychological and social aspects of quality of life (WHOQoL-BREF)
	33 HCS	Gender and age perception video	Sustained Attention to Response Task (SART)		
Banati et al. [10]	*40 MS*	Baron-Cohen Faces test and Eyes Test	IQ (WAIS -R/7) verbal IQ and performance IQ	Eyes test and faces test significantly impaired lower scores in faux pas test, empathy Quotient only in EDSS (2.5-4.5) group	*NO*
	37 RR; 3 SP	Faux pas test			But with duration disease with eyes and faces test and empathy and with EDSS >2 with faux pas test and empathy
	35 HCS	Empathy Quotient Baron-Cohen			
Ouellet et al. [19]	41 MS	Happé Strange Stories	Complete battery: attention "bells test"; memory (Rey's Auditory Verbal Memory Test), working memory (PASAT, Digit Span subtest WAIS-III), executive functions (Trails A &B, verbal fluency, Zoo Map Test, Mazes subtest; Stroop Test), reasoning (similarities subtest, Card Sorting Test; Three-minute reasoning test; Picture arrangement subtest (WAIS III))	Significant results in all tests but faux pas test only in cognitively impaired MS patients (36 patients)	*YES*
	22 RR; 13 SP; 5 PP; 1 indeterminate form	Faux pas test			Attention "bells test"; memory (Rey's Auditory Verbal Memory Test) and working memory (Digit Span subtest WAIS-III)
	20 HCS	Video Conversations and Insinuations (C&I)			

(continued)

Table 15.1 (continued)

Study	Samples and MS types	Social cognition tests	Neurocognitive tests	Results	Correlation with cognitive impairment or other
Jehna et al. [5]	*20 MS*	Computerized test for facial emotion	Processing speed (faces processing speed)	No significant results	*YES*
	12 CIS; 7 RR; 1 SP				Lower scores of faces processing speed
	23 HCS				
Henry et al. [6]	27 MS (forms not precised)	Ekman-Friesen Faces	SEFCI : Screening examination for cognitive impairment	Significative difference in eyes test difference only found for anger and fear in the faces test	*YES*
	30 HCS	Baron-Cohen Eyes Test	Verbal fluency		Verbal fluency and processing speed
Kleeberg et al. [27]	*20 MS*	Gambling Task	Behavioral Assessment of the Dysexecutive Syndrome (BADS)	Slower learning than HCS (persisting in making disadvantageous choices)	*NO*
	16 HCS	Emotional reactivity by a skin conductance response	Trail Making Test		No correlations with executive tests
			Dysexecutive Quotient		Correlations with anxiety
			Hospital Anxiety and Depression Scale		
Beatty et al. [16]	47 MS	Aprosodia Battery (primary emotional states)	MMSE	Significative difference; greatest impairment for patients with severe physical disability	*NO*
	19 HCS		Boston Naming Test		With BDI, hearing loss, and neurocognition
			Western Aphasia Battery		
			Repeatable Battery for the Assessment of Neuropsychological Status (RBANS)		
			Beck Depression Inventory		

Benedict et al. [20]	34 MS EDSS 1–8,5 14 HCS	Hogan Empathy Scale (HES)	Language (Token test, Boston Naming Test), spatial processing (Line Orientation Test and Complex Figure Test), memory (California Verbal Learning Test and Brief Visuospatial Memory Test), attention (Trail Making Test part B; PASAT), executive function (Wisconsin Card Sorting test and Bookland Category Test)	Reduction in empathy in MS patients	*YES* With executive tests
Beatty et al. [11]	21 RR-MS	Benton Facial Recognition Test	MMSE	Poor performance in judging emotional expressions but also less accurate in discriminating neutral faces (significantly lower scores on the BFRT)	NO
		Ekman and Friesen Test	BDI		No correlations with BDI Positive correlations between scores on the BFRT (performance on the BFRT accounted for 43.6 % of the variance in performance on the affective judgment task)

Abbreviations: *MS* multiple sclerosis, *HCS* healthy control subjects, *RR* relapse-remitting, *SP* secondary progressive, *PP* primary progressive, *EDSS* expanded disability status scale

In relation with the two last studies [3, 4]: the reduced communication between key emotional regions and the increased activation of brain regions implicated in emotional treatment can be interpreted as a compensatory mechanism aimed to adapt the clinical expression of emotional symptoms in MS [3].

These observations fit into the concept of functional brain reorganization with the progression of MS [28].

Structural MRI Study

In addition to the studies described above investigating lesion mapping together with fMRI in which no difference were observed in morphological measures, another study assessed the relationship of morphological MRI parameters with SC in MS. This study [29] used quantitative MRI methods in patients with MS to investigate the impact of white matter lesion load and cortical atrophy on emotion recognition and mental state attributions. Emotion recognition test (Baron-Cohen Faces Test) and ToM tasks (Baron-Cohen eyes test and faux pas test) were administered in 49 MS patients and 24 healthy subjects. Then MRI images (T1- and T2-weighted three-dimensional brain) at 3 T were acquired from all patients with MS and 18 healthy controls. MS patients performed significantly poorer in the faces test and in the eyes test but not in the faux pas test. The MRI study showed that, after correction for the confounding factors (gender, EDSS, anxiety, and depression), both poor faces test and eyes test performances correlated with total T1-weighted lesion load and regional T1-lesion load of association fiber tracts interconnecting cortical regions related to visual processing (splenium of corpus callosum in particular) and emotion processing (genu of corpus callosum, right inferior longitudinal fasciculus, inferior fronto-occipital fasciculus, and left and right uncinate fasciculus) . No correlation was found with total T2-weighted lesion load. This study also showed that both poor faces test and eyes test performances correlated with cortical thinning of the *right fusiform face area.* Faces test showed correlation with thickness of the *right entorhinal cortex*. Eyes test performance showed correlation with thickness of the *left temporal pole* and *the right frontal eye field.* Although interesting, these results are preliminary, and other studies are needed for understanding the relationship between SC impairment and the disease process in MS.

References

1. Filley CM. White matter and behavioral neurology. Ann N Y Acad Sci. 2005;1064:162–83.
2. Krause M, Wendt J, Dressel A, Berneiser J, Kessler C, Hamm AO, et al. Prefrontal function associated with impaired emotion recognition in patients with multiple sclerosis. Behav Brain Res. 2009;205:280–5.

3. Passamonti L, Cerasa A, Liguori M, Gioia MC, Valentino P, Nistico R, et al. Neurobiological mechanisms underlying emotional processing in relapsing-remitting multiple sclerosis. Brain. 2009;132:3380–91.
4. Jehna M, Langkammer C, Wallner-Blazek M, Neuper C, Loitfelder M, Ropele S, et al. Cognitively preserved MS patients demonstrate functional differences in processing neutral and emotional faces. Brain Imaging Behav. 2011;5:241–51.
5. Jehna M, Neuper C, Petrovic K, Wallner-Blazek M, Schmidt R, Fuchs S, et al. An exploratory study on emotion recognition in patients with a clinically isolated syndrome and multiple sclerosis. Clin Neurol Neurosurg. 2010;112:482–4.
6. Henry JD, Phillips LH, Beatty WW, Mcdonald S, Longley WA, Joscelyne A, et al. Evidence for deficits in facial affect recognition and theory of mind in multiple sclerosis. J Int Neuropsychol Soc. 2009;15:277.
7. Henry A, Tourbah A, Chaunu M-P, Rumbach L, Montreuil M, Bakchine S. Social cognition impairments in relapsing-remitting multiple sclerosis. J Int Neuropsychol Soc. 2011;17:1122–31.
8. Berneiser J, Wendt J, Grothe M, Kessler C, Hamm AO, Dressel A. Impaired recognition of emotional facial expressions in patients with multiple sclerosis. Mult Scler Relat Disord. 2014;3:482–8.
9. Phillips LH, Henry JD, Scott C, Summers F, Whyte M, Cook M. Specific impairments of emotion perception in multiple sclerosis. Neuropsychology. 2011;25:131–6.
10. Banati M, Sandor J, Mike A, Illes E, Bors L, Feldmann A, et al. Social cognition and theory of mind in patients with relapsing-remitting multiple sclerosis. Eur J Neurol. 2010;17:426–33.
11. Beatty WW, Goodkin DE, Weir WS, Staton RD, Monson N, Beatty PA. Affective judgments by patients with Parkinson's disease or chronic progressive multiple sclerosis. Bull Psychon Soc. 1989;27:361–4.
12. Prochnow D, Donell J, Schäfer R, Jörgens S, Hartung HP, Franz M, et al. Alexithymia and impaired facial affect recognition in multiple sclerosis. J Neurol. 2011;258:1683–8.
13. Ekman P, Friesen W. Pictures of facial affect. Palo Alto, CA: Consulting Psychologist Press; 1976.
14. Baron-Cohen S, Wheelwright S. Jolliffe, Therese. Is there a "language of the eyes"? Evidence from normal adults, and adults with autism or Asperger syndrome. Vis Cogn. 1997;4:311–31.
15. Bowers D, Blonder LX, Heilman KM. The Florida affect battery, manual (revised). Gainesville: Center for Neuropsychological Studies, University of Florida; 2001.
16. Beatty WW, Orbelo DM, Sorocco KH, Ross ED. Comprehension of affective prosody in multiple sclerosis. Mult Scler. 2003;9:148–53.
17. Kraemer M, Herold M, Uekermann J, Kis B, Daum I, Wiltfang J, Berlit P, Diehl RR, Abdel-Hamid M. Perception of affective prosody in patients at an early stage of relapsing-remitting multiple sclerosis. J Neuropsychol. 2013;1:91–106.
18. Kraemer M, Herold M, Uekermann J, Kis B, Wiltfang J, Daum I, et al. Theory of mind and empathy in patients at an early stage of relapsing remitting multiple sclerosis. Clin Neurol Neurosurg. 2012;115:1016–22.
19. Ouellet J, Scherzer PB, Rouleau I, Métras P, Bertrand-Gauvin C, Djerroud N, et al. Assessment of social cognition in patients with multiple sclerosis. J Int Neuropsychol Soc. 2010;16:287.
20. Benedict RH, Priore RL, Miller C, Munschauer F, Jacobs L. Personality disorder in multiple sclerosis correlates with cognitive impairment. J Neuropsychiatry Clin Neurosci. 2001;13:70–6.
21. Roca M, Manes F, Gleichgerrcht E, Ibanez A, Gonzalez de Toledo ME, Marenco V, Bruno D, Torralva T, Sinay V. Cognitive but not affective theory of mind deficits in mild relapsing-remitting multiple sclerosis. Cogn Behav Neurol. 2014;1:25–30.
22. Dziobek I, Fleck S, Kalbe E, Rogers K, Hassenstab J, Brand M, et al. Introducing MASC: a movie for the assessment of social cognition. J Autism Dev Disord. 2006;36:623–36.

23. Pottgen J, Dziobek I, Reh S, Heesen C, Gold SM. Impaired social cognition in multiple sclerosis. J Neurol Neurosurg Psychiatry. 2013;84:523–8.
24. Langer-Gould A, Zhang JL, Chung J, et al. Incidence of acquired CNS demyelinating syndromes in a multiethnic cohort of children. Neurology. 2011;77:1143–8.
25. Charvet L, Cleary R, Vazquez K, Belman A, Krupp L, on behalf of the US Network for Pediatric MS. Social cognition in pediatric-onset multiple sclerosis (MS). Mult Scler J. 2014;82.
26. Bechara A, Damasio AR, Damasio H, Anderson SW. Insensitivity to future consequences following damage to human pre- frontal cortex. Cognition. 1994;50:7–15.
27. Kleeberg J, Bruggimann L, Annoni J-M, van Melle G, Bogousslavsky J, Schluep M. Altered decision-making in multiple sclerosis: a sign of impaired emotional reactivity? Ann Neurol. 2004;56:787–95.
28. Sumowski JF, Wylie GR, Deluca J, Chiaravalloti N. Intellectual enrichment is linked to cerebral efficiency in multiple sclerosis: functional magnetic resonance imaging evidence for cognitive reserve. Brain. 2009;133(2):362–74.
29. Mike A, Strammer E, Aradi M, Orsi G, Perlaki G, Hajnal A, et al. Disconnection mechanism and regional cortical atrophy contribute to impaired processing of facial expressions and theory of mind in multiple sclerosis: a structural MRI study. PLoS ONE. 2013;8, e82422.

Chapter 16
Cognitive Impairment in Multiple Sclerosis

Aurélie Ruet

Abstract Cognitive impairment (CI) is important to be detected in patients living with MS due to several following reasons. First, even if CI is often underestimated by patients and physicians, patients with MS are frequently cognitively impaired, and cognitive deficits could be observed in different stages and phenotypes of MS. Information processing speed has been proposed to be the main cognitive domain impaired in patients with MS. It appears crucial to take cognition into account in the clinical practice and to perform neuropsychological assessment with dedicated tools. Concretely, CI could affect daily, familial, social, and vocational activities and alter the health-related quality of life of patients with MS. The pathophysiology of CI is still not completely elucidated, and this research field gains interest. Both focal and diffuse white and gray matter damage participate in explaining CI in MS. At the early stage of the disease, CI could be used as a prognostic marker and could contribute in defining the severity of the pathology. Consequently, detecting CI could influence the therapeutic strategy in MS and studies investigating specific treatment are in progress.

Keywords Cognition • Neuropsychological battery • Information processing speed • Episodic memory • Executive function • Prognostic • Cognitive compensation • Cognitive reserve • Cognitive remediation

Introduction

The nature, frequency, severity, and evolution of cognitive impairment (CI) seen in patients with multiple sclerosis (MS) will be explained in the first part of this chapter. Then, the neuropsychological (NP) batteries used in MS will be described and each NP test will be detailed. In the third part, the consequences of CI will be

A. Ruet, MD, PhD (✉)
Centre Hospitalier Universitaire Pellegrin, Bordeaux, France

Department of Neurology, Université de Bordeaux, Bordeaux 33076, France
e-mail: aurelie.ruet@chu-bordeaux.fr

B. Brochet (ed.), *Neuropsychiatric Symptoms of Inflammatory Demyelinating Diseases*, Neuropsychiatric Symptoms of Neurological Disease,
DOI 10.1007/978-3-319-18464-7_16

addressed. Concerning the pathophysiology of CI, imaging and histopathological data will be reported in order to illustrate anatomical substrates underlying CI in MS. Cognitive compensation and cognitive reserve will be approached in order to explain the clinico-radiological paradox and heterogeneity seen in patients with MS. Based on these correlates, the prognostic value of CI in MS will be demonstrated in the fifth part. Finally, therapeutic options will be discussed for managing CI in patients with MS.

Nature, Frequency, Severity, and Evolution of Cognitive Deficits

Cognitive dysfunction in patients with MS has long been underestimated both by patients and physicians in part due to the fact that cognitive deficits are invisible compared to motor or cerebellar symptoms, for instance. This topic has progressively gained interest in research and in clinical practice, and there is now increasing evidence that CI is common in MS [1, 2].

Nature of Cognitive Deficits

Information processing speed (IPS) is commonly reduced in patients with MS. There are some controversial data concerning the respective contribution of IPS and working memory on cognitive functioning. One approach is to consider that impairment in IPS could affect primarily the functioning of the other cognitive domains. Thus, patients with MS could perform normally if they have enough time at least in the beginning of the disease. Some studies have supported this theory suggesting that IPS impairment is a central and key cognitive defect in this disease [3, 4]. Another approach is to consider the mediating role of working memory that has been recently proposed in a study performed in patients with early relapsing-remitting MS (RRMS) [5]. Besides this important deficit in IPS, episodic memory is frequently impaired in MS [6, 7]. In a mixed sample of patients with MS, impairment in verbal and visuospatial episodic memories has been reported [7]. Poor performances were found at both the immediate and delayed recall suggesting impairment in the coding of the information. Impairment of executive functions is also an important cognitive deficit occurring during the disease with a negative impact [7].

Frequency of Cognitive Deficits

It has been recognized that CI is frequent in MS and could be identified in all types and stages of MS [1, 2, 8]. The frequency rates of CI in patients with MS could vary from 35 to 70 % at both early and late stages of clinically definite MS (CDMS) [6, 8].

On one hand, a comprehensive NP battery was administered to 100 community-based heterogeneous patients with MS, and 43 % of CI was detected in this pivotal study reported by Rao et al. [6]. One the other hand, previous university-based medical centers have reported that cognitive deficits were present in 54–65 % in patients with MS [9–11]. Recent studies have focused on more homogeneous sample of patients with MS. Thus, in a cohort of 44 early relapsing-remitting MS (RRMS), CI was detected in 45 % of patients within six months after MS diagnosis (defined by at least two abnormal NP tests below the fifth percentile compared to matched healthy controls (HCs)) [12]. In the same stage and using the same definition, an Italian multicentric study has reported 34.9 % of CI in a large cohort of more than 500 early RRMS patients [13]. In the same sample using a more stringent definition for CI (at least three abnormal NP tests below the fifth percentile compared to matched HCs), only 19.5 % of patients were classified as having CI. In contrast, at later stages of the disease but with mild disability assessed by the Expanded Disability Status Scale (EDSS), CI was observed in 45 % of a group of 163 patients who have been so-called benign MS (BMS) defined by a score of EDSS less or equal to 3.0 after at least 15 years of disease duration [14]. In fact, the real proportion of BMS patients could be overestimated through the lack of systematic cognitive assessment in MS in practice.

It is noteworthy that the frequency of CI reported in studies including patients with MS is basically heterogeneous. This is mainly due to methodological aspects. Indeed, the estimation of CI could vary in relation to the sample composition and could depend on the norms used for the interpretation of the results (published normative data or own sample of HCs matched to the studied patients for age, sex, and educational level). Moreover, the determination of CI depends on the method and the chosen definition used for classifying patients with or without CI. In fact, this comes from a lack of consensus on how to define CI in MS. Thus, the questions remain concerning the minimal numbers of abnormal NP tests or cognitive domains before classifying a patient as cognitively impaired. Another approach is to use Z-scores with the following formula for each NP score: "MS patient's score - mean value of their own matched HCs group)/SD of the matched HCs." Then, a chosen cut-off could be applied to Z-score per NP test in order to define a cognitively impaired patient for a given NP test. Besides, there is no strong consensus on the cut-off for defining an abnormal performance. The data are not homogeneous across the studies and could vary between 1 standard deviation (SD) to 2 SD when comparing the scores or Z-scores of patients to matched HCs. Considering a threshold of 1.64 SD (equivalent to the fifth percentile) could be a good compromise. This important question has been addressed in an interesting paper comparing the criteria of CI in MS studies according to inclusion criteria of patients (early versus late stages of MS) [15]. Three classification strategies have been individualized among 20 approaches used for classifying CI in MS and were applied differently depending on the stage of MS. One strategy is based on the number of abnormal NP tests, another on the determination of a composite score, and the last is a combination of the first two. Even if most of the researchers applied the first strategy, they used different cut-off for defining an abnormal score for each NP test. Nevertheless, it appears that the cut-off on about 20 % of abnormal tests with a score below the fifth percentile is used in most of the cases. One of the conclusions is that the choice of the classification appears to be driven by the sample of patients (early versus late stage of MS). In studies done at the early stage of the disease, a more liberal definition is mainly chosen, whereas a more stringent and conservative definition is applied at later stage of the pathology.

The relationship between the frequency of CI and disease duration has been questioned. After the first clinical event suggestive of MS called clinically isolated syndrome (CIS), there is increasing evidence that cognitive deficits could be present even if they could be detected in a lower frequency than those observed in RRMS (from 25 to 30 %). Additionally, the deficits are more focused in CIS than in later stage of MS [10, 16–21], and the most impaired cognitive domains are IPS, working memory, attention, and verbal fluency. Moreover, at a preclinical stage suggestive of MS called radiologically isolated syndrome (RIS), the same pattern of cognitive deficits has been observed as previously described in one third of the sample (from 27.6 [22] to 30.8 % [23].

In contrast to RRMS, little information is available concerning cognitive dysfunction in progressive MS patients [24–30]. In one study comparing CIS, RRMS, and progressive MS divided by primary and secondary progressive MS (PPMS and SPMS, respectively), a continuum has been demonstrated in terms of frequency of cognitively impaired patients taking into account the scores of each NP test included in the battery [29]. These data suggest that there is an increase of CI from CIS to RRMS to SPMS.

In contrast, the actual frequency and the nature of CI in patients with PPMS are not fully established due to some methodological limitations of studies including heterogeneous samples of patients with MS. Indeed, patients with RRMS and those with PPMS are frequently different in terms of demographics findings such as age and gender, so appropriate control groups are needed for correct matching a priori. One study has specifically taken these differences into account by including more than 400 HCs in order to match adequately patients and controls for age, sex, and educational level [30]. It has been demonstrated that patients with PPMS had more diffuse CI than those with RRMS form. IPS was the most frequently impaired cognitive domain in both PPMS and RRMS patients, and the two cognitive domains, which differed between these two types of MS, were verbal episodic memory and executive function with respect to the frequency.

Severity of Cognitive Impairment

Few studies have directly compared the severity of CI in different types of MS [24–30]. In the study comparing 415 HCs, 60 RRMS patients, and 41 PPMS patients, one important finding was the difference of CI in terms of severity between these two types of MS [30]. Patients with PPMS had not only more diffuse CI but also more severe cognitive deficits than patients with RRMS especially in verbal episodic memory and working memory. Notably, patients with PPMS had more pronounced CI than patients with RRMS, even after controlling for physical disability, as assessed using the EDSS score, with the same mean disease duration.

Evolution of Cognition in MS

Whereas there are a lot of cross-sectional studies on cognition in MS, few studies had a longitudinal design that could investigate the progression of cognitive deficits in patients with MS. One should be cautious in the interpretation of the results in that type of studies due to inter-patient variability. The follow-up period varies in range from 1 to 18 years [31–38]. The course of cognitive performance in patients with MS is partly contradictory, as some studies have reported the preservation of cognitive functioning, whereas others have observed a mild to moderate cognitive decline over time in MS [39]. In fact, methodological factors often limit the direct comparison of the results, such as the difference in the composition of studied sample, the length of the follow-up period, and the definition chosen for cognitive decline over time. In one 3-year follow-up study, patients with MS were divided into two groups – a group of cognitively preserved (CP) and a group of cognitively impaired patients at baseline – with the same level of physical disability [32]. The patients from the first group remained cognitively stable in the majority of cases, except for one third of patients who exhibited slight deterioration. In contrast, more than two thirds of the patients considered impaired at baseline presented a cognitive decline in many NP tests. These findings suggest that early cognitive decline could predict further widespread and progressive deterioration, whereas patients with intact cognitive performances might remain stable. The relative short-term of follow-up could explain the absence of cognitive decline in the first group of patients. In a 10-year longitudinal study of 45 MS patients, cognitive deterioration was reported in all patients, even in patients without initial CI [33]. During the first 7 years after MS diagnosis, 40.9 % of cognitively impaired patients and 59.1 % of CP patients showed deterioration in memory domains, whereas almost one third of patients (22.7 %) – including both patients with and without CI – presented IPS deterioration [40]. One recent study has reported the cognitive performances of patients included in one phase III clinical trial of intramuscular interferon beta 1a [38]. One advantage of this study is the long period of follow-up since the last assessment was performed 18 years after the inclusion. A cognitive deterioration has been observed and it concerns mainly IPS domain. Interestingly, the decline over time of IPS was found more frequently in the unimpaired patients than the impaired group of patients at baseline. Looking at the early stage of the disease, it has been reported that the proportion of cognitively impaired patients could almost double in the years following the CIS (from 29 % at the CIS stage to 54 % 5 years later) [41]. In one-year follow-up study, the occurrence of isolated cognitive relapses (ICRs) was associated with poor cognitive performance suggesting ICRs as a factor for cognitive decline in MS [42]. The ICRs were defined as a transient reduction of the Symbol Digit Modalities Test (SDMT) [43] score of at least four points during the relapse in comparison to pre- and post-relapse assessment. Notably, ICRs were not reported by patients who did not feel any change either in cognition, mood, or fatigue and were detected only by objective evaluation.

How to Assess Cognitive Function

One challenging question is how to assess cognitive function in patients with MS in clinical practice and in research activities. The gold standard consists of the administration of a comprehensive NP battery performed by a qualified practitioner (neuropsychologist, neurologist). Thus, the most commonly used NP battery in MS is the Brief Repeatable Battery of Neuropsychological Tests (BRB-N) proposed by Rao et al. [44], which includes tests of attention, IPS, episodic verbal and visuospatial memory, and verbal fluency (Tables 16.1 and 16.2). A French NP battery has been proposed after modifying some NP tests and adding others in order to explore executive functions in particular [45] (Table 16.1). In 2002, a group of experts proposed a new battery called the Minimal Assessment of Cognitive Function in MS (MACFIMS) based on a consensus approach [46] (Table 16.1). The aim of that battery is to cover five cognitive domains commonly impaired in MS such as IPS/working memory, learning and memory, executive function, visual-spatial processing, and language. In

Table 16.1 Neuropsychological tests included in neuropsychological batteries used in multiple sclerosis [44–46]

	BRB-N [44]	BCcogSEP [45]	MACFIMS [46]
Information processing speed	SDMT	WAIS	SDMT
Working memory	PASAT 3 s	PASAT	PASAT 3 s
		Numeral backward span test	PASAT 2 s
Verbal episodic memory	SRT	Modified SRT	CVLT-II
Visuospatial episodic memory	SPART (10/36)	SPART	BVMT-R
		Numeral forward span test	
Executive functions	WLG	WLG	COWAT
Language, verbal fluency		Opposite orders, Go/No-Go, letter/numbers sequences	D-KEFS sorting test
Others			
Visual perception/spatial processing			JLOT

BRB-N Brief Repeatable Battery of Neuropsychological Tests [44], BCcogSEP [45], *MACFIMS* Minimal Assessment of Cognitive Function in Multiple Sclerosis [46], *SDMT* Symbol Digit Modalities Test, *WAIS* Wechsler Adult Intelligence Scale, *PASAT 3 s* Paced Auditory Serial Addition Test 3.0 s, *PASAT 2 s* Paced Auditory Serial Addition Test 2.0 s, *SRT* Selective Reminding Test, *CVLT-II* California Verbal Learning Test-Second Edition, *SPART (10/36)* Spatial Recall Test, *BVMT-R* Brief Visuospatial Memory Test-Revised, *WLG 90* Word List Generation Test, *COWAT* Controlled Oral Word Association Test, *D-KEFS sorting test* Delis-Kaplan Executive Function System Sorting Test, *JLOT* Judgment of Line Orientation Test parallel, it is worth to mention that confounding factors like fatigue, depression, and anxiety must be assessed as they could influence cognitive performance.

Another option to assess cognition is the administration of self-questionnaires. Thus, one auto-questionnaire called the MS Neuropsychological Screening Questionnaire (MSNQ) has been proposed for patients and informants [47]. Unluckily, the cognitive self-report complaints do not reflect cognitive test performance in MS, but are more likely associated with depressive symptoms [47–49]. Nevertheless, fulfilling this type of questionnaires by informants could be helpful as it has been considered more reliable than self-reports fulfilled directly by patients with MS [47, 49].

One limitation of the use of comprehensive NP is that its administration is not feasible everywhere in clinical practice and it is time consuming. So, the issue has been to determine which relevant NP tests could be used minimally for detecting cognitive dysfunction in MS and for selecting patients who require additional evaluation from an expert. The SDMT [43] has been proposed as a good candidate for detecting CI in comparison to other NP tests in early RRMS patients [48] and in a mixed sample of patients with MS (both RRMS and SPMS patients) [50]. This test described in Table 16.2 is part of both the BRB-N and MACFIMS batteries. Notably, it is associated with a good reliability in several assessments [51, 52]. Thus, this IPS test has been chosen to be part of the Brief International Cognitive Assessment for Multiple Sclerosis (BICAMS) which consists of the minimal cognitive evaluation required for patients with MS [53] (Table 16.3). Nevertheless, one weakness of the SDMT is its practice effect, and a computerized screening cognitive test (CSCT) [54], detailed in Table 16.2, has been proposed for limiting this. The CSCT was associated with a good accuracy for assessing IPS in patients with MS in comparison to other IPS tests included in the test of attentional performance (TAP) [57]. In addition to the SDMT, it has been recommended by this group of experts to include the California Verbal Learning Test-Second Edition [58] to assess episodic verbal memory and the Brief Visuospatial Memory Test-Revised [59] to explore episodic visuospatial memory as memory dysfunction occurs frequently in MS too (Tables 16.2 and 16.3). The application of this brief cognitive assessment is ongoing in international research studies in MS.

Consequences of Cognitive Impairment

Cognitive impairment could affect different aspects in the lives of persons with MS. There are some direct and indirect consequences in terms of daily activities, social function, leisure activities, and interpersonal relationships with family, partners, and friends [1, 2, 60]. Moreover, cognitively impaired patients were more unemployed than cognitively unimpaired patients in several studies [60–62]. Importantly, early cognitive status, independently to physical disability, contributed to the vocational status change in a cohort of patients included after the diagnosis of

Table 16.2 Neuropsychological tests used for assessing the different cognitive domains impaired in multiple sclerosis [44, 46, 54–57]

Cognitive domains and functions	Neuropsychological tests	Description	Score	Advantage	Weakness
Information processing speed (visual)	Oral SDMT	Nine digits/symbols, oral substitution test during 90 s	Number of accurate answers	Very sensitive test, short test	Practice effect
		A unique key showing the association of symbols with digits is provided, and it is similar for each test session			
	CSCT	Nine digits/symbols, oral substitution computerized test during 90 s	Number of accurate answers	Very sensitive test	Need to be validated in multicentric study
		The sequences of symbols and digits of the key are automatically generated for each session of training and testing and are different from one session to another		Short test	
				Weak practice effect	
	WAIS-R	Nonverbal digits/symbols, substitution tasks	Number of accurate answers		Written test
	Digit symbol				Less sensitive than the SDMT
					Could be affected by hand deficiencies
Information processing speed (auditory) and working memory	PASAT	Oral and auditory test	Number of correct sums	Two alternative forms	Important practice effect
	Version 2 or 3 s	61 numbers are given orally every 2 or 3 s and the subjects should add the number they just heard with the number they heard before			Stressful
					Exploration of different functions (working memory, IPS, inhibition, mental arithmetic)

	Numeral backward span test	Oral test	Number of correct sequences of numbers	Working memory could be analyzed independently of IPS	
		The subject should repeat a sequence of numbers in reverse			
Verbal episodic memory	SRT	Oral test	Immediate and delayed recall scores: number of corrected words	Two alternative forms	Important practice effect
		A list of 12 unrelated words is read aloud to the subject who is asked to repeat as many words as possible in any sequence. The examiner then provides any words not recalled, after which the subject could try to recall the entire list. This test is repeated five times, followed by a 20- to 25-min interval, after which the subject should recall and replicate the list again			Only short- and long-term memory are investigated
					No exploration of indexing
	Modified SRT	Oral test	Immediate and delayed recall scores: number of corrected words	16 words of the BcCog	
	CVLT	Oral test	Immediate and delayed recall scores: number of corrected words	Learning task More comprehensive test of memory: interference could be investigated	
		Learning test of 16 words with five learning trials and interference learning task before the delayed recall after 20- to 25-min interval			
	Numeral forward span test	Oral test	Number of correct sequences of numbers		Low sensibility of this test
		The subject should repeat a sequence of numbers in the correct order			

(continued)

Table 16.2 (continued)

Cognitive domains and functions	Neuropsychological tests	Description	Score	Advantage	Weakness
Visual/spatial episodic memory	10/36 SPART	Memory test using a 6 X 6 checkerboard with 10 pieces placed in specific locations. After 10 s, the subject should replicate the pattern on a blank checkerboard. The test is repeated 3 times, followed by a 20- to 25-min interval, and then the subject should recall and replicate the pattern again	Immediate and delayed recall scores: number of correct location of the pieces		Low sensibility of this test
	BVMT-R	A piece of paper with six visual designs is presented	Immediate and delayed recall scores: number of correct drawn design	Sensitive test	Could be affected by hand deficiencies
		After 10 s, the subject is given a blank sheet of paper and should draw each of the designs in the correct location		Six alternative forms	
		After a 20- to 25-min interval, the subject should recall and draw as many designs as possible			
Executive functions					
Verbal fluency	WLG/COWAT	Oral test	Number of adequate nouns		Could be affected by reduced IPS
		Verbal fluency is tested (a maximum of nouns of animals starting with letter "P" should be given) in 90 s			
Sorting test	WCST	Executive test	Number of correct series		Long test
		Card sorting test with different rules			
Sorting test	D-KEF Sorting Test	Executive test	Number of correct series		Long test
		Card sorting test with different rules			

Inhibition	Stroop	Oral test	Number of correct denominations	Interference task	
		The subject should read color nouns written with an ink in a different color			
	Go-No-Go	Executive test	Reaction times		Low sensitivity of this test
		The subject should perform an action given certain stimuli (e.g., press a button – Go) and inhibit that action under a different set of stimuli (e.g., not press that same button – No-Go)			
Flexibility	TMT-A/TMT-B	Executive test	Number of correct series	TMT-B investigates mental flexibility	
		The subject should link letters and numbers according to a predefined rule			
		Two versions: TMT-A and TMT-B			
Attention	TAP	Computerized battery of different aspect of attention such as visual and auditory divided attention, alertness without and with warning, visual scanning with and without target	Number of accurate answers	Reaction times and number of accurate answers could be analyzed separately	Long

*SDMT** Symbol Digit Modalities Test, *CSCT* Computerized Screening Cognitive Test [54], *WAIS-R Digit Symbol* Wechsler Adult Intelligence Scale-Revised Digit Symbol, *PASAT** Paced Auditory Serial Addition Test, *SRT** Selective Reminding Test, *CVLT** California Verbal Learning Test, *10/36 SPART** Spatial Recall Test, *BVMT-R** Brief Visuospatial Memory Test-Revised, *WLG** Word List Generation Test, *COWAT** Controlled Oral Word Association Test, *WCST* Wisconsin Card Sorting Test [55], *D-KEFS** Delis-Kaplan Executive Function System Sorting Test, *TMT* Trail Making Test [56], *TAP* Test of Attentional Performance [57]. * Neuropsychological tests included in the BRB-N, BCcogSEP, and/or MACFIMS batteries [44, 46]

Table 16.3 Proposition for minimal cognitive assessment for multiple sclerosis (BICAMS) [53]

Cognitive domain	Neuropsychological test
Information processing speed	SDMT
Verbal episodic memory	CVLT-II: first five recalls
Visuospatial episodic memory	BVMT-R: first three recalls

BICAMS Brief International Cognitive Assessment for Multiple Sclerosis [53], *SDMT* Symbol Digit Modalities Test, *CVLT-II* California Verbal Learning Test-Second Edition, *BVMT-R* Brief Visuospatial Memory Test-Revised

MS and followed during seven years [62]. In particular, IPS impairment could predict this change, and cognitive deterioration was associated with both the vocational status at the end of the follow-up and its change over the first seven years after the diagnosis.

There is a negative impact on mood too and CI could interfere in self-esteem feeling and copying strategy. In general, CI could alter life satisfaction and the health-related quality of life [60, 62–67]. Driving capacities could be compromised depending on the extent and the severity of CI. In terms of the general treatment of the disease, the presence of CI does modify medical decisions and medication adherence. The management of CI and rehabilitation programs are further detailed in part VI of this chapter.

Pathophysiology of Cognitive Impairment

The pathological substrate of CI in patients with MS is not completely understood. Structural and functional imaging and histopathological studies have provided data suggesting the role of both focal and diffuse brain damage within and outside MS lesions in white and gray matter (WM and GM, respectively) [Review in 68–70].

The first approach is to consider simple imaging parameters such as the distribution, amount, and the extent of focal WM lesions. White matter lesion volume has been found greater in cognitively impaired than in CP patients with MS in many studies [68, 70], but there are only mild to moderate correlation with CI. These modest associations between WM lesions and CI in MS could be explained by the fact that T2 hyperintensities reflect heterogeneous pathologic substrates, including edema, inflammation, demyelination, remyelination, gliosis, axonal loss, and there is a lack of pathological specificity. More importantly, specific locations have been highlighted, and lesions in corpus callosum have been associated with CI in patients with MS [71]. Moreover, some clinical and imaging studies have suggested the role of the cerebellum in CI and in particular in IPS impairment in MS [72–75]. Secondly, it appears interesting to focus on diffuse brain damage and in particular to study the so-called normal-appearing white matter or brain tissue (NAWM and NABT, respectively). In a cross-sectional study, diffuse brain damage assessed by magnetization transfer imaging (MTI) was associated with early CI in patients recently diagnosed with RRMS [12]. These results were replicated in other studies and especially in sample including patients after the first clinical demyelinating event suggestive of MS [76]. Cognitive impairment could be the consequence of brain disconnection due to these abnormalities located in WM tracts. Diffusion tensor imaging (DTI) protocols have allowed to study different metrics including fractional anisotropy in the whole WM skeleton using a tract-based spatial statistic analysis [77, 78] or in specific WM tracts [79] showing the relative contribution of lesional and non-lesional WM in cognitive performance in patients with MS. Several functional MRI (fMRI) studies have also provided interesting findings in patients with MS without CI and with CI and illustrated cortical reorganization that is different according to the stage of MS [68–70]. Brain compensatory mechanisms have been found at early stage of the disease [74, 80, 81], and functional disconnection may affect these mechanisms needed to overcome focal and diffuse structural damage occurring during the disease. There are only few longitudinal studies that included early RRMS patients with several cognitive and MRI evaluations with a long-term follow-up. In one 7-year follow-up study, MRI parameters reflecting the extent and the severity of the diffuse damage in NABT and the net consequence of the diffuse brain damage assessed by atrophy measurements (whole brain and central atrophy) more strongly predicted CI in RRMS patients than visible lesions in the WM [40].

Besides WM, there is increasing interest concerning the damage within the GM for explaining CI in MS [82]. Cortical lesion volume has been found to be higher in cognitively impaired than CP patients with MS [83]. Once again, lesions in specific locations have been considered clinically relevant and were associated with CI in patients with MS. In particular, the regions of interest are deep GM structures such as the thalamus and other basal ganglia and the hippocampus [68–70]. Moreover, brain atrophy appears as a better predictor of cognitive deterioration in patients with MS than WM lesion load [68]. In particular, GM atrophy might play a significant role in the physiopathology of CI in MS, and both cortical and subcortical atrophy have been significantly correlated to CI in patients with MS [68, 82]. Some studies have investigated the role of thalamic atrophy in CI in patients with MS and this topic gains interest [70, 82]. Moreover, a few studies have focused on the hippocampus showing the role of its atrophy mainly in memory impairment in patients with MS [70, 84].

Finally, structural and functional approaches could be combined in order to better explore cognitive functions in patients with MS. A functional disconnection between GM structures at least, partially secondary to damage located in specific WM areas, has been suggested as one of the most important mechanisms leading to CI in MS. A promising method could be to investigate resting-state connectivity. In early MS patients, both structural damage and resting-state functional connectivity changes in brain networks have been investigated [85]. Interestingly, when comparing the different effect sizes of MRI metrics, the highest value was found among the functional connectivity measurements. Moreover, atrophy in one specific area, namely, the posterior cingulate cortex (PCC), was the only predictor of the functional correlation between the medial prefrontal cortex and the PCC. Moreover, the presence of brain and cognitive reserve could attenuate the negative effect of the cumulative brain damage on cognitive performance in patients with MS [86–88]. An interesting longitudinal study including patients after the first clinical demyelinating event (CIS) was performed to investigate the correlates of the evolution of cognitive scores with the change of MRI parameters within 2-years of follow-up [89]. Surprisingly, no significant differences were observed between baseline cognitive status and both baseline and change of MRI metrics in this CIS cohort. One of the explanations could be the presence of cognitive reserve present at this very early stage of the disease.

Few studies have focused only on patients with PPMS. Focal and diffuse WM damage and GM pathology have been reported as significant predictors of cognitive performance in IPS, attention, and executive function in a 5-year follow-up study including 31 patients with PPMS [90]. Additionally, in an immunohistochemical study of postmortem brains of 26 patients with PPMS, a generalized diffuse meningeal inflammation was reported [91]. This confined inflammation might play a significant role in the pathogenesis of cortical GM lesions and contribute to the clinical disability in these patients.

Prognostic Factor

Physical disability and CI could occur independently from each other during the course of the disease, and patients could present CI even before the manifestation of physical symptoms. Aforementioned, patients who are so-called BMS could have CI despite of a low EDSS supporting the need to detect cognitive deficits for evaluating the severity of the disease. Notably, it has been proposed a modification of the definition of BMS in order to include cognitive assessment [92]. The relationship between physical disability and cognition has been questioned in MS. Significant correlations between the EDSS score and cognitive test performances have been reported [93, 94]. Even if modest relationships are typically observed between CI and physical disability in MS, the majority of these results primarily concern the measurement of IPS [94–97]. These data highlight the prognostic value of IPS impairment that is considered as a central defect in MS. In a 7-year longitudinal study, the cognitive deterioration was correlated with MRI parameters reflecting mainly the initial brain diffuse axonal injury and its early change within the first two years [40]. These results support the role of early central atrophy in CI in patients with RRMS and in particular its correlation with IPS decline in early MS. The early identification of IPS impairment could be a relevant marker of early central atrophy that has been used for predicting the progression of the disability assessed through changes in EDSS [98].

Management of Cognitive Impairment

Medications: Disease-Modifying Drugs and Symptomatic Treatment

Aforementioned, cognitive status should be included in treatment decisions independently of physical disability as it represents a marker for disease severity and progression. Nevertheless, the historical clinical trials did not take into account these data in defining the efficacy of treatments in MS. Cognitive functions have been evaluated mainly in post hoc analysis of the first clinical trials of disease-modifying drugs in MS. Few studies have chosen cognitive outcome as a primary endpoint. Cognitive secondary outcome measures of randomized controlled trials or their extension have been reported [99]. For instance, a positive effect of interferon beta 1b subcutaneous has been demonstrated in patients included after a CIS [100]. Another randomized clinical trial was performed for evaluating the effect on cognitive function of different types of interferon beta (Avonex, Rebif, and Betaferon) in newly diagnosed RRMS with one-year of follow-up [101]. In accordance with some previous studies focusing on the effect on interferon beta in MS [102, 103], the results suggest a positive effect on these disease-modifying drugs in preventing cognitive deterioration in MS. Encouragingly, cognitive performances have been also improved during an observational open-label study testing one monoclonal antibody in RRMS patients [104]. Moreover, fingolimod was tested in lipopolysaccharide (LPS) model in rats in order to explore the link between immune activation and cognition [105]. Indeed, the LPS was used as an agent inducing microglial cell activation and brain inflammation. Interestingly, a protective effect of fingolimod was demonstrated at different experimental levels (functional, histological, and transcriptional steps) suggesting its application in treating memory impairment in neuroinflammatory conditions.

Moreover, several symptomatic drugs have been tested to improve cognition in patients with MS, such as anticholinesterasics (donepezil, rivastigmine) and channel blockers [99]. However, no drug has shown positive results in large randomized controlled trials. Some positive results have been reported on short-term follow-up with L-amphétamine [99]. In conclusion, these studies provide insufficient data for prescribing symptomatic treatment for preventing and treating CI in patients with MS.

Cognitive Rehabilitation and Remediation

There is a lack of well-designed research studies investigating the effectiveness of cognitive rehabilitation programs in patients with MS [106, 107]. As the impairment of IPS is a key deficit in MS and has a prognostic value in this disease, its early detection and management seem to be clinically relevant and justify putting some efforts to investigate the impact of specific cognitive rehabilitation and remediation programs. Moreover, managing episodic memory is also a challenge of this type of programs and some specific studies are in progress. Besides, it is clinically relevant to focus on ecological validity of this type of rehabilitation.

Conclusion

Cognitive impairment is common in MS and could be seen in each type and stage of the disease. It affects primarily information processing speed, and episodic memory is frequently impaired too. CI has a negative impact on daily activities and in particular on vocational status of patients living with MS. Even if there is a high variability, cognitive functions tend to deteriorate over time as cumulative brain damage occurs. There is increasing evidence that CI could be due to a disconnection syndrome relative to the accumulated focal and diffuse brain damage within the white and gray matter structures. Educational level, leisure activities, and intelligence quotient contribute to cognitive reserve and have the potential to attenuate the consequences of cognitive deficits at least at the beginning of the pathology. The presence of brain compensatory mechanisms supported the development of rehabilitation and cognitive remediation programs. Longitudinal studies with long follow-up including clinical, neuropsychological, and imaging assessments are still needed to better understand the pathophysiology of cognitive impairment in both active and non-active patients with MS. One of the remaining challenge is the treatment of cognitive impairment in patients with MS, and works are in progress.

References

1. Langdon DW. Cognition in multiple sclerosis. Curr Opin Neurol. 2011;24:244–9.
2. Chiaravalloti ND, DeLuca J. Cognitive impairment in multiple sclerosis. Lancet Neurol. 2008;7:1139–51.
3. DeLuca J, Chelune GJ, Tulsky DS, Lengenfelder J, Chiaravalloti ND. Is speed of processing or working memory the primary information processing deficit in multiple sclerosis? J Clin Exp Neuropsychol. 2004;26:550–62.
4. Forn C, Belenguer A, Parcet-Ibars MA, Avila C. Information-processing speed is the primary deficit underlying the poor performance of multiple sclerosis patients in the Paced Auditory Serial Addition Test (PASAT). J Clin Exp Neuropsychol. 2008;13:1–8.
5. Berrigan LI, Lefevre JA, Rees LM, Berard J, Freedman MS, Walker LA. Cognition in early relapsing-remitting multiple sclerosis: consequences may be relative to working memory. J Int Neuropsychol Soc. 2013;19:938–49.
6. Rao SM, Leo GJ, Bernardin L, Unverzagt F. Cognitive dysfunction in multiple sclerosis. I. Frequency, patterns, and prediction. Neurology. 1991;41:685–91.
7. Benedict RH, Cookfair D, Gavett R, et al. Validity of the minimal assessment of cognitive function in multiple sclerosis (MACFIMS). J Int Neuropsychol Soc. 2006;12:549–58.

8. Brochet B. Prevalence, profile and functional impact of cognitive impairment in multiple sclerosis. In: Amato MP, editor. Cognitive impairment in multiple sclerosis. Milano: Elsevier; 2011. p. 1–8.
9. Peyser JM, Edwards KR, Poser CM, Filskov SB. Cognitive function in patients with multiple sclerosis. Arch Neurol. 1980;37:577–9.
10. Lyon-Caen O, Jouvent R, Hauser S, et al. Cognitive function in recent-onset demyelinating diseases. Arch Neurol. 1986;43:1138–41.
11. Truelle JL, Palisson E, Le Gall D, Stip E, Derouesne C. Intellectual and mood disorders in multiple sclerosis. Rev Neurol (Paris). 1987;143:595–601.
12. Deloire MS, Salort E, Bonnet M, et al. Cognitive impairment as marker of diffuse brain abnormalities in early relapsing-remitting multiple sclerosis. J Neurol Neurosurg Psychiatry. 2005;76:519–26.
13. Patti F, Amato M, Trojano M, et al. Cognitive impairment and its relation with disease measures in mildly disabled patients with relapsing–remitting multiple sclerosis: Baseline results from the Cognitive Impairment in Multiple Sclerosis (COGIMUS) study. Mult Scler. 2009;15:779–88.
14. Amato MP, Zipoli V, Goretti B, et al. Benign multiple sclerosis: cognitive, psychological and social aspects in a clinical cohort. J Neurol. 2006;253:1054–9.
15. Fischer M, Kunkel A, Bublak P, Faiss JH. How reliable is the classification of cognitive impairment across different criteria in early and late stages of multiple sclerosis? J Neurol Sci. 2014;343:91–9.
16. Feinstein A, Kartsounis LD, Miller DH, Youl BD, Ron MA. Clinically isolated lesions of the type seen in multiple sclerosis: a cognitive, psychiatric, and MRI follow up study. J Neurol Neurosurg Psychiatry. 1992;55:869–76.
17. Callanan MM, Logsdail SJ, Ron MA, Warrington EK. Cognitive impairment in patients with clinically isolated lesions of the type seen in multiple sclerosis. A psychometric and MRI study. Brain. 1989;112:361–74.
18. Pelosi L, Geesken JM, Holly M, Hayward M, Blumhardt LD. Working memory impairment in early multiple sclerosis. Evidence from an event-related potential study of patients with clinically isolated myelopathy. Brain. 1997;120:2039–58.
19. Achiron A, Barak Y. Cognitive impairment in probable multiple sclerosis. J Neurol Neurosurg Psychiatry. 2003;74:443–6.
20. Feuillet L, Reuter F, Audoin B, et al. Early cognitive impairment in patients with clinically isolated syndrome suggestive of multiple sclerosis. Mult Scler. 2007;13:124–7.
21. Zipoli V, Goretti B, Hakiki B, et al. Cognitive impairment predicts conversion to multiple sclerosis in clinically isolated syndromes. Mult Scler. 2010;16:62–7.
22. Amato MP, Hakiki B, Goretti B, et al. Association of MRI metrics and cognitive impairment in radiologically isolated syndromes. Neurology. 2012;78:309–14.
23. Lebrun C, Blanc F, Brassat D, Zephir H, J Seze J, CFSEP. Cognitive function in radiologically isolated syndrome. Mult Scler. 2010;16:919–25.
24. Comi G, Filippi M, Martinelli V, et al. Brain magnetic resonance imaging correlates of cognitive impairment in primary and secondary progressive multiple sclerosis. J Neurol Sci. 1995;132:222–7.
25. Foong J, Rozewicz L, Chong WK, Thompson AJ, Miller DH, Ron MA. A comparison of neuropsychological deficits in primary and secondary progressive multiple sclerosis. J Neurol. 2000;247:97–101.
26. Gaudino EA, Chiaravalloti ND, DeLuca J, Diamond BJ. A comparison of memory performance in relapsing-remitting, primary progressive and secondary progressive, multiple sclerosis. Neuropsychiat Neuropsychol Behav Neurol. 2001;14:32–44.
27. Huijbregts SC, Kalkers NF, de Sonneville LM, de Groot V, Reuling IE, Polman CH. Differences in cognitive impairment of relapsing remitting, secondary, and primary progressive MS. Neurology. 2004;63:335–9.
28. Wachowius U, Talley M, Silver N, Heinze HJ, Sailer M. Cognitive impairment in primary and secondary progressive multiple sclerosis. J Clin Exp Neuropsychol. 2005;27:65–77.

29. Potagas C, Giogkaraki E, Koutsis G, et al. Cognitive impairment in different MS subtypes and clinically isolated syndromes. J Neurol Sci. 2008;267:100–6.
30. Ruet A, Deloire M, Charré-Morin J, Hamel D, Brochet B. Cognitive impairment differs between primary progressive and relapsing-remitting MS. Neurology. 2013;80:1501–8.
31. Jennekens-Schinkel A, Laboyrie PM, Lanser JB, van der Velde EA. Cognition in patients with multiple sclerosis after four years. J Neurol Sci. 1990;99:229–47.
32. Kujala P, Portin R, Ruutiainen J. The progress of cognitive decline in multiple sclerosis A controlled 3-year follow-up. Brain. 1997;120:289–97.
33. Amato MP, Ponziani G, Siracusa G, Sorbi S. Cognitive dysfunction in early-onset multiple sclerosis: a reappraisal after 10 years. Arch Neurol. 2001;58:1602–6.
34. Rosti E, Hämäläinen P, Koivisto K, Hokkanen L. One-year follow-up study of relapsing-remitting MS patients' cognitive performances: Paced Auditory Serial Addition Test's susceptibility to change. J Int Neuropsychol Soc. 2007;13:791–8.
35. Denney DR, Lynch SG, Parmenter BA. A 3-year longitudinal study of cognitive impairment in patients with primary progressive multiple sclerosis: speed matters. J Neurol Sci. 2008;267:129–36.
36. Duque B, Sepulcre J, Bejarano B, Samaranch L, Pastor P, Villoslada P. Memory decline evolves independently of disease activity in MS. Mult Scler. 2008;14:947–53.
37. Amato MP, Portaccio E, Goretti B, et al. Relevance of cognitive deterioration in early relapsing–remitting MS: a 3-year follow-up study. Mult Scler. 2010;16:1474–82.
38. Strober LB, Rao SM, Lee JC, Fischer E, Rudick R. Cognitive impairment in multiple sclerosis: An 18 year follow-up study. Mult Scler Relat Disord. 2014;3:473–81.
39. Portaccio E, Amato MP. Natural history. In: Amato MP, editor. Cognitive impairment in multiple sclerosis. Milano: Elsevier; 2011. p. 29–36.
40. Deloire MS, Ruet A, Hamel D, Bonnet M, Dousset V, Brochet B. MRI predictors of cognitive outcome in early multiple sclerosis. Neurology. 2011;76:1161–7.
41. Reuter F, Zaaraoui W, Crespy L, et al. Frequency of cognitive impairment dramatically increases during the first 5 years of multiple sclerosis. J Neurol Neurosurg Psychiatry. 2011;82:1157–9.
42. Pardini M, Uccelli A, Grafman J, Yaldizli Ö, Mancardi G, Roccatagliata L. Isolated cognitive relapses in multiple sclerosis. J Neurol Neurosurg Psychiatry. 2014;85:1035–7.
43. Smith A. Symbol Digit Modalities Test (SDMT) manual (revised). Los Angeles: Western Psychological Services; 1982.
44. Rao SM, The Cognitive Function Study Group of the National Multiple Sclerosis Society. A manual for the brief repeatable battery of neuropsychological tests in multiple sclerosis. Milwaukee: Medical College of Wisconsin; 1990.
45. Dujardin K, Sockeel P, Cabaret M, De Sèze J, Vermersch P. BCcogSEP: a French test battery evaluating cognitive functions in multiple sclerosis. Rev Neurol (Paris). 2004;160:51–62.
46. Benedict RH, Fischer JS, Archibald C, et al. Minimal neuropsychological assessment of MS patients: a consensus approach. Clin Neuropsychol. 2002;16:381–97.
47. Benedict RH, Munschauer F, Linn R, et al. Screening for multiple sclerosis cognitive impairment using a self-administered 15-item questionnaire. Mult Scler. 2003;9:95–101.
48. Deloire MS, Bonnet MC, Salort E, et al. How to detect cognitive dysfunction at early stages of multiple sclerosis? Mult Scler. 2006;12:445–52.
49. Benedict RH, Cox D, Thompson LL, Foley F, Weinstock-Guttman B, Munschauer F. Reliable screening for neuropsychological impairment in multiple sclerosis. Mult Scler. 2004;10:675–8.
50. Parmenter BA, Weinstock-Guttman B, Garg N, Munschauer F, Benedict RH. Screening for cognitive impairment in multiple sclerosis using the Symbol digit Modalities Test. Mult Scler. 2007;13:52–7.
51. Benedict RH, Duquin JA, Jurgensen S, et al. Repeated assessment of neuropsychological deficits in multiple sclerosis using the Symbol Digit Modalities Test and the MS Neuropsychological Screening Questionnaire. Mult Scler. 2008;14:940–6.

52. Benedict RH, Smerbeck A, Parikh R, Rodgers J, Cadavid D, Erlanger D. Reliability and equivalence of alternate forms for the Symbol Digit Modalities Test: implications for multiple sclerosis clinical trials. Mult Scler. 2012;18:1320–5.
53. Langdon DW, Amato MP, Boringa J, et al. Recommendations for a Brief International Cognitive Assessment for Multiple Sclerosis (BICAMS). Mult Scler. 2012;18:891–8.
54. Ruet A, Deloire MS, Charré-Morin J, Hamel D, Brochet B. A new computerised cognitive test for the detection of information processing speed impairment in multiple sclerosis. Mult Scler. 2013;19:1665–72.
55. Weigl E. On the psychology of so-called process of abstraction. J Abnorm Soc Psychol. 1941;36:3–33.
56. Rossini ED, Karl MA. The Trail Makin Test A and B: a technical note on structural nonequivalence. Percept Mot Skills. 1994;78:625–6.
57. Zimmermann P, Fimm B. 2.1 Tests d'évaluation de l'attention. Würzelen: Psytest; 2009.
58. Delis DC, Kramer JH, Kaplan E, Ober BA. California verbal learning test manual: second edition, adult version. San Antonio: Psychological Corporation; 2000.
59. Benedict RH. Brief visuospatial memory test-revised. Professional manual. Odessa: Psychological Assessment Resources, Inc.; 1997.
60. Rao SM, Leo GJ, Ellington L, Nauertz T, Bernardin L, Unverzagt F. Cognitive dysfunction in multiple sclerosis. II Impact on employment and social functioning. Neurology. 1991;41:692–6.
61. Morrow SA, Drake A, Zivadinov R, Munschauer F, Weinstock-Guttman B, Benedict RH. Predicting loss of employment over three years in multiple sclerosis: clinically meaningful cognitive decline. Clin Neuropsychol. 2010;24:1131–45.
62. Ruet A, Deloire M, Hamel D, Ouallet JC, Petry K, Brochet B. Cognitive impairment, health-related quality of life and vocational status at early stages of multiple sclerosis: a 7-year longitudinal study. J Neurol. 2013;260:776–84.
63. Benito-León J, Morales JM, Rivera-Navarro J. Health-related quality of life and its relationship to cognitive and emotional functioning in multiple sclerosis patients. Eur J Neurol. 2002;9:497–502.
64. Mitchell AJ, Benito-León J, Morales González JM, Rivera-Navarro J. Quality of life and its assessment in multiple sclerosis: integrating physical and psychological components of well-being. Lancet Neurol. 2005;4:556–66.
65. Clavelou P, Auclair C, Taithe F, Gerbaud L. Quality of life in multiple sclerosis. Rev Neurol (Paris). 2009;165 Suppl 4:S123–8.
66. Fernández O, Baumstarck-Barrau K, Simeoni MC, Auquier P. MusiQoL study group. Patient characteristics and determinants of quality of life in an international population with multiple sclerosis: assessment using the MusiQoL and SF-36 questionnaires. Mult Scler. 2011;17:1238–49.
67. Baumstarck-Barrau K, Simeoni MC, Reuter F, et al. Cognitive function and quality of life in multiple sclerosis patients: a cross-sectional study. BMC Neurol. 2011;2:11–7.
68. Filippi M, Rocca MA, Benedict RH, et al. The contribution of MRI in assessing cognitive impairment in multiple sclerosis. Neurology. 2010;75:2121–8.
69. DeLuca GC, Yates RL, Beale H, Morrow SA. Cognitive impairment in multiple sclerosis: clinical. Radiol Pathol Insight Brain Pathol. 2015;25:79–98.
70. Rocca MA, Amato MP, De Stefano N, MAGNIMS Study Group, et al. Clinical and imaging assessment of cognitive dysfunction in multiple sclerosis. Lancet Neurol. 2015;14:302–17.
71. Rossi F, Giorgio A, Battalini M, et al. Relevance of brain lesion location to cognition in relapsing multiple sclerosis. Plos One. 2012;7, e44826.
72. Ruet A, Hamel D, Deloire MS, Charré-Morin J, Saubusse A, Brochet B. Information processing speed impairment and cerebellar dysfunction in relapsing-remitting multiple sclerosis. J Neurol Sci. 2014;347:246–50.
73. Cerasa A, Valentino P, Chiriaco C, et al. MR imaging and cognitive correlates of relapsing-remitting multiple sclerosis patients with cerebellar symptoms. J Neurol. 2013;260:1358–66.

74. Bonnet MC, Dilharreguy B, Allard M, Deloire MS, Petry KG, Brochet B. Differential cerebellar and cortical involvement according to various attentional load: role of educational level. Hum Brain Mapp. 2009;30:1133–43.
75. Rocca MA, Bonnet MC, Meani A, et al. Differential cerebellar functional interactions during an interference task across multiple sclerosis phenotypes. Radiology. 2012;265:864–73.
76. Faiss JH, Dähne D, Baum K, et al. Reduced magnetisation transfer ratio in cognitively impaired patients at the very early stage of multiple sclerosis: a prospective, multicenter, cross-sectional study. BMJ Open. 2014;4:e004409. doi:10.1136/bmjopen-2013-04409.
77. Dineen RA, Vilisaar J, Hlinka J, et al. Disconnection as a mechanism for cognitive dysfunction in multiple sclerosis. Brain. 2009;132:239–49.
78. Roosendaal SD, Geurts JJ, Vrenken H, et al. Regional DTI differences in multiple sclerosis patients. Neuroimage. 2009;44:1397–403.
79. Mesaros S, Rocca MA, Kacar K, et al. Diffusion tensor MRI tractography and cognitive impairment in multiple sclerosis. Neurology. 2012;78:969–75.
80. Audoin B, Au Duong MV, Ranjeva JP, et al. Magnetic resonance study of the influence of tissue damage and cortical reorganization on PASAT performance at the earliest stage of multiple sclerosis. Hum Brain Mapp. 2005;24:216–28.
81. Bonnet MC, Allard M, Dilharreguy B, Deloire M, Petry KG, Brochet B. Cognitive compensation failure in multiple sclerosis. Neurology. 2010;75:1241–8.
82. Hulst HE, Geurts JJ. Gray matter imaging in multiple sclerosis: what have we learned? BMC Neurol. 2011;11:153.
83. Calabrese M, Agosta F, Rinaldi F, et al. Cortical lesions and atrophy associated with cognitive impairment in relapsing-remitting multiple sclerosis. Arch Neurol. 2009;66:1144–50.
84. Sicotte NL, Kern KC, Giesser BS, et al. Regional hippocampal atrophy in multiple sclerosis. Brain. 2008;131:1134–41.
85. Louapre C, Perlbarg V, Garcia-Lorenzo D, et al. Brain networks disconnection in early multiple sclerosis cognitive deficits: an anatomofunctional study. Hum Brain Mapp. 2014;35:4706–17.
86. Sumowski JF, Chiaravalloti N, DeLuca J. Cognitive reserve protects against cognitive dysfunction in multiple sclerosis. J Clin Exp Neuropsychol. 2009;31:913–26.
87. Sumowski JF, Wylie GR, Deluca J, Chiaravalloti N. Intellectual enrichment is linked to cerebral efficiency in multiple sclerosis: functional magnetic resonance imaging evidence for cognitive reserve. Brain. 2010;133:362–74.
88. Arnett PA, Brochet B. How can cognitive reserve in multiple sclerosis inform clinical care? Neurology. 2013;80:1724–5.
89. Uher T, Blahova-Dusankova J, Horakova D, et al. Longitudinal MRI and neuropsychological assessment of patients with clinically isolated syndrome. J Neurol. 2014;261:1735–44.
90. Penny S, Khaleeli Z, Cipolotti L, Thompson A, Ron M. Early imaging predicts later cognitive impairment in primary progressive multiple sclerosis. Neurology. 2010;74:545–52.
91. Choi SR, Howell OW, Carassiti D, et al. Meningeal inflammation plays a role in the pathology of primary progressive multiple sclerosis. Brain. 2012;1–13.
92. Rovaris M, Barkhof F, Calabrese M, et al. MRI features of benign multiple sclerosis: toward a new definition of this disease phenotype. Neurology. 2009;72:1693–701.
93. Nocentini U, Pasqualetti P, Bonavita S, et al. Cognitive dysfunction in patients with relapsing-remitting multiple sclerosis. Mult Scler. 2006;12:77–87.
94. Deloire M, Ruet A, Hamel D, Bonnet M, Brochet B. Early cognitive impairment in multiple sclerosis predicts disability outcome several years later. Mult Scler. 2010;16:581–7.
95. Hohol MJ, Guttmann CR, Orav J, et al. Serial neuropsychological assessment and magnetic resonance imaging analysis in multiple sclerosis. Arch Neurol. 1997;54:1018–25.
96. De Sonneville LM, Boringa JB, Reuling IE, Lazeron RH, Adèr HJ, Polman CH. Information processing characteristics in subtypes of multiple sclerosis. Neuropsychologia. 2002;40:1751–65.
97. Lynch SG, Parmenter BA, Denney DR. The association between cognitive impairment and physical disability in multiple sclerosis. Mult Scler. 2005;11:469–76.

98. Lukas C, Minneboo A, de Groot V, et al. Early central atrophy rate predicts 5 year clinical outcome in multiple sclerosis. J Neurol Neurosurg Psychiatry. 2010;81:1351–6.
99. Amato MP, Langdon D, Montalban X, et al. Treatment of cognitive impairment in multiple sclerosis: position paper. J Neurol. 2013;260(6):1452–68.
100. Penner IK, Stemper B, Calabrese P, et al. Effects of interferon beta-1b on cognitive performance in patients with a first event suggestive of multiple sclerosis. Mult Scler. 2012;18(10):1466–71.
101. Mokhber N, Azarpazhhoh A, Orouji E, et al. Cognitive dysfunction in patients with multiple sclerosis treated with different types of interferon beta: a randomized clinical trial. J Neurol Sci. 2014;342:16–20.
102. Fischer JS, Priore RL, Jacobs LD, Cookfair DL, Rudick RA, Herndon RM, et al. Neuropsychological effects of interferon beta-1 a in relapsing multiple sclerosis. Multiple Sclerosis Collaborative Research group. Ann Neurol. 2000;48:885–92.
103. Barak Y, Achiron A. Effect of interferon-beta-1 b on cognitive functions in multiple sclerosis. Eur Neurol. 2002;47:11–4.
104. Iaffaldano P, Viterbo RG, Paolicelli D, et al. Impact of natalizumab on cognitive performances and fatigue in relapsing multiple sclerosis: a prospective, open-label, two years observational study. PLoS One. 2012;7, e35843.
105. Omidbakhsh R, Rajabli B, Nasoohi S, et al. Fingolimod affects gene expression profile associated with LPS-induced memory impairment. Exp Brain Res. 2014;232:3687–96.
106. O'Brien AR, Chiaravalloti N, Goverover Y, Deluca J. Evidenced-based cognitive rehabilitation for persons with multiple sclerosis: a review of the literature. Arch Phys Med Rehabil. 2008;89:761–9.
107. Rosti-Otajärvi EM, Hämäläinen PI. Neuropsychological rehabilitation for multiple sclerosis. Cochrane Database Syst Rev. 2011;9, CD009131.

Chapter 17
Neuropsychiatry of Neuromyelitis Optica

Frédéric Blanc

Abstract Neuromyelitis optica (NMO), also called Devic's disease, is a central nervous system inflammatory disease characterized by optic neuritis and longitudinally extensive acute transverse myelitis but also with brain involvement. Cognitive impairment is present in 54–57 % of patients. The main cognitive deficits are in long-term memory, speed of information processing, attention, and executive functions. The neural basis of cognitive troubles in NMO seems to be the whole WM but particularly the corpus callosum and the superior longitudinal fascicles. Depression and fatigue are frequent behavioral symptoms of NMO. Guidelines for the treatment of neuropsychological symptoms of NMO are needed: when clinical trials will be conducted in NMO patients, a neuropsychological evaluation will be necessary.

Keywords Neuromyelitis optica • Neuropsychiatry • Neural basis of cognitive impairment • Cognitive impairment • Multiple sclerosis • Brain atrophy • MRI • DTI

Introduction

Neuromyelitis optica (NMO), also called Devic's disease, is a central nervous system (CNS) inflammatory disease characterized by optic neuritis (ON) and longitudinally extensive acute transverse myelitis (ATM). First described by Eugene Devic and Fernand Gault in the nineteenth century as a distinct disease, NMO was then classified as a subtype of multiple sclerosis (MS) [1]. The discovery of an autoantibody called NMO-IgG that targets aquaporin-4 (AQP4) at the beginning of the twenty-first century has transformed NMO in a clearly different CNS pathology [2].

F. Blanc, MD, PhD, HdR (✉)
CMRR (Memory Resources and Research Centre), University Hospital of Strasbourg, 1, avenue Molière, 67000 Strasbourg, France

University of Strasbourg and CNRS, ICube Laboratory UMR 7357 and FMTS (Fédération de Médecine Translationnelle de Strasbourg), Team IMIS/Neurocrypto, Strasbourg, France
e-mail: frederic.blanc@chru-strasbourg.fr

B. Brochet (ed.), *Neuropsychiatric Symptoms of Inflammatory Demyelinating Diseases*, Neuropsychiatric Symptoms of Neurological Disease,
DOI 10.1007/978-3-319-18464-7_17

Epidemiological and population-based studies suggest that the prevalence of NMO varies among countries: 0.52/100,000 in Cuba, 4.2/100,000 in French West Indies [3], 0.72/100,000 in Mexico [4], 1.96/100,000 in the United Kingdom [5], and 4.4/100,000 in Denmark [6]. Such differences are probably due to the genetic background but also the improvement of treatment and decreasing mortality of NMO patients [3].

The classical presentation of NMO patients is a relapsing disease with attacks of optic neuritis (ON), ATM, or both in 90 % of cases [7]. The monophasic course or progressive course is less common [8]. The two types of relapses ON and myelitis are really disabling, and remission is poorer than in MS, particularly if untreated.

The diagnostic criteria for NMO were revised in 2006 after the detection of NMO-IgG [9]. In addition to the two clinical events of ON and acute myelitis, a diagnosis of NMO requires these two of the three following supportive criteria be fulfilled:

- Contiguous spinal cord magnetic resonance imaging (MRI) lesion extending over three or more vertebral segments
- Brain MRI not meeting diagnostic criteria of MS according to Paty [10]
- NMO-IgG seropositive status

Fourteen percent of NMO patients can have an initial short transverse myelitis (less than three vertebral segments) [11]. A longitudinally extensive ATM is the more frequent, but a short one does not exclude NMO. We will discuss thereafter the presence/absence of brain lesions ON MRI. NMO-IgG status depends on the detection method used: thus, 10–50 % of patients with NMO are negative for NMO-IgG [12]. Insufficient assay sensitivity is the main cause of AQP4-IgG seronegativity, as shown in various comparative studies [13]. These criteria have been recently revised but not published (Wingerchuk et al. 2014, oral presentation American Academy of Neurology, S63.001, "revised diagnostic criteria for neuromyelitis optica spectrum disorders"): these criteria have been expanded to include 6 different core characteristics: optic neuritis, acute myelitis, area postrema syndrome (nausea, vomiting, and hiccups), other brainstem syndromes, symptomatic narcolepsy or acute diencephalic syndrome with MRI findings, and symptomatic cerebral syndrome with MRI findings. Antibody-positive patients need to show at least one of these core characteristics. Antibody-negative patients need to show at least two of the core characteristics with the following requirements: (1) At least one of the core symptoms must be optic neuritis, myelitis, or area postrema syndrome; (2) the core characteristics must be disseminated in space; and (3) MRI findings must distinguish NMO from MS, or other demyelinating disorders.

Patients with NMO have an extensive loss of AQP4 and a decreased astrocyte concentration in acute and chronic NMO lesions of the spinal cord and the optic nerves [14]. Brain lesions of NMO patients are localized at sites of high aquaporin 4 expression [15]. Pathology findings argue also for larger tissue damages including brainstem and brain [16]. Such lesions in the brain of NMO patients led us to ask the question of cognitive impairment and behavioral modifications in NMO patients.

Cognitive Impairment in NMO

Because of the predominance of the lesions in the optic nerves and the spinal cord, the research of cognitive troubles in NMO was late taken into account. However, cognitive impairment in NMO is common. Thus, we have for the first time in 2008 demonstrated that NMO patients have cognitive dysfunctions [17]. This finding was subsequently confirmed by others [17–22]. Cognitive impairment is present in more than half of NMO patients: 57 % of NMO patients for Vanotti et al. [19] and also for Saji et al. [20] and 54 % in our cohort [23].

The main cognitive deficits were found in long-term memory, speed of information processing, attention, and executive functions. Whatever the studies, no clear differences were found between the cognitive pattern of NMO and MS. It means that the cognitive pattern in NMO patients was a "subcortical" cognitive impairment, including a decreased speed of treatment of information (DSST, PASAT), executive function impairment (PASAT, fluencies), attention impairment (PASAT, forward and backward digit span), and memory impairment (SRT, 10/36). The term "subcortical" was first used by K. Wilson when he described in 1912 patients with Wilson's disease with cognitive deficits different from other dementias and then for patients with degenerative extrapyramidal disorders [24]. This term is also used for inflammatory diseases. This term "subcortical" is particularly coherent with NMO patients where the cognitive pattern and the atrophy pattern seem to be linked to the WM involvement (see infra).

Using the emotional morphing task, Cardona et al. have demonstrated that NMO patients have also difficulties to recognize negative emotions (disgust, anger, and fear), in comparison to controls [22]. Such results could participate to behavioral modifications in NMO patients.

Behavioral Aspects of NMO

Although very few papers describe behavioral aspects of NMO, the main behavioral symptom described in NMO is depression [25]. In 2004, a first case of major depression concomitant to a relapse was described in a NMO case [26]. Using the 15 items geriatric depression scale (GDS), Kawahara has demonstrated that NMO patients are depressed in more than 35 % of cases [25]. These results were confirmed by Chanson et al., using a more specific scale (EHD for "Echelle d'Humeur Dépressive" in French and depressive mood scale in English) [27]. This 11-item French questionnaire has been specifically designed and validated for the assessment of depression in MS. Interestingly, it seems that the best predictor of depression is the importance of the handicap measured by EDSS score [27].

Fatigue has been also frequently described with NMO. It has been first described associated with hypothalamic lesions and endocrinopathies [28]. Symptoms associated with hypothalamus are various including fatigue, hypothermia, hyperphagia, obesity, and symptoms associated with each axis such as the CRH-ACTH axis or the TRH-TSH axis, but also hyperprolactinemia with amenorrhea and galactorrhea [28]. Fatigue and NMO have been also described to be associated with high level of creatine kinase in three Japanese cases [29]. Chanson et al. have demonstrated that the scores for all dimensions of fatigue were lower in NMO than in MS, but this difference reached the level of statistical significance only for the psychological dimension [27]. Finally, it is of interest to note that fatigue and depression can affect cognitive functions in NMO patients as in other inflammatory diseases [21].

Neural Basis of Neuropsychological Modifications in NMO

Brain Lesions in NMO

Brain involvement in NMO patients seems to be more frequent than the first descriptions of the disease. Brain MRI at the beginning of the disease is usually normal. However, after a disease course of several years, non-MS-like lesions were found in 50 % of patients, whereas MS-like lesions were present in only 10 % of cases [30]. Logically, brain lesions in NMO are localized at sites of high AQP4 expression [15]. Using T1-3D MRI images, we have found a decreased WM volume in the frontal and parietal lobes (including the superior longitudinal fascicle), corpus callosum, cerebellum, brainstem, and optic chiasm compared to healthy control subjects, but no GM atrophy [23]. Using diffusion tensor imaging, brain tissue abnormalities have been found in normal appearing gray matter (NAGM) and normal-appearing white matter (NAWM) [31]. Furthermore, using magnetization transfer (MT) MRI, Rocca et al. showed reduced MT ratio of the NAGM [32]. Magnetic resonance (MR) spectroscopy showed no abnormality in brain NMO, including gray matter (GM) and white matter (WM) [33–35]. However, in these studies, the analysis was on the centrum ovale [33] or other specific regions of the brain. To our best knowledge, no study has tested the whole brain using MR spectroscopy in NMO.

Neuropathologically, Popescu et al. did not find any demyelination of the GM of the brain including cerebellum, and more pathological data are needed on the brain WM, even if case of cerebellum involvement has been described [36, 37]. Popescu et al. have found also a preservation of aquaporin-4 (AQP4) in cerebral cortex of patients with NMO. In the same way, Saji et al. demonstrated no cortical demyelination, but they found neuronal loss in cortical layers II, III, and IV, with reaction of aquaporin-4 (AQP4)-negative astrocytes in layer I, massive activated microglia in layer II, and meningeal inflammation [20]. Interestingly, this cortical degeneration could explain the cognitive impairment also found in these Japanese patients [20].

Neural Basis of Cognitive Impairment in NMO

Because of the existence of few brain lesions in NMO patients, one could think that NMO patients with cognitive impairment would have more WM lesions than patients without. However, no correlation between brain lesions and cognitive impairment has been found in any studies [17, 18, 23]. On the other hand, we have demonstrated that NMO patients with cognitive impairment compared to NMO patients without any have a large decreased WM volume including brainstem, cerebellum, corticospinal tracts, and also the important fascicles of the brain such as corpus callosum, superior longitudinal fascicle, and inferior longitudinal fascicle [23]. Moreover, we demonstrated correlations between cognitive tests performance and white matter volume. First, we found an association between low performance for immediate and delayed spatial memory and a reduced volume of the optic chiasm, the corpus callosum, limbic lobe including parahippocampal gyri, and fronto-parieto-occipital regions including the superior longitudinal fascicle. Visual impairment could explain difficulties in performing a visual memory task. Chronic disconnection of the corpus callosum by surgery is known to be responsible for moderate memory impairment, particularly topographical memory [38]. The parahippocampal gyri are regions well known to be involved in memory, particularly in spatial memory (where stream) [39]. Finally, the frontoparietal network, the superior longitudinal fascicle, and the parietal cortex are also of importance for memory, particularly spatial working memory [40, 41].

Reduced WM in the corpus callosum and the pons was correlated with poor performances on PASAT. Norepinephrine-synthesizing neurons that send diffuse projection from a part of the pons, the locus coeruleus, have a major role in attention, particularly in focused attention and the ability to redirect attention [42, 43]. Attention is of high importance to succeed in performing the PASAT. Anterior callosal abnormalities are reported to be correlated with impaired PASAT performance in MS [44]. In the same way, impaired PASAT was associated with numerous little frontal WM regions involved in the working memory [45]. A verbal memory test (BCcog-SRT) was found to be correlated logically to regions of importance for episodic memory (thalamus, fornix, hippocampus, frontal lobe) and for recognizing words (lingual gyrus) [46]. Digit span was found to be correlated with perisylvian atrophy, as previously described [47]. DSST, a speed writing test, was found to be correlated with the precentral gyrus, which is also the primary motor cortex, indispensable to do this test.

In the same way, He et al. have demonstrated using DTI significantly correlations between corpus callosum, frontal regions, and cognitive tests concerning verbal memory and speed of information processing [48]. Thus, with PASAT, they found correlations with corpus callosum (fraction of anisotropy and mean diffusivity). These data are coherent with ours showing the importance of the corpus callosum and the superior longitudinal fascicle in NMO for cognitive aspects.

Treatment

Current NMO treatments include general immunosuppressive agents such as oral azathioprine, oral prednisone, oral mycophenolate mofetil, or oral methotrexate. B-cell depletion using intravenous rituximab is of interest particularly for patients with pejorative evolution [49]. For relapses, intravenous methylprednisolone and plasma exchange are the treatments of reference [49]. However, no treatments have been tested for cognitive and behavioral aspects of NMO. Moreover, no controlled clinical trials in NMO patients have been ever conducted to date even for the classical aspects of NMO, i.e., NO and ATM [50].

Conclusion

NMO is an inflammatory disease of the central nervous system not only responsible for NO and ATM, but also brain involvement including GM and WM. That is the reason why cognitive impairment in NMO is logical. The neural basis of cognitive troubles in NMO seems to be the whole WM but particularly the corpus callosum and the superior longitudinal fascicles. Depression and fatigue are frequent behavioral symptoms of NMO. There are no guidelines for the treatment of neuropsychological symptoms of NMO: when clinical trials will be conducted in NMO patients, a neuropsychological evaluation will be necessary.

References

1. Gault F. De la Neuromyélite Optique aiguë. Thèse à la faculté de Médecine et de Pharmacie de Lyon 981; 1894.
2. Lennon VA, Wingerchuk DM, Kryzer TJ, Pittock SJ, Lucchinetti CF, et al. A serum autoantibody marker of neuromyelitis optica: distinction from multiple sclerosis. Lancet. 2004;364:2106–12.
3. Cabre P, Gonzalez-Quevedo A, Lannuzel A, Bonnan M, Merle H, et al. Descriptive epidemiology of neuromyelitis optica in the Caribbean basin. Rev Neurol (Paris). 2009;165:676–83.
4. Rivera JF, Kurtzke JF, Booth VJ, Corona VT. Characteristics of Devic's disease (neuromyelitis optica) in Mexico. J Neurol. 2008;255:710–5.
5. Cossburn M, Tackley G, Baker K, Ingram G, Burtonwood M, et al. The prevalence of neuromyelitis optica in South East Wales. Eur J Neurol. 2012;19:655–9.
6. Asgari N, Lillevang ST, Skejoe HP, Falah M, Stenager E, et al. A population-based study of neuromyelitis optica in Caucasians. Neurology. 2011;76:1589–95.
7. Jarius S, Ruprecht K, Wildemann B, Kuempfel T, Ringelstein M, et al. Contrasting disease patterns in seropositive and seronegative neuromyelitis optica: a multicentre study of 175 patients. J Neuroinflammation. 2012;9:4.
8. Collongues N, Marignier R, Zephir H, Papeix C, Blanc F, et al. Neuromyelitis optica in France: a multicenter study of 125 patients. Neurology. 2010;74:736–42.

9. Wingerchuk DM, Lennon VA, Pittock SJ, Lucchinetti CF, Weinshenker BG. Revised diagnostic criteria for neuromyelitis optica. Neurology. 2006;66:1485–9.
10. Paty DW, Oger JJ, Kastrukoff LF, Hashimoto SA, Hooge JP, et al. MRI in the diagnosis of MS: a prospective study with comparison of clinical evaluation, evoked potentials, oligoclonal banding, and CT. Neurology. 1988;38:180–5.
11. Flanagan EP, Weinshenker BG, Krecke KN, Lennon VA, Lucchinetti CF, et al. Short myelitis lesions in Aquaporin-4-IgG-positive neuromyelitis optica spectrum disorders. JAMA Neurol. 2015;72:81–7.
12. Jarius S, Wildemann B. Aquaporin-4 antibodies (NMO-IgG) as a serological marker of neuromyelitis optica: a critical review of the literature. Brain Pathol. 2013;23:661–83.
13. Marignier R, Bernard-Valnet R, Giraudon P, Collongues N, Papeix C, et al. Aquaporin-4 antibody-negative neuromyelitis optica: distinct assay sensitivity-dependent entity. Neurology. 2013;80:2194–200.
14. Misu T, Fujihara K, Kakita A, Konno H, Nakamura M, et al. Loss of aquaporin 4 in lesions of neuromyelitis optica: distinction from multiple sclerosis. Brain. 2007;130:1224–34.
15. Pittock SJ, Weinshenker BG, Lucchinetti CF, Wingerchuk DM, Corboy JR, et al. Neuromyelitis optica brain lesions localized at sites of high aquaporin 4 expression. Arch Neurol. 2006;63:964–8.
16. Nakamura M, Endo M, Murakami K, Konno H, Fujihara K, et al. An autopsied case of neuromyelitis optica with a large cavitary cerebral lesion. Mult Scler. 2005;11:735–8.
17. Blanc F, Zephir H, Lebrun C, Labauge P, Castelnovo G, et al. Cognitive functions in neuromyelitis optica. Arch Neurol. 2008;65:84–8.
18. Saji E, Toyoshima Y, Yanagawa K, Nishizawa M, Kawachi I. Neuropsychiatric presentation of neuromyelitis optica spectrum disorders. Neurology. 2010;74:A169.
19. Vanotti S, Cores EV, Eizaguirre B, Melamud L, Rey R, et al. Cognitive performance of neuromyelitis optica patients: comparison with multiple sclerosis. Arq Neuropsiquiatr. 2013;71:357–61.
20. Saji E, Arakawa M, Yanagawa K, Toyoshima Y, Yokoseki A, et al. Cognitive impairment and cortical degeneration in neuromyelitis optica. Ann Neurol. 2013;73:65–76.
21. He D, Chen X, Zhao D, Zhou H. Cognitive function, depression, fatigue, and activities of daily living in patients with neuromyelitis optica after acute relapse. Int J Neurosci. 2011;121:677–83.
22. Cardona JF, Sinay V, Amoruso L, Hesse E, Manes F, et al. The impact of neuromyelitis optica on the recognition of emotional facial expressions: a preliminary report. Soc Neurosci. 2014;9:633–8.
23. Blanc F, Noblet V, Jung B, Rousseau F, Renard F, et al. White matter atrophy and cognitive dysfunctions in neuromyelitis optica. PLoS One. 2012;7, e33878.
24. Cummings JL. Subcortical dementia. Neuropsychology, neuropsychiatry, and pathophysiology. Br J Psychiatry. 1986;149:682–97.
25. Kawahara Y, Ikeda M, Deguchi K, Hishikawa N, Kono S, et al. Cognitive and affective assessments of multiple sclerosis (MS) and neuromyelitis optica (NMO) patients utilizing computerized touch panel-type screening tests. Intern Med. 2014;53:2281–90.
26. Borden A, Kulkarni C, Krieger D, Bhalerao S. Depression and Devic's syndrome. Am J Psychiatr. 2004;161:1128–36.
27. Chanson JB, Zephir H, Collongues N, Outteryck O, Blanc F, et al. Evaluation of health-related quality of life, fatigue and depression in neuromyelitis optica. Eur J Neurol. 2011;18:836–41.
28. Vernant JC, Cabre P, Smadja D, Merle H, Caubarrere I, et al. Recurrent optic neuromyelitis with endocrinopathies: a new syndrome. Neurology. 1997;48:58–64.
29. Suzuki N, Takahashi T, Aoki M, Misu T, Konohana S, et al. Neuromyelitis optica preceded by hyperCKemia episode. Neurology. 2010;74:1543–5.
30. Pittock SJ, Lennon VA, Krecke K, Wingerchuk DM, Lucchinetti CF, et al. Brain abnormalities in neuromyelitis optica. Arch Neurol. 2006;63:390–6.

31. Yu CS, Lin FC, Li KC, Jiang TZ, Zhu CZ, et al. Diffusion tensor imaging in the assessment of normal-appearing brain tissue damage in relapsing neuromyelitis optica. AJNR Am J Neuroradiol. 2006;27:1009–15.
32. Rocca MA, Agosta F, Mezzapesa DM, Martinelli V, Salvi F, et al. Magnetization transfer and diffusion tensor MRI show gray matter damage in neuromyelitis optica. Neurology. 2004;62:476–8.
33. de Seze J, Blanc F, Kremer S, Collongues N, Fleury M, et al. Magnetic resonance spectroscopy evaluation in patients with neuromyelitis optica. J Neurol Neurosurg Psychiatry. 2010;81:409–11.
34. Aboul-Enein F, Krssak M, Hoftberger R, Prayer D, Kristoferitsch W. Diffuse white matter damage is absent in neuromyelitis optica. AJNR Am J Neuroradiol. 2010;31:76–9.
35. Bichuetti DB, Rivero RL, de Oliveira EM, Oliveira DM, de Souza NA, et al. White matter spectroscopy in neuromyelitis optica: a case control study. J Neurol. 2008;255:1895–9.
36. Popescu BF, Parisi JE, Cabrera-Gomez JA, Newell K, Mandler RN, et al. Absence of cortical demyelination in neuromyelitis optica. Neurology. 2010;75:2103–9.
37. Chalumeau-Lemoine L, Chretien F, Gaelle Si Larbi A, Brugieres P, Gray F, et al. Devic disease with brainstem lesions. Arch Neurol. 2006;63:591–3.
38. Zaidel DW. The case for a relationship between human memory, hippocampus and corpus callosum. Biol Res. 1995;28:51–7.
39. Eichenbaum H, Lipton PA. Towards a functional organization of the medial temporal lobe memory system: role of the parahippocampal and medial entorhinal cortical areas. Hippocampus. 2008;18:1314–24.
40. Vestergaard M, Madsen KS, Baare WF, Skimminge A, Ejersbo LR, et al. White matter microstructure in superior longitudinal fasciculus associated with spatial working memory performance in children. J Cogn Neurosci. 2011;23:2135–46.
41. Cabeza R, Ciaramelli E, Olson IR, Moscovitch M. The parietal cortex and episodic memory: an attentional account. Nat Rev Neurosci. 2008;9:613–25.
42. Benarroch EE. The locus ceruleus norepinephrine system: functional organization and potential clinical significance. Neurology. 2009;73:1699–704.
43. Sara SJ. The locus coeruleus and noradrenergic modulation of cognition. Nat Rev Neurosci. 2009;10:211–23.
44. Ozturk A, Smith SA, Gordon-Lipkin EM, Harrison DM, Shiee N, et al. MRI of the corpus callosum in multiple sclerosis: association with disability. Mult Scler. 2010;16:166–77.
45. Bledowski C, Kaiser J, Rahm B. Basic operations in working memory: contributions from functional imaging studies. Behav Brain Res. 2010;214:172–9.
46. Tulving E. Episodic memory: from mind to brain. Annu Rev Psychol. 2002;53:1–25.
47. Koenigs M, Acheson DJ, Barbey AK, Solomon J, Postle BR, et al. Areas of left perisylvian cortex mediate auditory-verbal short-term memory. Neuropsychologia. 2011;49:3612–9.
48. He D, Wu Q, Chen X, Zhao D, Gong Q, et al. Cognitive impairment and whole brain diffusion in patients with neuromyelitis optica after acute relapse. Brain Cogn. 2011;77:80–8.
49. Jarius S, Wildemann B, Paul F. Neuromyelitis optica: clinical features, immunopathogenesis and treatment. Clin Exp Immunol. 2014;176:149–64.
50. Papadopoulos MC, Bennett JL, Verkman AS. Treatment of neuromyelitis optica: state-of-the-art and emerging therapies. Nat Rev Neurol. 2014;10:493–506.

Chapter 18
Dementia in Multiple Sclerosis

Gilles Defer and Pierre Branger

Abstract Dementia in multiple sclerosis is not a well-defined condition as it may refer to different clinical situations. In addition, as the term has now disappeared from the last version of the Diagnostic and Statistical Manual of Mental Disorders (DSM-5), dementia may be now defined as a major neurocognitive disorder leading to significant alteration of social and/or physical independence that could be explained from a pathophysiological point of view as a cortico-subcortical multiple disconnection syndrome. Today, there was no informative prospective study permitting to establish the prevalence of dementia in MS. However, according to the literature, major neurocognitive disorder significantly limiting daily life activities, social relationships, or patient's ability to work may be higher than previously suspected in MS patients, especially in the progressive forms of the disease. Sometimes, early rapid and severe cognitive impairment may be the initial or the main manifestation of the disease. Patient's cognitive profile in case of major neurocognitive disorder does not differ from what is known about cognitive impairment in MS including major alteration of information processing speed. Prospective multicenter studies would be useful to better define the frequency and time of occurrence of dementia in this disease. Other main challenges for the future should be the identification of modifiable risk factors for severe cognitive impairment and the goal of reducing cognitive disability progression through new therapeutic procedures.

Keywords Dementia • Cognitive impairment • Cognitive reserve • Mental disorders • Intellectual disability • Multiple sclerosis

G. Defer, MD (✉) • P. Branger, MD
Department of Neurology, University Hospital Center of Caen,
Avenue de la Côte de Nacre, CS 30001, 14033 Caen Cedex 9, France
e-mail: defer-gi@chu-caen.fr; branger-p@chu-caen.fr

B. Brochet (ed.), *Neuropsychiatric Symptoms of Inflammatory Demyelinating Diseases*, Neuropsychiatric Symptoms of Neurological Disease,
DOI 10.1007/978-3-319-18464-7_18

Introduction

Psychiatric and cognitive manifestations of multiple sclerosis (MS) are largely described in details in other chapters of this book. However, concerning cognitive impairment, and beside the identification of the altered cognitive domains for one patient, one important question the neurologist may have to face is the estimation of the severity of this alteration and its consequences in daily life activities, social relationships, or patient's ability to work. Indeed, this cognitive disability adds to physical disability to greatly affect patient's quality of life [1] and represents a hard therapeutic challenge as, today, there is no drug able to significantly and permanently improve cognitive impairment of MS patients [2, 3]. One peculiar issue, related to severe or even very severe cognitive dysfunction, is the patient may early have or develop after years of disease a dementia. Usually dementia in MS is not only a rare situation but in addition not a well-defined clinical picture. Indeed, most of the reported cases of dementia in MS are individual case reports or small series [4–7], and there was no informative prospective study reporting the incidence or prevalence of dementia in MS. The main explanation of this lack of data probably comes from the absence of recognized published diagnosis criteria. When reading of the last version of the Diagnostic and Statistical Manual of Mental Disorders (DSM-5) published in 2013 [8], it is surprising to see that the term dementia is now leaved but included under the newly named entity *Major Neurocognitive Disorder*. Therefore, MS, associated or not with behavioral symptoms (which is also an important symptom in this disease), found its place as an example in the paragraph giving criteria and coding clues for *Neurocognitive Disorder due to Another Medical Condition* as Alzheimer and Lewy body diseases, frontotemporal disorders, Parkinson's, Huntington's, or vascular diseases, etc.

According to these remarks, this chapter will deal successively the question on the definition of dementia in MS and its cognitive profile according to the underlying pathophysiology. The frequency and main data issued from the literature, including cases with histological confirmation, will be discussed in the second part.

Concept of Dementia

Definition

Initially, the concept of dementia described by Esquirol [9] was the acquired character of cognitive decline before proposing a distinction between acute and chronic dementia. Acute dementia was renamed later delirium and characterized by the DSM-IV (*Diagnostic and Statistical Manual of Mental Disorders, text revised* [10]) as a disturbance of consciousness and a change in cognition that develop over a

Table 18.1 Diagnostic criteria of major neurocognitive disorder in DSM-5 [8]

(A) Evidence of significant cognitive decline from a previous level of performance in one or more cognitive domains (complex attention, executive function, learning and memory, language, perceptual-motor, or social cognition) based on:
1. Concern of the individual, a knowledgeable informant, or the clinician that there has been a significant decline in cognitive function
2. A substantial impairment in cognitive performance, preferably documented by standardized neuropsychological testing or, in its absence, another quantified clinical assessment
(B) The cognitive deficits interfere with independence in everyday activities (i.e., at a minimum, requiring assistance with complex instrumental activities of daily living such as paying bills or managing medications)
(C) The cognitive deficits do not occur exclusively in the context of a delirium
(D) The cognitive deficits are not better explained by another mental disorder (e.g., major depressive disorder, schizophrenia)

Table 18.2 Diagnostic criteria of major or mild neurocognitive disorder due to another medical condition in DSM-5 [8]

(A) The criteria are met for major or mild neurocognitive disorder
(B) There is evidence from the history, physical examination, or laboratory findings that the neurocognitive disorder is the pathophysiological consequence of another medical condition
(C) The cognitive deficits are not better explained by another mental disorder or another specific neurocognitive disorder (e.g., Alzheimer's disease, HIV infection)

short period of time, usually hours to days, and tend to fluctuate during the course of the day. In this manual, dementia was characterized by multiple cognitive deficits including memory impairment and at least one of the following cognitive disturbances: aphasia, apraxia, agnosia, or disturbance in executive functioning, causing significant decline from a previous level of functioning [10]. In the last published version of this classification of mental disorders (DSM-5 [8]), the term delirium was kept, whereas the term dementia was changed for mild (insufficient to interfere with independence) or major neurocognitive disorders (dementia in DSM-IV). The other main change is the disappearance of the classically required presence of "memory impairment" for all dementias. It has been recognized that memory impairment is not the first domain to be affected in all of the other diseases that cause a neurocognitive disorder, as in frontotemporal lobar disorder where language could be affected first. Diagnostic criteria of major neurocognitive disorders in DSM-5 are reported in Table 18.1. Therefore, each classical disease or syndrome (as Alzheimer's disease, vascular disorders, or head trauma) responsible of neurocognitive disorders has its own, mild or major, diagnosis criteria, whereas MS has no specific criteria and has then to be classified in the part entitled neurocognitive disorders due to another medical condition (Table 18.2). According to that, it seems from clinical practice that a significant number of patients with MS may fulfill these criteria with sufficient decline in memory and executive abilities causing an irreversible source of social and/or professional disabilities.

Cortical Versus Subcortical Dementia in Multiple Sclerosis, Myth, or Reality

Distinction between cortical and subcortical dementia is still debated but had allowed to classify, since many years, most of neurocognitive disorders. The concept of cortical dementia, whose prototype is Alzheimer's disease, is linked to a foreground memory impairment, with at least one associated cognitive disorder such as aphasia, apraxia, or agnosia, resulting in alteration of the social and/or physical independence. The concept of subcortical dementia [11] is less consensual and commonly used to describe very different patterns of cognitive dysfunction in diseases such as Huntington's disease, progressive supranuclear palsy, Parkinson's disease, or HIV infection. Bradyphrenia and dysexecutive disorders are predominant and can affect visuospatial and memory abilities without aphaso-apraxo-agnosia syndrome. Sometimes, cognitive deficits may be directly related to lesions in the subcortical gray matter and white matter but also indirectly related through dysfunction of cortico-subcortical frontal circuits explaining the use of "sub-corticofrontal dementia" expression in place of subcortical dementia [12]. A proposal of classical recognized differences between cortical and subcortical dementia is presented in Table 18.3 [13]. However, this distinction is now controversial because the anatomical damaged systems underlying to

Table 18.3 Neuropsychological characteristics of cortical and subcortical dementias [13]

Dementia	Cortical	Subcortical
Global cognitive efficiency	Very disturbed	+/– Preserved
Episodic memory	In the foreground	In the background
Encoding	Disturbed	Preserved
Free recall	Disturbed	Disturbed
Cued recall	Disturbed	+/– Preserved
Recognition	Disturbed	Preserved
Implicit memory	Disturbed	Preserved
Priming effect	Preserved	Disturbed
Procedural memory		
Executive functions	In the background	In the foreground
Bradyphrenia	Rarely observed	Present
Oral/written language	Lack of word, dysorthography	Often preserved
Speech	Long preserved	Dysarthria
Praxis	Disturbed	Preserved
Gnosis	Disturbed	Preserved
Mood	+/– Preserved	Disturbed

the different observed subcortical dementia cognitive patterns are of great variability. In this context, it can be allowed that the cognitive profile of MS is somewhere between these two concepts. Indeed, the disruption of communicating inside networks as a consequence of white matter damage could constitute the anatomical substrate of this cognitive impairment [14]. Calabrese et al. [15] suggested the term "multiple disconnection" to describe the variety of neuropsychological deficits encountered in this demyelinating disease (different from those observed in typical cortical or subcortical dementia), which may result from damages to temporal lobes, diencephalic memory systems, and to fronto-striatal circuits. This concept is now clearly supported by different studies [16–19]. Louapre et al. compared 15 cognitively impaired RRMS with abnormal performances in at least 3 tests to 20 cognitively preserved RRMS and 20 age-matched healthy controls. Cognitively impaired patients had higher white matter lesion load, and more severe atrophy in gray matter regions highly connected to networks involved in cognition. In addition, the authors showed that functional connectivity of attentional and default mode networks was decreased in the cognitively impaired group compared to the cognitively preserved group. The disconnection of these networks may deprive the brain of compensatory mechanisms required to face widespread structural damage [20]. Another multicenter study comparing fMRI scans in 42 RRMS (in whom 47 % were considered cognitively impaired) and 52 sex-matched healthy controls when performing the N-back task found that cognitively preserved patients had increased recruitment of the right dorsolateral prefrontal cortex, whereas cognitively impaired MS patients had reduced activations of several areas in the fronto-parieto-temporal lobes. These data suggest that preserved efficient frontal network is associated with a better cognitive profile in MS patients [21].

Beside relationship with lesion burden [22, 23], measurement of global or regional brain atrophy seems also to be particularly sensitive to cognitive status. Change in brain volume may be evaluated using different techniques such as the brain parenchymal volume (BPF), structural image evaluation using normalization of atrophy (SIENA), or the voxel-based morphometry (VBM) approach. Several studies have related brain volume change to cognitive alteration in MS cognitive dysfunction [24–28]. Cortical atrophy may contribute to the development of cognitive impairment in MS [29] and dominates the pathological process as MS progresses [30], even if Benedict et al. [31] concluded in a cross-sectional study concerning 82 patients that central and cortical atrophy participate equally to the development of cognitive dysfunction in MS. Whatever the underlying mechanisms driving cognitive dysfunction in MS, some patients are able to withstand without no or weak cognitive impairment despite significant white/gray matter lesions and/or cerebral atrophy suggesting of efficient cognitive reserve. The maximal lifetime brain growth [32], hereditary, and intellectual enrichment [33] are described as factors contributing to sources of cognitive reserve. These sources of reserve may help identify patients at greatest risk for cognitive decline and may be targeted for early-intervention cognitive rehabilitation [34].

Dementia in Multiple Sclerosis

Frequency

As previously pointed out, to date there is no informative prospective study able to give the prevalence of dementia in MS. Again frequency of major cognitive impairment in MS population is difficult to provide from the literature because of lack of published validated diagnosis criteria. However, some studies had tried to approximate this question. Through different articles, Rao et al. [35–37] reported that 40–65 % of MS patients may have cognitive impairment during their disease, with consequences on daily life [38, 39], and that 20–30 % may have severe dementia. In fact, this assertion is based on a small study where 28/44 patients with progressive MS were shown to have memory disorders and among them 9 had "broader cognitive alteration." However [40], nothing was specified regarding the impact on daily life activities of cognitive impairment in these patients which is a mandatory criteria for dementia.

Scales objectifying loss of independence (Minimal Record Disability [41]) due to cognitive impairment were incorporated in two studies [42, 43]. Rodriguez et al. [43] explored in Olmsted County, Minnesota, a cohort of 179 cases of definite or probable MS and found few patients (3.7 %) with severe decrease in cognition requiring supervision. Using the mental functional subscore of the EDSS, Midgard et al. [42] (from Møre and Romsdal County, Norway) found that 4 % of their 124 patients had severe cognitive disorder, whereas when using the Incapacity Status Scale, 10.3 % were reported to have decrease in mentation severe enough to interfere with everyday activities. However, it is known that the EDSS mental subscore do not well reflect the nature and severity of cognitive alteration. In addition in these two studies, the nature of cognitive alterations was not studied or specified then challenging the diversity of cognitive impairment required for the diagnosis of dementia. Similar limitations concern another study [44] exploring a small cohort of 26 patients with progressive MS where a dementia scale testing impact on daily life activities (Blessed Dementia Scale [45]) was used and showed that 50 % of patients matched the criterion of "major cognitive dysfunction or dementia."

In a review devoted to MS, Benedict and Bobholz [46] reported high occurrence of dementia in a selected group of patients. Indeed, using data from a previous prospective study evaluating in a large cohort of MS patients using the MACFIMS cognitive battery [47], they search for patients meeting DSM-IV criteria for dementia. They found that 22 % of the 291 evaluated patients had low (<2.0) abnormal neuropsychological Z-score (calculated from the data of 56 healthy controls) for at least one memory test and most tests exploring other cognitive domains with impairment of vocational status. This group of MS-associated dementia has very similar demographic features as compared with other MS population in the literature except for a higher frequency (51 %) of secondary progressive disease and more behavioral disturbances.

In 2009, Staff et al. [48] reported a retrospective case series through the Mayo Clinic data retrieval system between 1996 and 2008. They plan to identify severe cognitive impairment as a primary neurological symptom in MS patients. Severe cognitive impairment was formally assessed by the Kokmen Short Test of Mental Status (38-point cognitive screening test assessing orientation, attention, learning and recall, calculation, abstraction, construction, and knowledge) which has a good sensitivity and specificity for patients younger than 50 years [49]. This screening procedure led to the identification of 172 MS patient with severe cognitive impairment among 549 having cognitive alterations which represent 30 % of patients presenting potentially a dementia. Unfortunately, the number of MS patients without cognitive impairment in the database and clinical details of these patients was not available.

Whatever the limitations of these two studies, they suggest from single center experience that among MS patients with cognitive alteration, dementia may be relatively high between 20 and 30 %.

Clinical Presentation of MS Cases Inaugurated by Dementia

Different case reports or small series have reported severe cognitive impairment or dementia initiating or with rapid progression early on in the disease.

In 1976, Young et al. [4], before imaging area (only one patient had brain scan), reported 5 cases of patients presenting behavioral and cognitive disorders with or without initial neurological manifestations, but getting finally a diagnosis of MS. The study mainly concerns the observation of cognitive impairment early on in the disease, especially because 2 patients had mental symptoms several years before neurological ones with severe alteration of some cognitive domains on psychological testing and behavioral disorder suggesting of dementia. In all patients, neuropsychological assessment found a decline in overall intellectual efficiency worsening at most evaluation.

Fontaine et al. [5] reported two cases of women with MS histological confirmation. The first patient aged 51 years developed a progressive pragmatic behavior with alteration of short-term memory and learning abilities with later on constructive and dressing apraxia. Initially, neurological evaluation did not show sensorimotor, gait disorder, cranial nerves, or cerebellar abnormalities. This patient died 4 years after the beginning of symptoms. Postmortem examination showed numerous plaques in the periventricular white matter with severe atrophy of corpus callosum. Plaques were also seen in the white matter of both hippocampus and the columns of the fornix that explains memory impairment. The second patient presented with behavioral disturbances starting during childhood and had to left school at 12. She had mystic hallucinations and was depressive. Cognitive testing showed severe intellectual and memory dysfunction; visuospatial processes were severely affected. MR imaging showed periventricular lesions. Stereotactic biopsy of a large left frontal plaque confirmed MS diagnosis.

In 2005, Leyhe et al. [6] reported 4 cases in patients older than 60 years who were referred to the memory clinic for diagnosis of dementia. All of them were found to have evidence of biological (positive oligoclonal bands on CSF examination) and MR imaging chronic inflammatory CNS process compatible with the diagnosis of MS, mainly primary progressive (3/4). Only one had progressive gait disturbances associated with major cognitive alterations, whereas others had mainly sensitive symptoms. This series shows that MS could be considered as a differential diagnosis of dementia in older patients especially in case of white matter abnormalities on MR imaging. In addition, it has been recently showed that Alzheimer disease (AD) may coexist with MS dementia and that in patients presenting complicated cognitive presentation, ^{18}F-fluorodeoxyglucose PET imaging and CSF-AD biomarkers may be of diagnostic value [50].

Another MS case with dementia was reported by Stoquart-Elsankari et al. [7]. The observation concerned a 48-year-old woman presenting with cognitive slowing and hemiparesis who had dramatic dementia evolution. Motor symptoms responded well to steroids treatment, but cognitive abilities continue over 6 months to severely deteriorate with cerebral MRI showing diffuse hyperintensities of the white matter especially in frontal regions despite immunosuppressive therapy with mitoxantrone.

In 2003, Zarei et al. [51] reported 6 patients initially presenting with an undiagnosed progressive dementia syndrome with prominent amnesia often accompanied by classic cortical features including dysphasia, dysgraphia, or dyslexia. Mood disturbance was ubiquitous, and in three patients, there was a long history of preceding severe depression. Thereafter, all patients developed on follow-up, physical signs, and marked disabilities. MR imaging and ancillary investigations establish the diagnosis of MS. These cases led the authors to propose the concept of cortical variant of MS supported by 17 other similar older cases found in the literature. This concept was more developed in another paper [52] where the authors underline a potential relationship between radiological and neuropathological demonstrated high frequency of cortical lesions and neurobehavioral symptoms, including depression, amnesia, or other distinct cortical syndromes. Most of the reported patients had depression, and 40 % presented amnesia or psychiatric symptoms (personality change, aggression, perseveration, circumstantiality, or inappropriate hunger). Indeed, behavioral symptoms and related impairments are now well described and more frequently observed than previously suspected [53], even sometimes as an initial manifestation of disease. However, these cases, some of them having mainly psychiatric presentation, cannot be restricted to "a cortical disease" as accepted MS pathophysiological processes supporting cognitive alteration are now well linked to a multiple disconnection or brain networks alterations including basal ganglia [54, 55].

Presentation and related clinical features of severe cognitive alteration have been well approached by the already discussed study of Staff et al. [48]. In this study, the authors identified among 172 patients with severe cognitive impairment, after having applied numerous exclusion criteria using extensive biological examinations, 23 patients (representing 4 % of the whole initial examined cohort of 549 patients) in whom severe cognitive impairment was the primary neurological symptom and who had no significant MS-related impairment in other neurological domains or alternative diagnosis for cognitive dysfunction. In 9 patients, cognitive dysfunction occurred in an attack-related subacute fulminant presentation with five suffering of single severe cognitive attack without full resolution. All had a relapsing-remitting form of the disease. In the 14 other patients, cognitive alteration evolved in progressive fashion leading to significant disability in the context of a primary or secondary progressive disease form. Among the whole cohort, 65 % exhibited psychiatric symptoms, 57 % had mild cerebellar ataxia, and 39 % had cortical symptoms and signs (seizure, aphasia, apraxia), whereas 70 % had brain atrophy on MR imaging. Interestingly, 67 % had a history of smoking and 93 % of which was active at the time of disease onset.

Discussion/Conclusion

Today, there is no doubt that dementia in MS may occur during the course of the disease. Whatever the terminology used (now major neurocognitive disorder interfering with independence according to DMS-5), this disabling disorder needs to be better identified in MS patients. Cognitive profile of demented MS patients does not seem to differ from cognitive alterations classically observed in this disease including major alteration of information processing speed or verbal and visual memory impairment [46]. Regarding clinical presentation, the clinician may have to encounter two main situations. The first one, probably the most frequent, may be a less or more rapid worsening of cognitive alteration during disease course mainly in progressive forms but not only as severe cognitive impairment may be observed in nonclinically active relapsing-remitting MS where progression of major neurocognitive disorder is the only cause of disability (personal observations). This profile is in line with the results of different longitudinal studies which demonstrated that in a defined cohort, the percentage of patients with major neurocognitive disorder continues to increase overtime and that preexisting alterations of cognition may favor this evolution [56–61].

Secondly, dementia may start as an acute or subacute picture of encephalopathy possibly associated with other symptoms without full resolution [7, 48]. At that time, a differential diagnosis with acute disseminated encephalomyelitis (ADEM) may be difficult needing immediate MR imaging and CSF examination [7]. In fact, this picture may be linked to a severe cognitive relapse which needs to be actively treated with steroid as any usual motor or visual relapses. The concept of isolated cognitive relapse has recently emerged from the literature and may be applied to this situation [62–64]. Moreover, in case of poor resolution, disease-modifying therapy would be quickly introduced even in case of lack or minor physical symptoms or disability, in order to prevent progressive worsening of cognitive disability [65, 66]. Finally, the third picture including significant psychiatric disorders, especially depression, that occur before or are associated with cognitive alteration with or without cortical symptoms may be rarer and probably need more time and clinical and paraclinical investigations to be definitively linked to MS.

According to the literature, it seems that in MS, major neurocognitive disorder sufficient to interfere with independence may occur at a higher rate (between 20 and 30 %) than previously suspected, especially in the progressive form of the disease, keeping in mind lack of convincing informative prospective study on that question. Then, one of the main challenges in the research field of cognition in MS would be to set multicenter prospective studies having as a primary goal to better define the prevalence of major neurocognitive disorder impairing daily life activities, according to DSM-5, especially in the progressive forms of the disease. Inclusion criteria would probably consider patients with at least cognitive alterations higher than 2SD compared to healthy controls or normative data for 3 or more cognitive domains including IPS and episodic memory. In addition, pooling clinical and conventional/nonconventional MR imaging data from MS cases with early and severe cognitive impairment as the initial and main manifestation of the disease would be useful to understand the underlying pathophysiological aspects responsible of this cognitive disability. As Staff et al. [48] pointed out active smoking in most of their patients with major neurocognitive disorder, such study would be necessarily completed by simultaneous search of potential risk factors as some modifiable ones have been already identified as potentially modifying the relationship dementia/dependence in elderly patients [67, 68]. Finally and because there is today no specific therapy which had clearly demonstrated an efficacy for cognitive impairment in MS, the development of new therapeutic strategies and/or new treatments is needed to help these patients reduce or prevent cognitive disability.

References

1. Mitchell AJ, Benito-León J, González J-MM, Rivera-Navarro J. Quality of life and its assessment in multiple sclerosis: integrating physical and psychological components of wellbeing. Lancet Neurol. 2005;4(9):556–66.
2. Defer G [Cognitive effects of symptomatic and disease treatments] in Neuropsychologie de la Sclérose en plaques – Defer G, Brochet B, Pelletier J, Editions Masson; 2010. p 181–188.
3. Amato MP, Langdon D, Montalban X, Benedict RHB, DeLuca J, Krupp LB, et al. Treatment of cognitive impairment in multiple sclerosis: position paper. J Neurol. 2013;260(6):1452–68.
4. Young AC, Saunders J, Ponsford JR. Mental change as an early feature of multiple sclerosis. J Neurol Neurosurg Psychiatry. 1976;39(10):1008–13.
5. Fontaine B, Seilhean D, Tourbah A, Daumas-Duport C, Duyckaerts C, Benoit N, et al. Dementia in two histologically confirmed cases of multiple sclerosis: one case with isolated dementia and one case associated with psychiatric symptoms. J Neurol Neurosurg Psychiatry. 1994;57(3):353–9.
6. Leyhe T, Laske C, Buchkremer G, Wormstall H, Wiendl H. Dementia as a primary symptom in late onset multiple sclerosis. Case series and review of the literature. Nervenarzt. 2005;76(6):748–55.
7. Stoquart-Elsankari S, Périn B, Lehmann P, Gondry-Jouet C, Godefroy O. Cognitive forms of multiple sclerosis: report of a dementia case. Clin Neurol Neurosurg. 2010;112(3):258–60.
8. American Psychiatric Association. Diagnostic and statistical manual of mental disorders. 5th ed. Arlington: American Psychiatric Association; 2013.

9. Esquirol E. Traité des maladies mentales considérées sous le rapport médical, hygiénique et médicolégal. Paris: J.-B. Baillière; 1838.
10. American Psychiatric Association. Diagnostic and statistical manual of mental disorders. 4th ed., text rev. Washington, DC: American Psychiatric Association; 2000.
11. Albert ML, Feldman RG, Willis AL. The « subcortical dementia » of progressive supranuclear palsy. J Neurol Neurosurg Psychiatry. 1974;37(2):121–30.
12. Dubois B, Boller F, Pillon B, Agid Y. Cognitive deficit in Parkinson's disease. In: Boller F, Grafman J, editors. Handbook of neuropsychology. Amsterdam: Elsevier; 1991. p. 195–239.
13. Defer G, Daniel F. Démence in Neuropsychologie de la Sclérose en plaques- Defer G, Brochet B, Pelletier J, Editions Masson; 2010 p. 135–9.
14. Audoin B, Au Duong MV, Malikova I, Confort-Gouny S, Ibarrola D, Cozzone PJ, et al. Functional magnetic resonance imaging and cognition at the very early stage of MS. J Neurol Sci. 2006;245(1–2):87–91.
15. Calabrese P. Neuropsychology of multiple sclerosis–an overview. J Neurol. 2006;253 Suppl 1:I10–5.
16. Mevel K, Grassiot B, Chételat G, Defer G, Desgranges B, Eustache F. The default mode network: cognitive role and pathological disturbances. Rev Neurol (Paris). 2010;166(11):859–72.
17. Morgen K, Sammer G, Courtney SM, Wolters T, Melchior H, Blecker CR, et al. Distinct mechanisms of altered brain activation in patients with multiple sclerosis. Neuroimage. 2007;37(3):937–46.
18. Rocca MA, Pagani E, Absinta M, Valsasina P, Falini A, Scotti G, et al. Altered functional and structural connectivities in patients with MS: a 3-T study. Neurology. 2007;69(23):2136–45.
19. Rocca MA, Valsasina P, Absinta M, Riccitelli G, Rodegher ME, Misci P, et al. Default-mode network dysfunction and cognitive impairment in progressive MS. Neurology. 2010;74(16):1252–9.
20. Louapre C, Perlbarg V, García-Lorenzo D, Urbanski M, Benali H, Assouad R, et al. Brain networks disconnection in early multiple sclerosis cognitive deficits: an anatomofunctional study. Hum Brain Mapp. 2014;35(9):4706–17.
21. Rocca MA, Valsasina P, Hulst HE, Abdel-Aziz K, Enzinger C, Gallo A, et al. Functional correlates of cognitive dysfunction in multiple sclerosis: a multicenter fMRI Study. Hum Brain Mapp. 2014;35(12):5799–814.
22. Sperling RA, Guttmann CR, Hohol MJ, Warfield SK, Jakab M, Parente M, et al. Regional magnetic resonance imaging lesion burden and cognitive function in multiple sclerosis: a longitudinal study. Arch Neurol. 2001;58(1):115–21.
23. Lazeron RH, Langdon DW, Filippi M, van Waesberghe JH, Stevenson VL, Boringa JB, et al. Neuropsychological impairment in multiple sclerosis patients: the role of (juxta)cortical lesion on FLAIR. Mult Scler. 2000;6(4):280–5.
24. Bastianello S, Giugni E, Amato MP, Tola M-R, Trojano M, Galletti S, et al. Changes in magnetic resonance imaging disease measures over 3 years in mildly disabled patients with relapsing-remitting multiple sclerosis receiving interferon β-1a in the COGnitive Impairment in MUltiple Sclerosis (COGIMUS) study. BMC Neurol. 2011;11:125.
25. Bermel RA, Bakshi R. The measurement and clinical relevance of brain atrophy in multiple sclerosis. Lancet Neurol. 2006;5(2):158–70.
26. Morgen K, Sammer G, Courtney SM, Wolters T, Melchior H, Blecker CR, et al. Evidence for a direct association between cortical atrophy and cognitive impairment in relapsing-remitting MS. Neuroimage. 2006;30(3):891–8.
27. Shiee N, Bazin P-L, Zackowski KM, Farrell SK, Harrison DM, Newsome SD, et al. Revisiting brain atrophy and its relationship to disability in multiple sclerosis. PLoS One. 2012;7(5), e37049.
28. Lazeron RHC, Boringa JB, Schouten M, Uitdehaag BMJ, Bergers E, Lindeboom J, et al. Brain atrophy and lesion load as explaining parameters for cognitive impairment in multiple sclerosis. Mult Scler. 2005;11(5):524–31.

29. Amato MP, Bartolozzi ML, Zipoli V, Portaccio E, Mortilla M, Guidi L, et al. Neocortical volume decrease in relapsing-remitting MS patients with mild cognitive impairment. Neurology. 2004;63(1):89–93.
30. Fisher E, Lee J-C, Nakamura K, Rudick RA. Gray matter atrophy in multiple sclerosis: a longitudinal study. Ann Neurol. 2008;64(3):255–65.
31. Benedict RHB, Bruce JM, Dwyer MG, Abdelrahman N, Hussein S, Weinstock-Guttman B, et al. Neocortical atrophy, third ventricular width, and cognitive dysfunction in multiple sclerosis. Arch Neurol. 2006;63(9):1301–6.
32. Satz P. Brain reserve capacity on symptom onset after brain injury: a formulation and review of evidence for threshold theory. Neuropsychology. 1993;7(3):273–95.
33. Stern Y. What is cognitive reserve? Theory and research application of the reserve concept. J Int Neuropsychol Soc. 2002;8(3):448–60.
34. Sumowski JF, Leavitt VM. Cognitive reserve in multiple sclerosis. Mult Scler. 2013;19(9):1122–7.
35. Rao SM, Hammeke TA, McQuillen MP, Khatri BO, Lloyd D. Memory disturbance in chronic progressive multiple sclerosis. Arch Neurol. 1984;41(6):625–31.
36. Rao SM, Leo GJ, Bernardin L, Unverzagt F. Cognitive dysfunction in multiple sclerosis. I. Frequency, patterns, and prediction. Neurology. 1991;41(5):685–91.
37. Rao SM. Neuropsychology of multiple sclerosis. Curr Opin Neurol. 1995;8(3):216–20.
38. Rao SM, Leo GJ, Ellington L, Nauertz T, Bernardin L, Unverzagt F. Cognitive dysfunction in multiple sclerosis. II Impact on employment and social functioning. Neurology. 1991;41(5):692–6.
39. Patti F, Amato MP, Trojano M, Bastianello S, Tola MR, Goretti B, et al. Cognitive impairment and its relation with disease measures in mildly disabled patients with relapsing–remitting multiple sclerosis: baseline results from the Cognitive Impairment in Multiple Sclerosis (COGIMUS) study. Mult Scler. 2009;15(7):779–88.
40. Rao SM. White matter disease and dementia. Brain Cogn. 1996;31(2):250–68.
41. Slater RJ, LaRocca NG, Scheinberg LC. Development and testing of a minimal record of disability in multiple sclerosis. Ann NY Acad Sci. 1984;436:453–68.
42. Midgard R, Riise T, Nyland H. Impairment, disability, and handicap in multiple sclerosis. A cross-sectional study in an incident cohort in Møre and Romsdal County, Norway. J Neurol. 1996;243(4):337–44.
43. Rodriguez M, Siva A, Ward J, Stolp-Smith K, O'Brien P, Kurland L. Impairment, disability, and handicap in multiple sclerosis: a population-based study in Olmsted County, Minnesota. Neurology. 1994;44(1):28–33.
44. Patti F, Di Stefano M, De Pascalis D, Ciancio MR, De Bernardis E, Nicoletti F, et al. May there exist specific MRI findings predictive of dementia in multiple sclerosis patients? Funct Neurol. 1995;10(2):83–90.
45. Blessed G, Tomlinson BE, Roth M. The association between quantitative measures of dementia and of senile change in the cerebral grey matter of elderly subjects. Br J Psychiatry. 1968;114(512):797–811.
46. Benedict RHB, Bobholz JH. Multiple sclerosis. Semin Neurol. 2007;27(1):78–85.
47. Benedict RHB, Cookfair D, Gavett R, Gunther M, Munschauer F, Garg N, et al. Validity of the minimal assessment of cognitive function in multiple sclerosis (MACFIMS). J Int Neuropsychol Soc. 2006;12(4):549–58.
48. Staff NP, Lucchinetti CF, Keegan BM. Multiple sclerosis with predominant, severe cognitive impairment. Arch Neurol. 2009;66(9):1139–43.
49. Kokmen E, Smith GE, Petersen RC, Tangalos E, Ivnik RC. The short test of mental status. Correlations with standardized psychometric testing. Arch Neurol. 1991;48(7):725–8.
50. Flanagan EP, Knopman DS, Keegan BM. Dementia in MS complicated by coexistent Alzheimer disease: Diagnosis premortem and postmortem. Neurol Clin Pract. 2014;4(3):226–30.
51. Zarei M, Chandran S, Compston A, Hodges J. Cognitive presentation of multiple sclerosis: evidence for a cortical variant. J Neurol Neurosurg Psychiatry. 2003;74(7):872–7.

52. Zarei M. Clinical characteristics of cortical multiple sclerosis. J Neurol Sci. 2006;245(1–2):53–8.
53. Rosti-Otajärvi E, Hämäläinen P. Behavioural symptoms and impairments in multiple sclerosis: a systematic review and meta-analysis. Mult Scler. 2013;19(1):31–45.
54. Calabrese M, Rinaldi F, Grossi P, Mattisi I, Bernardi V, Favaretto A, et al. Basal ganglia and frontal/parietal cortical atrophy is associated with fatigue in relapsing-remitting multiple sclerosis. Mult Scler. 2010;16(10):1220–8.
55. Menéndez-González M, Salas-Pacheco JM, Arias-Carrión O. The yearly rate of Relative Thalamic Atrophy (yrRTA): a simple 2D/3D method for estimating deep gray matter atrophy in Multiple Sclerosis. Front Aging Neurosci. 2014;6:219.
56. Kujala P, Portin R, Ruutiainen J. The progress of cognitive decline in multiple sclerosis. A controlled 3-year follow-up. Brain. 1997;120(Pt 2):289–97.
57. Amato MP, Ponziani G, Siracusa G, Sorbi S. Cognitive dysfunction in early-onset multiple sclerosis: a reappraisal after 10 years. Arch Neurol. 2001;58(10):1602–6.
58. Rosti E, Hämäläinen P, Koivisto K, Hokkanen L. One-year follow-up study of relapsing-remitting MS patients' cognitive performances: Paced Auditory Serial Addition Test's susceptibility to change. J Int Neuropsychol Soc. 2007;13(5):791–8.
59. Bergendal G, Fredrikson S, Almkvist O. Selective decline in information processing in subgroups of multiple sclerosis: an 8-year longitudinal study. Eur Neurol. 2007;57(4):193–202.
60. Duque B, Sepulcre J, Bejarano B, Samaranch L, Pastor P, Villoslada P. Memory decline evolves independently of disease activity in MS. Mult Scler. 2008;14(7):947–53.
61. Achiron A, Chapman J, Magalashvili D, Dolev M, Lavie M, Bercovich E, et al. Modeling of cognitive impairment by disease duration in multiple sclerosis: a cross-sectional study. PLoS One. 2013;8(8), e71058.
62. Pardini M, Uccelli A, Grafman J, Yaldizli Ö, Mancardi G, Roccatagliata L. Isolated cognitive relapses in multiple sclerosis. J Neurol Neurosurg Psychiatry. 2014;85(9):1035–7.
63. Larner AJ, Young CA. Acute amnesia in MS revisited. Int MS J. 2009;16(3):102–4.
64. Coebergh JAF, Roosendaal SD, Polman CH, Geurts JJ, van Woerkom TCAM. Acute severe memory impairment as a presenting symptom of multiple sclerosis: a clinical case study with 3D double inversion recovery MR imaging. Mult Scler. 2010;16(12):1521–4.
65. Zéphir H, de Seze J, Dujardin K, Dubois G, Cabaret M, Bouillaguet S, et al. One-year cyclophosphamide treatment combined with methylprednisolone improves cognitive dysfunction in progressive forms of multiple sclerosis. Mult Scler. 2005;11(3):360–3.
66. Zéphir H, de Sèze J, Dujardin K, Dubois G, Cabaret M, Bouillaguet S, et al. Cognitive impact of mitoxantrone and methylprednisolone in multiple sclerosis: an open label study. Rev Neurol (Paris). 2008;164(1):47–52.
67. Rist PM, Capistrant BD, Wu Q, Marden JR, Glymour MM. Dementia and dependence: do modifiable risk factors delay disability? Neurology. 2014;82(17):1543–50.
68. Beydoun MA, Beydoun HA, Gamaldo AA, Teel A, Zonderman AB, Wang Y. Epidemiologic studies of modifiable factors associated with cognition and dementia: systematic review and meta-analysis. BMC Public Health. 2014;14:643.

Chapter 19
Depression, Anxiety, and Cognitive Functioning in Multiple Sclerosis

Jean Pelletier, Audrey Rico, and Bertrand Audoin

Abstract Patients with multiple sclerosis (MS) often show depressive and anxiety symptoms as well as cognitive disorders. However, it is difficult for several reasons to determine how depression and anxiety may contribute to these patients' cognitive impairments. On the one hand, other contributory factors such as fatigue may be involved; and on the other hand, the methods currently used to assess depression and anxiety are subject to bias of various kinds, which makes it difficult to determine whether and to what extent these disorders are responsible for the expression of cognitive symptoms. Lastly, little is known so far about the physiopathological mechanisms underlying these disorders.

Anxiety seems to have only minor effects on MS patients' cognitive functioning. It is generally recognized that the anxiety induced by patients' difficulty in adapting to the disease and the painful emotional feelings it generates are triggered by psychosocial pressures without affecting patients' cognitive performances.

Although the effects of depressive disorders on cognitive functioning are still a matter of debate, it is nevertheless possible that these disorders may influence the cognitive problems associated with MS. It is also possible that the presence of focal or diffuse brain lesions might at least partly explain the onset and the aggravation of depressive symptoms and cognitive impairments in these patients.

At some particular stages in the disease (namely, the early and the progressive phase), it is particularly important to note the presence of anxiety and depressive symptoms and their repercussions on the patients' cognitive deficits, such as those affecting the performance of attentional and executive tasks. In view of the high prevalence of latter deficits, means of assessing and treating them are urgently required in order to be able to improve the patients' quality of life.

Keywords Anxiety • Depression • Cognitive functioning • MRI

J. Pelletier, M.D., Ph.D. (✉) • A. Rico, M.D., Ph.D. • B. Audoin, M.D., Ph.D.
Pôle de Neurosciences Cliniques, Service de Neurologie, APHM, Hôpital de la Timone, Marseille, France

CNRS, CRMBM CEMEREM UMR 7739, Aix-Marseille Université, Marseille, France
e-mail: jean.pelletier@ap-hm.fr; audrey.rico@ap-hm.fr; bertrand.audoin@ap-hm.fr

B. Brochet (ed.), *Neuropsychiatric Symptoms of Inflammatory Demyelinating Diseases*, Neuropsychiatric Symptoms of Neurological Disease,
DOI 10.1007/978-3-319-18464-7_19

Introduction

The fact that anxiety, depressive symptoms, and cognitive deficits are often associated with multiple sclerosis (MS) no longer needs to be proved, since these disorders are known to be extremely frequent in MS patients and to seriously affect their quality of life. In addition, the impact of these symptoms on the everyday lives of MS patients and their families would certainly justify the introduction of systematic screening so that appropriate treatment can be initiated. However, the physiopathological mechanisms underlying these symptoms are rather complex because they may involve focal or global brain lesions as well as patients' reactions to the disease. These intricate mechanisms can therefore make it difficult for MS patients to meet their socioprofessional and familial obligations. The repercussions of the disease are just as serious, if not more so (because they are not clearly visible) than those resulting from the patients' physical and sensory disabilities.

Although it is essential to document how the presence of brain lesions contributes to depression, anxiety, and cognitive disorders, the other contributory parameters possibly involved also need to be studied in order to be able to fully understand and treat these disorders, namely:

- The nature of the disease, which frequently occurs quite suddenly in fairly young individuals at a strategic time in their lives when they tend to be investing heavily in their emotional and professional future
- The unpredictable evolution of the disease
- The characteristics of the disease at specific moments (at disclosure of the diagnosis, when the patients' disabilities worsen, when side effects occur, when the treatment has to be changed)
- The presence of side effects such as fatigue and chronic pain
- The patient's basic personality
- The quality of the social support provided and the modes of coping used to deal with the potentially destabilizing situations resulting from the disease

The Impact of Depression on MS Patients' Cognitive Functioning

Among the various factors possibly responsible for depression in MS patients, it is still very difficult to determine the extent to which patients' motor and sensory problems, their cognitive impairments, fatigue, and their individual personality traits may prevent them from adapting to the disease. And conversely, the possible effects of depression on these patients' cognitive disorders are still a matter of debate. The prevalence of depressive symptoms in MS (50 %) has been found to be practically identical to the prevalence of cognitive disorders [1].

The authors of several studies failed to find any clear-cut relationship between these patients' cognitive deficits and depression [1–4]. Since no significant links were found to exist between the severity of the cognitive impairments and the level of depression, it was assumed for several years that depressive symptoms have no effect on MS patients' cognitive dysfunction. This conclusion seems rather surprising, however, especially in view of the fact that cognitive disturbances are known to be generally associated with severe depressive states [5–7]. It is therefore surely be rather paradoxical to believe that although cognitive disorders can occur in depressed subjects, depressive symptoms have no effect on MS patients' cognitive abilities. The method used to assess the cognitive deficits identified in MS patients may partly explain why this paradoxical conclusion was reached: the cognitive skills which are the most strongly disturbed by depressive symptoms may be those requiring fast information processing, working memory, and executive functions [8–14]. These skills, which mobilize attentional and executive abilities in particular, are often impaired in MS patients, especially in the early stage of the disease [15, 16]. The occurrence of these specific cognitive disorders may therefore be largely attributable to the presence of depressive symptoms, especially at the onset of the disease and at specific stages during its evolution (in patients with a progressive pattern of evolution). It therefore seems to be necessary to assess the impact of depressive symptoms on cognitive deficits of this kind before we are entitled to conclude that depressive symptoms have no effect on MS patients' cognitive performances.

On the other hand, it is necessary to take the various evolutive forms of the disease into account before deciding whether or not there exists a link between cognitive impairments and depressive symptoms and to pursue further longitudinal research in order to determine how depressive symptoms may affect cognitive disorders with time. Previous studies have shown that depressive and cognitive disorders both occur more frequently in the progressive forms of the disease, especially in the secondary progressive form, than in the remitting form [17, 18]. There are two possible explanations, which are not mutually exclusive, for this finding. In the first place, the lesional load detected by MRI may be heavier in the secondary progressive form, in which greater physical and psychological stress may be generated by the newly evolutive pattern of the disease. In addition, the severity of the disease in terms of the cognitive deficits due to the slowing of the information processing speed (as assessed using the PASAT test) was found to be significantly positively correlated with depression [19]. A significant correlation was also found to exist by other authors between depressive disorders and the subjective perception of a cognitive deficit [20, 21], and a link was found to exist between depression and cognitive disorders as well as with uncertainty about the future, loss of hope, and defensive strategies for coping with emotion. It was established in another study that the depression scores increased with the aggravation of the disease [22]. Although MS patients' impaired functional parameters were found to be significantly associated with their depressive state regardless of age, educational level, and the duration of the disease, their cognitive performances were more strongly correlated with their depressive symptoms. Another study has shown that depressive MS patients' cognitive deficits were clearly correlated with depressive moods and a negative self-image, but not with the neurovegetative symptoms of depression [10]. In addition, the existence of a relationship between depression and cognitive disorders, especially in tasks involving information processing speed, executive functions, and working memory, was established particularly clearly in severely depressed patients [10, 23–26]. Only a few longitudinal studies have focused on the effects of depression on cognitive impairments [11, 27–33]. Some of them have clearly shown that the aggravation of patients' cognitive impairments was significantly linked to increasing levels of depression [18, 27–32]. In particular, the patients' mean-term and long-term performances (4 years and 10 years after the onset, respectively) deteriorated in parallel with the aggravation of their levels of depression [29, 30]. In another study, it was reported that the level of depression sometimes affected the cognitive performances tested using SDMT and PASAT tests in the early stages of the disease (up to 2 years after the onset) but that this relationship depended mainly on the patients' level of disability (in terms of their EDSS scores) [33].

In short, assessing the impact of depressive disorders on MS patients' cognitive impairments is a difficult task despite the fact that symptoms of both kinds occur so frequently in these patients. It seems likely that although depressive disorders may accentuate some cognitive deficits at the onset and during the progressive phase of the disease, they seem to have less impact than the physical markers of disability (EDSS) [33]. However, it would no doubt be worth assessing MS patients' levels of depression and taking this information into account when evaluating their cognitive problems.

The Impact of Anxiety on MS Patients' Cognitive Functioning

Few studies have focused so far on the prevalence of anxiety and its correlations with other aspects of MS such as the cognitive impairments involved. Although the prevalence of anxiety in MS has been rated at 40 % [1, 34–41], it is difficult at present to determine its effects on the occurrence of cognitive disorders and their aggravation.

However, the few data available seem to suggest that anxiety does not significantly affect MS patients' cognitive performances [35]. In particular, it was recently reported that the level of anxiety was higher in the group of patients whose cognitive performances showed no deterioration than in the group in which they deteriorated [40]. However, as found to occur in the case of depressive disorders, the periods corresponding to the onset of the disease and the aggravation of the patients' physical disabilities may induce greater anxiety and thus affect some specific cognitive functions (especially those mobilizing attentional and executive abilities).

Here again, it seems to be worth recommending that MS patients should be screened for anxiety and tested using neuropsychological methods in order to assess the levels involved.

The Influence of Brain Damage on Depression, Anxiety, and Cognitive Functioning

Since little is known so far about the morphological substrates responsible for cognitive disturbances, anxiety, and depression, it is difficult to explain the links between these three parameters.

Few studies have dealt with the relationships between anxiety and the presence of brain lesions. In one study, no correlations were found to exist between the level of anxiety and the anomalies identified by MRI: the latter authors put forward the idea that anxiety may be a reaction to the psychosocial pressures to which patients are subjected [38]. Anxiety would therefore be due to difficulty in adapting to the disease and the painful emotions it generates rather than to the severity and/or the topographical distribution of the brain lesions sustained. No studies have been published so far to our knowledge on the effects of diffuse lesions in the white or gray matter on anxiety. Nor is any information available at present as to whether global or focal atrophy is liable to contribute to the occurrence of anxiety.

The neuroanatomical regions involved in depression in MS patients have not yet been clearly identified, and it is difficult to apply suitable methods because of the lack of neuropathological models for depression. Some authors have observed the existence of a significant link between the volume of patients' brain lesions, their distribution, and the level of depression [42–48]. In particular, lesions located in the frontal and temporal regions (especially in the left anterior part) are thought to be more strongly correlated with the level of depression and that of the cognitive impairments [38, 42, 46]. Likewise, a significant link has been found to exist between the level of depression and the degree of atrophy observed either in overall or in specific brain regions such as the corpus callosum, the hippocampus, the forceps minor, and the amygdala [49–54]. This link may be involved in some of the cognitive impairments associated with MS. Recently developed nonconventional MRI methods have yielded more detailed information about the substrates liable to account for the links possibly existing between depressive symptoms and cognitive disorders [55–64]. It is worth noting that gray matter lesions seem to be associated with patients' depression and cognitive impairments much more than white matter lesions [47, 56, 58, 60, 63]. Diffuse gray matter lesions are therefore promising candidate markers of depressive and cognitive disorders [47, 56, 59, 61]. Lastly, the existence of functional disturbances in terms of impaired neuronal connectivity and the emergence of functional reorganization processes may mediate some specific types of depression (as well as fatigue) and cognitive impairments (such as those resulting from information processing speed deficits) [65, 66]. A neurobiological process involving impaired connections between the prefrontal regions and the amygdala and compensatory functional mechanisms preventing the clinical expression of these impairments may, for instance, at least partly account for the anxiety and depression often associated with MS [67].

Conclusion

Although the exact effects of symptoms of anxiety and depression on cognitive dysfunction still remain to be determined, in view of the frequency with which these symptoms occur in MS patients, it seems to be worth recommending that they should be taken into account when assessing MS patients' cognitive status and treated whenever they are found to be present.

It seems to be particularly important to check the presence of anxiety and depression concomitantly with that of cognitive disorders during specific phases of the disease (such as the early and progressive phases). The presence of symptoms of depression and anxiety is liable in fact to influence the interpretation of the cognitive disturbances identified.

Accordingly, in view of the potentially aggravating effects of anxiety and depression on MS patients' cognitive deficits, if these symptoms are found to be present, suitable treatment should be prescribed (including anxiolytics, antidepressants, and psychotherapy). This treatment should improve the patients' quality of life by helping to prevent the subsequent aggravation of their somatic and cognitive symptoms and the resulting effects on the quality of their social and relational lives.

References

1. Siegert RJ, Abernathy DA. Depression in multiple sclerosis: a review. J Neurol Neurosurg Psychiatry. 2005;76:469–75.
2. Rao SM. Neuropsychology of multiple sclerosis. Curr Opin Neurol. 1995;8:216–20.
3. Rao SM. Neuropsychology of multiple sclerosis: a critical review. J Clin Exp Neuropsychol. 1986;8:503–42.
4. Brassington JC, Marsh NV. Neuropsychological aspects of multiple sclerosis. Neuropsychol Rev. 1998;8:43–77.
5. Shenal BV, Harrison DW, Demaree HA. The neuropsychology of depression: a literature review and preliminary model. Neuropsychol Rev. 2003;13:33–42.
6. Elliott R. The neuropsychological profile in unipolar depression. Trends Cogn Sci. 1998;2:447–54.
7. Hartlage S, Alloy LB, Vazquez C, et al. Automatic and effortful processing in depression. Psychol Bull. 1993;113:247–78.
8. Arnett PA, Higginson CI, Voss WD, et al. Depressed mood in multiple sclerosis: relationship to capacity-demanding memory and attentional functioning. Neuropsychology. 1999;13:434–46.
9. Arnett PA, Higginson CI, Voss WD, et al. Depression in multiple sclerosis: relationship to working memory capacity. Neuropsychology. 1999;13:546–56.
10. Arnett PA, Higginson CI, Randolph JJ. Depression in multiple sclerosis: relationship to planning ability. J Int Neuropsychol Soc. 2001;7:665–74.
11. Landro NI, Celius EGHS. Depressive symptoms account for deficient information processing speed but not for impaired working memory in early phase multiple sclerosis (MS). J Neurol Sci. 2004;217:211–6.
12. Demaree HA, Gaudino E, DeLuca J. The relationship between depressive symptoms and cognitive dysfunction in multiple sclerosis. Cogn Neuropsychiatry. 2003;8:161–71.

13. Demaree HA, DeLuca J, Gaudino E, et al. Speed of information processing as a key deficit in multiple sclerosis. J Neurol Neurosurg Psychiatry. 1999;67:661–3.
14. DeLuca J, Gaudino EA, Diamond BJ, et al. Acquisition and storage deficits in multiple sclerosis. J Clin Exp Neuropsychol. 1998;20:376–90.
15. Feuillet L, Reuter F, Audoin B, et al. Early cognitive impairment in patients with clinically isolated syndrome suggestive of multiple sclerosis. Mult Scler. 2007;13:124–7.
16. Achiron A, Barak Y. Cognitive impairment in probable multiple sclerosis. J Neurol Neurosurg Psychiatry. 2003;74:443–6.
17. Feinstein A, Kartsounis LD, Miller D, et al. Clinically isolated lesions of the type seen in multiple sclerosis: a cognitive, psychiatric, and MRI follow up study. J Neurol Neurosurg Psychiatry. 1992;55:869–76.
18. Piras MR, Magnano I, Canu EDG, et al. Longitudinal study of cognitive dysfunction in multiple sclerosis: neuropsychological, neuroradiological, and neurophysiological findings. J Neurol Neurosurg Psychiatry. 2003;74:878–85.
19. Camp SJ, Stevenson VL, Thompson AJ, et al. A longitudinal study of cognition in primary progressive multiple sclerosis. Brain. 2005;128:2891–8.
20. Shawaryn MA, Schiaffino KM, LaRocca NG, et al. Determinants of health-related quality of life in multiple sclerosis: the role of illness intrusiveness. Mult Scler. 2002;8:310–8.
21. Maor Y, Olmer L, Mozes B. The relation between objective and subjective impairment in cognitive function among multiple sclerosis patients-the role of depression. Mult Scler. 2001;7:131–5.
22. Lynch SG, Kroencke DC, Denney DR. The relationship between disability and depression in multiple sclerosis: the role of uncertainty, coping and hope. Mult Scler. 2001;7(6):411–6.
23. Chwastiak L, Ehde D, Gibbons L, et al. Depressive symptoms and severity of illness in multiple sclerosis: epidemiologic study of a large community sample. Am J Psychiatry. 2002;159:1862–8.
24. Ghaffar O, Feinstein A. The neuropsychiatry of multiple sclerosis: a review of recent developments. Curr Opin Psychiatry. 2007;20(3):278–85.
25. Julian LJ, Arnett PA. Relationships among anxiety, depression, and executive functioning in multiple sclerosis. Clin Neuropsychol. 2009;23(5):794–804.
26. Barwick FH, Arnett PA. Relationship between global cognitive decline and depressive symptoms in multiple sclerosis. Clin Neuropsychol. 2011;25(2):193–209.
27. Patti F, Amato MP, Trojano M, et al. Cognitive impairment and its relation with disease measures in mildly disabled patients with relapsing-remitting multiple sclerosis: baseline results from the Cognitive Impairment in Multiple Sclerosis (COGIMUS) study. Mult Scler. 2009;15(7):779–88.
28. Amato MP, Ponziani G, Pracucci G, et al. Cognitive impairment in early-onset multiple sclerosis. Arch Neurol. 1995;52:168–72.
29. Amato MP, Ponziani G, Siracuse G, Sorbi S. Cognitive dysfunction in early-onset multiple sclerosis: a reappraisal after 10 years. Arch Neurol. 2001;58:1602–6.
30. Christodoulou C, Melville P, Scherl WF, et al. Negative affect predicts subsequent cognitive change in multiple sclerosis. J Int Neuropsychol Soc. 2009;15:53–61.
31. Denney DR, Lynch SG, Parmenter BA. A 3-year longitudinal study of cognitive impairment in patients with primary progressive multiple sclerosis: speed matters. J Neurol Sci. 2008;267:129–36.
32. Siepman TA, Janssens AC, de Koning I, et al. The role of disability and depression in cognitive functioning within 2 years after multiple sclerosis diagnosis. J Neurol. 2008;255(6):910–6.
33. Janssens AC, Van Doorn PA, de Boer JB, et al. Perception of prognostic risk in patients with multiple sclerosis: the relationship with anxiety, depression and disease-related distress. J Clin Epidemiol. 2004;57(2):180–6.
34. Galeazzi GM, Ferrari S, Giaroli G, et al. Psychiatric disorders and depression in multiple sclerosis outpatients: impact of disability and interferon beta therapy. Neurol Sci. 2005;26:255–62.

35. Janssens AC, van Doorn PA, de Boer JB, et al. Anxiety and depression influence the relation between disability status and quality of life in multiple sclerosis. Mult Scler. 2003;9:397–403.
36. Feinstein A, O'Connor P, Gray T, Feinstein K. The effects of anxiety on psychiatric morbidity in patients with multiple sclerosis. Mult Scler. 1999;5:323–6.
37. Janssens ACJW, van Doorn PA, de Boer JB, et al. Impact of recently diagnosed multiple sclerosis on quality of life, anxiety, depression and distress of patients and partners. Acta Neurol Scand. 2003;108:389–95.
38. Zorzon M, de Masi R, Nasuelli D, et al. Depression and anxiety in multiple sclerosis. A clinical and MRI study in 95 subjects. J Neurol. 2001;248:416–21.
39. Korostil M, Feinstein A. Anxiety disorders and their clinical correlates in multiple sclerosis patients. Mult Scler. 2007;13:67–72.
40. Rogers JM, Panegyres PK. Cognitive impairment in multiple sclerosis: evidence-based analysis and recommendations. J Clin Neurosci. 2007;14(10):919–27.
41. Karadayi H, Arisoy O, Altunrende B, et al. The relationship of cognitive impairment with neurological and psychiatric variables in multiple sclerosis patients. Int J Psychiatry Clin Pract. 2014;18(1):45–51.
42. Feinstein A. Neuropsychiatric syndromes associated with multiple sclerosis. J Neurol. 2007;254(2):73–6.
43. Bakshi R, Czarnecki D, Shaikh ZA, et al. Brain MRI lesions and atrophy are related to depression in multiple sclerosis. Neuroreport. 2000;11(6):1153–8.
44. Pujol J, Bello J, Deus J, Marti-Vilalta JL, Capdevila A. Lesions in the left arcuate fasciculus region and depressive symptoms in multiple sclerosis. Neurology. 1997;49(4):1105–10.
45. Tiemann L, Penner IK, Haupts M, Schlegel U, Calabrese P. Cognitive decline in multiple sclerosis: impact of topographic lesion distribution on differential cognitive deficit patterns. Mult Scler. 2009;15:1164–74.
46. Feinstein A, Roy P, Lobaugh N, et al. Structural brain abnormalities in multiple sclerosis patients with major depression. Neurology. 2004;62:586–90.
47. Gobbi C, Rocca MA, Riccitelli G, et al. Influence of the topography of brain damage on depression and fatigue in patients with multiple sclerosis. Mult Scler. 2014;20(2):192–201.
48. Amato MP, Hakiki B, Goretti B, et al. Association of MRI metrics and cognitive impairment in radiologically isolated syndromes. Neurology. 2012;78(5):309–14.
49. Benedict RH, Carone DA, Bakshi R. Correlating brain atrophy with cognitive dysfunction, mood disturbances, and personality disorder in multiple sclerosis. J Neuroimaging. 2004;14(S3):36S–45.
50. Hildebrandt H, Hahn HK, Kraus JA, et al. Memory performance in multiple sclerosis patients correlates with central brain atrophy. Mult Scler. 2006;12(4):428–36.
51. Hildebrandt H, Eling P. A longitudinal study on fatigue, depression, and their relation to neurocognition in multiple sclerosis. J Clin Exp Neuropsychol. 2014;36(4):410–7.
52. Gold SM, O'Connor MF, Gill R, et al. Detection of altered hippocampal morphology in multiple sclerosis-associated depression using automated surface mesh modeling. Hum Brain Mapp. 2014;35(1):30–7.
53. Yaldizli Ö, Penner IK, Frontzek K, et al. The relationship between total and regional corpus callosum atrophy, cognitive impairment and fatigue in multiple sclerosis patients. Mult Scler. 2014;20(3):356–64.
54. Gobbi C, Rocca M, Pagani E, et al. Forceps minor damage and co-occurrence of depression and fatigue in multiple sclerosis. Mult Scler. 2014;20(12):1633–40.
55. Feinstein A, O'Connor P, Akbar N, et al. Diffusion tensor imaging abnormalities in depressed multiple sclerosis patients. Mult Scler. 2010;16:189–96.
56. Riccitelli G, Rocca MA, Pagani E, et al. Mapping regional grey and white matter atrophy in relapsing–remitting multiple sclerosis. Mult Scler. 2012;18:1027–37.

57. Hulst HE, Gehring K, Uitdehaag BM, Visser LH, et al. Indicators for cognitive performance and subjective cognitive complaints in multiple sclerosis: a role for advanced MRI? Mult Scler. 2013;20(8):1131–4.
58. Nielsen AS, Kinkel RP, Madigan N, Tinelli E, Benner T, Mainero C. Contribution of cortical lesion subtypes at 7 T MRI to physical and cognitive performance in MS. Neurology. 2013;81(7):641–9.
59. Benedict RH, Schwartz CE, Duberstein P, et al. Influence of personality on the relationship between grey matter volume and neuropsychiatric symptoms in multiple sclerosis. Psychosom Med. 2013;75(3):253–61.
60. Papadopoulou A, Müller-Lenke N, Naegelin Y, et al. Contribution of cortical and white matter lesions to cognitive impairment in multiple sclerosis. Mult Scler. 2013;19(10):1290–6.
61. Llufriu S, Martinez-Heras E, Fortea J, et al. Cognitive functions in multiple sclerosis: impact of gray matter integrity. Mult Scler. 2014;20(4):424–32.
62. Faiss JH, Dähne D, Baum K, et al. Reduced magnetisation transfer ratio in cognitively impaired patients at the very early stage of multiple sclerosis: a prospective, multicenter, cross-sectional study. BMJ Open. 2014;4(4), e004409.
63. Shen Y, Bai L, Gao Y, et al. Depressive symptoms in multiple sclerosis from an in vivo study with TBSS. Biomed Res Int. 2014;2014:148465.
64. Cavallari M, Ceccarelli A, Wang GY, et al. Microstructural changes in the striatum and their impact on motor and neuropsychological performance in patients with multiple sclerosis. PLoS One. 2014;9(7), e101199.
65. Tona F, Petsas N, Sbardella E, et al. Multiple sclerosis: altered thalamic resting-state functional connectivity and its effect on cognitive function. Radiology. 2014;271(3):814–21.
66. Rocca MA, Valsasina P, Hulst HE, et al. Functional correlates of cognitive dysfunction in multiple sclerosis: a multicenter fMRI Study. Hum Brain Mapp. 2014;35(12):5799–814.
67. Passamonti L, Cerasa A, Liguori M, et al. Neurobiological mechanisms underlying emotional processing in relapsing-remitting multiple sclerosis. Brain. 2009;132(12):3380–91.

Index

B. Brochet (ed.), *Neuropsychiatric Symptoms of Inflammatory Demyelinating Diseases*, Neuropsychiatric Symptoms of Neurological Disease,
DOI 10.1007/978-3-319-18464-7

The manufacturer's authorised representative in the EU is Springer Nature Customer Service Centre GmbH, Europaplatz 3, 69115 Heidelberg, Germany. If you have any concerns regarding our products, please contact ProductSafety@springernature.com

Printed and bound by CPI Group (UK) Ltd, Croydon, CR0 4YY
15/07/2026
02167631-0004